Algorithms in Real Algebraic Geometry

Edited by
DENIS S. ARNON
Xerox Corporation, Palo Alto,
California
BRUNO BUCHBERGER
Johannes Kepler Universität,
Linz, Austria

Reprinted from the *Journal of Symbolic Computation*
Volume 5, Numbers 1 and 2, 1988

ACADEMIC PRESS · 1988

Harcourt Brace Jovanovich, Publishers

London   San Diego   New York   Boston   Sydney   Tokyo   Toronto

ACADEMIC PRESS LIMITED
24–28 Oval Road, London, NW1 7DX

United States Edition published by
ACADEMIC PRESS INC.
San Diego, CA 92101, U.S.A.

ISBN 0-12-063880-0

Printed in Great Britain by The Whitefriars Press Ltd, Tonbridge

Contents

Editorial 1

The Complexity of Linear Problems in Fields 3
V. WEISPFENNING

Real Quantifier Elimination is Doubly Exponential 29
J. H. DAVENPORT and J. HEINTZ

Solving Systems of Polynomial Inequalities in Subexponential Time . . . 37
D. YU. GRIGOR'EV and N. N. VOROBJOV (JR)

Complexity of Deciding Tarski Algebra 65
D. YU. GRIGOR'EV

Some Aspects of Complexity in Real Algebraic Geometry 109
J.-J. RISLER

Thom's Lemma, the Coding of Real Algebraic Numbers and the Computation of
the Topology of Semi-algebraic Sets 121
M. COSTE and M. F. ROY

Algebraic Decomposition of Regular Curves 131
S. ARNBORG and H. FENG

An Improved Projection Operation for Cylindrical Algebraic Decomposition of
Three-dimensional Space 141
S. McCALLUM

An Adjacency Algorithm for Cylindrical Algebraic Decompositions of Three-
dimensional Space 163
D. S. ARNON, G. E. COLLINS and S. McCALLUM

A Cluster-Based Cylindrical Algebraic Decomposition Algorithm 189
D. S. ARNON

A Polynomial-time Algorithm for the Topological Type of a Real Algebraic Curve 213
D. S. ARNON and S. McCALLUM

On Mechanical Quantifier Elimination for Elementary Algebra and Geometry . 237
D. S. ARNON and M. MIGNOTTE

Quantifier Elimination: Optimal Solution for Two Classical Examples . . . 261
D. LAZARD

A Bibliography of Quantifier Elimination for Real Closed Fields 267
D. S. ARNON

J. Symbolic Computation (1988) **5**, 1

Editorial

This special issue of the *Journal of Symbolic Computation* evolved from the nearly simultaneous submission of a number of papers treating algorithmic problems in real algebraic geometry to the journal. Additional papers were then solicited from researchers active in the field and a complete bibliography on the past research on quantifier elimination for real closed fields was compiled. As a result, the special issue is now the first collection of papers devoted to algorithms in real algebraic geometry. It is intended both to exhibit the present state of research and to facilitate access to the field for those who wish to enter. We hope that it stimulates further research in this challenging and promising area, where significant results have been achieved in recent years, but much more remains to be done. In particular, it is crucial that the geometric reasoning community be attracted to the new algebraic techniques and, conversely, that pure mathematicians working in real algebraic geometry become interested in the algorithmic problems inherent in advanced geometric reasoning applications.

It is the policy of the *Journal of Symbolic Computation* to stimulate, by special issues, this kind of interaction between research communities and to provoke new research activities. The *Journal of Symbolic Computation* will be proud to continue to serve as a forum for research in the algorithmic aspects of real algebraic geometry.

Dennis Arnon
Bruno Buchberger

J. Symbolic Computation (1988) **5**, 3–27

The Complexity of Linear Problems in Fields

VOLKER WEISPFENNING

University of Heidelberg, Mathematisches Institut,
Im Neuenheimer Feld 288, *D-6900 Heidelberg, FRG*

(*Received* 20 *November* 1984)

We consider linear problems in fields, ordered fields, discretely valued fields (with finite residue field or residue field of characteristic zero) and fields with finitely many independent orderings and discrete valuations. Most of the fields considered will be of characteristic zero. Formally, linear statements about these structures (with parameters) are given by formulas of the respective first-order language, in which all bound variables occur only linearly. We study symbolic algorithms (*linear elimination procedures*) that reduce linear formulas to linear formulas of a very simple form, i.e. quantifier-free linear formulas, and algorithms (*linear decision procedures*) that decide whether a given linear sentence holds in all structures of the given class. For all classes of fields considered, we find linear elimination procedures that run in double exponential space and time. As a consequence, we can show that for fields (with one or several discrete valuations), linear statements can be transferred from characteristic zero to prime characteristic p, provided p is double exponential in the length of the statement. (For similar bounds in the non-linear case, see Brown, 1978.) We find corresponding linear decision procedures in the Berman complexity classes $\bigcup_{c \in \mathbb{N}} STA(*, 2^{cn}, dn)$ for $d = 1, 2$. In particular, all these procedures run in exponential space. The technique employed is quantifier elimination via Skolem terms based on Ferrante & Rackoff (1975). Using ideas of Fischer & Rabin (1974), Berman (1977), Fürer (1982), we establish lower bounds for these problems showing that our upper bounds are essentially tight. For linear formulas with a bounded number of quantifiers all our algorithms run in polynomial time. For linear formulas of bounded quantifier alternation most of the algorithms run in time $2^{O(n^k)}$ for fixed k.

Introduction

Elementary statements about fields, ordered fields and valued fields play an important rôle in symbolic algebraic computations. On the one hand, their expressive power is strong enough to cover a good deal of commutative algebra, geometry and number theory, in particular most polynomial manipulations. On the other hand, one knows since the pioneering work of Tarski, A. Robinson and Ax–Kochen–Ershov, that these statements are amenable to computational methods, when considered in certain fields such as the reals, the complex numbers, the p-adics and certain power series fields.

Two methods are of prime importance in this connection:

(1) *Decision procedures*, i.e. symbolic manipulations for deciding the validity of formal elementary statements in certain classes of structures.
(2) *Quantifier elimination procedures*, i.e. symbolic manipulations reducing formal elementary statements with parameters equivalently in a class of structures to particularly simple such statements (quantifier-free formulas).

While a great number of recursive and even primitive recursive procedures of this kind have been established (cf. van den Dries, 1981; Macintyre *et al.*, 1983), only few of them, as e.g. for real algebra (cf. Collins, 1983; Ben-Or *et al.*, 1986), complex algebra (cf. Heintz, 1983), boolean algebra (cf. Kozen, 1980), are known to be elementary recursive, i.e. to run in time bounded by a finite iteration of the exponential function applied to the size of the input.

In this paper, we restrict our attention mostly to linear statements with parameters in fields, ordered fields, discretely valued fields and fields with several orderings and valuations. Roughly speaking, these are formulas of the respective elementary language, in which all essential variables, i.e. the variables x bound by a quantitier $\exists x$, $\forall x$, occur only linearly. This is, of course, a severe restriction in expressive power; on the other hand, it allows us to dispense with closure conditions such as real or p-adic closedness. The following two examples may illustrate the kind of problems that can be handled in this framework:

(1) Solvability of finite systems of linear equations, linear order inequalities, and linear p-adic divisibilities for variable prime p in the field $\mathbb{Q}$ of rationals.

(2) Solvability of finite systems of linear equations and in equations with specified orders of poles and zeros at finitely many places in the rational function field $\mathbb{Q}(t)$. Moreover, if t is specified as a transcendental real, linear order inequalities may be added.

(3) Elementary problems in computational geometry such as movability problems to the extent that they concern (not necessarily convex) polyhedra and translations (see section 6 for an example). In all examples, the parameters of the problem may be indeterminate.

The existence of quantifier elimination procedures for linear formulas (*linear elimination*, for short) has been studied in Macintyre *et al.*, 1983; Point, 1983; van den Dries, 1981. None of the procedures exhibited there is better than primitive recursive.

The purpose of this paper is to determine the exact complexity of linear elimination and the decision of linear sentences in the classes of fields mentioned above. Somewhat surprisingly, the outcome is essentially the same for all these classes of fields:

Except for the case of multiordered, multivalued fields, the decision problem for linear sentences in complete (under polynomial time reductions) for the Berman complexity class $\bigcup_{c \in \mathbb{N}} STA(*, 2^{cn}, n)$ (see Berman, 1977, 1980). So all these decision problems are computationally equivalent to the decision problem for real addition (see Berman, 1977) and for boolean algebras (see Kozen, 1980). In the exceptional case the problem is $\cup STA(*, 2^{cn}, n)$-hard and in $\cup STA(*, 2^{cn}, 2n)$. So in every case, the problem can be solved in exponential space and double exponential time.

We find linear elimination procedures running in double exponential space and time, and prove that any such procedure does, indeed, require double exponential space on infinitely many formulas. So the linear elimination problem for these classes of fields is of the same complexity as the quantifier elimination problem for torsion-free abelian groups and for algebraically closed fields (cf. section 5 and Heintz, 1983). Moreover, we show that any quantifier elimination procedure for the reals or p-adics requires double exponential space. So the quantifier elimination problem for arbitrary formulas and for linear formulas in the theory of real numbers are essentially of the same complexity.

We show in section 6 that the number of quantifiers (the *dimension*) of a linear formula is the prime source of computational complexity; the number of quantifier alternations is of secondary importance and the length of the formula enters only polynomially. Thus, for bounded dimension all our algorithms run in polynomial time; for bounded quantifier alternation many of them run in exponential time (with a polynomial exponent).

As a by-product, we find that the transfer of linear statements φ in fields and fields with one or several independent discrete valuations from characteristic zero to large positive characteristic p works for p double exponential in the length of φ, and does in fact require a lower bound of this size (see sections 2.7, 3.5, 4.3, 5.2' below).

The upper bounds are established by the method of quantifier elimination via Skolem terms, based on ideas of Ferrante & Rackoff (1975). The lower bounds use results of Fischer & Rabin (1974), Berman (1977, 1980) and Fürer (1982), in particular the construction of short linear formulas defining large finite sets.

The *plan of the paper* is as follows: Section 1 provides the logical background and presents the general method. Section 2 treats the case of fields and ordered fields, and section 3 the case of discretely valued fields. Section 4 combines the results of the previous sections with the appropriate approximation theorem for independent valuations and orders to treat the case of multivalued, multiordered fields. Section 5 establishes the lower bounds. Section 6 studies the modifications of these results for linear formulas with a bounded number of quantifiers or quantifier blocks.

1. The General Method

We begin with a short sketch of the logical background. A reader familiar with elementary logic may skip this paragraph.

We consider elementary languages L given by a finite set of constants and finitary operation and relation symbols. From these symbols together with an infinite supply V of variables $x, y, \ldots$, the equality sign "$=$", the logical symbols $\wedge, \vee, \neg, \exists, \forall$ and brackets $(,)$, the terms and formulas of L are built up: *terms* $t, t', \ldots$ are formed from constants and variables by superposition of operation symbols; *atomic formulas* are equations $(t = t')$ or formal relations $R(t_1, \ldots, t_n)$ between terms; arbitrary *formulas* $\varphi, \psi, \ldots$ are obtained from atomic formulas by closure under $\wedge, \vee, \neg$ and quantifiers $\exists x, \forall x$. $\varphi \to \psi$ and $\varphi \leftrightarrow \psi$ are abbreviations for $\neg \varphi \vee \psi$ and $(\varphi \to \psi) \wedge (\psi \to \varphi)$. An occurrence of a variable x in a formula φ is *bound*, if it is in the scope of a quantifier $\exists x$ or $\forall x$; otherwise, it is *free*. A formula containing no quantifier is *quantifier-free*. A formula containing no variable free is a *sentence*. A *theory* T in L is a set of L-sentences. An *L-structure* A is a non-empty set, where the constants, operation symbols and relation symbols of L are interpreted as elements, operations and relations of the appropriate arity. If $\varphi(\mathbf{x})$ is an L-formula with all free variables in the string $\mathbf{x}$, A is an L-structure and $\mathbf{a}$ is a string of elements of A matching $\mathbf{x}$, then $A \models \varphi(\mathbf{a})$ means "φ *holds in A for the parameters* $\mathbf{x} = \mathbf{a}$". For a sentence φ the parameters are deleted. A is a *model* of an L-theory T, if $A \models \varphi$ for all sentences $\varphi \in T$. A sentence φ is a *consequence* of T, $T \models \varphi$, if φ holds in all models of T. Let T be a theory in L and Φ a set of L-formulas. Then a *decision procedure* for Φ, T is a procedure that decides for any sentence $\varphi \in \Phi$ whether $T \models \varphi$ or not. A *quantifier elimination* for Φ, T is a procedure $\varphi \mapsto \varphi'$ assigning to any formula $\varphi \in \Phi$ a quantifier-free formula $\varphi' \in \Phi$ such that φ and φ' are T-equivalent. If in these definitions Φ is the set $Fo(L)$ of all L-formulas, then the reference to Φ is omitted.

Quantifier elimination procedures are an important tool in model theory; in particular, they reduce the decision problem for T to the decision of quantifier-free sentences with respect to T. The latter is solvable for many algebraic theories. Therefore, a great variety of techniques has been developed to prove the existence of quantifier elimination procedures (cf. Weispfenning, 1984). They yield—as a rule—only general recursive, at best primitive recursive procedures, and hence are far from being feasible. On the other hand, one knows from the work of Fischer & Rabin (1974) that the decision problem even for simple algebraic theories like real addition requires non-deterministic exponential time. The same technique can be applied to show that quantifier elimination for this theory requires double exponential space (see 5.1'). So the best time and space bound that can be expected in these matters is a finite iteration $2**2** \ldots 2**n$ of the exponential function $2**n = 2^n$. Functions computable on a Turing machine with such a time (or space) bound are called *elementary recursive*. Traditional quantifier elimination procedures violate such a bound for a rather trivial logical reason: They eliminate one quantifier at a time, and solve the quantifier elimination problem for a formula $\exists x(\varphi)$ with quantifier-free φ by reduction to the case that φ is simple enough to express a mathematically well-known problem of the respective theory T. This invariably involves an application of the distributive law for $\wedge, \vee$, in order to eliminate disjunctions from φ. But any such application may increase the length of the formula exponentially. So, for an unbounded number of quantifiers the length of the resulting formula surpasses any elementary recursive bound.

For some theories T, this problem can be circumvented in the following way: One disregards the internal structure of the formula $\exists x(\varphi)$, and tries instead to find a finite set of terms (depending on the free variables of this formula) to act as witnesses for the existence of an element x satisfying φ. This device dates back to the early work of Skolem. It was taken up by Cooper (1972) and Ferrante & Rackoff (1975) to find elementary recursive quantifier elimination procedures for Presburger arithmetic and for addition of reals with order (cf. also Weispfenning, 1986).

In the rest of this section, we present a uniform, abstract version of this method, suited for application to linear problems. A novel element—multiple substitution of Skolem terms—is introduced in order to cover the case of fields with several orderings and valuations in section 4. A corresponding method for finding elementary recursive *decision* procedures via finite sets of Henkin constants has received a uniform treatment in Ferrante & Rackoff (1979).

Let L be an elementary language and let T be a theory in L. Let $X \subset V$ be an infinite set of variables, Z a set of L-terms, θ a set of atomic L-formulas, and Φ the closure of θ under $\wedge, \vee, \neg$ and quantifiers $\exists x, \forall x$ with $x \in X$. Assume, moreover:

1.1 (i) All terms occurring in formulas $\vartheta \in \theta$ are in Z.
  (ii) There is a modified substitution procedure assigning to variables $x \in X$ and terms $t, t' \in Z$ a term $t(x//t') \in Z$ such that $T \models t(x//t') = t(x/t')$, where $t(x/t')$ denotes the term obtained from t by substituting t' for x in the usual sense.
  (iii) For $\vartheta \in \theta$, $x \in X$, $t' \in Z$, let $\vartheta(x//t')$ denote the expression resulting from ϑ by replacing every term t in ϑ by $t(x//t')$. Then $\vartheta(x//t')$ is a formula in θ.

If t is a term, φ is a formula, then $X(t), X(\varphi)$ denote the set of variables $x \in X$ occurring in t and φ respectively. If Ψ is a finite subset of θ, $x \in X$, S is a finite set of terms $t \in Z$ with $x \notin X(t)$, then we say, S *is a set of Skolem terms for* x, Ψ if

$$T \models \forall x \left(\bigvee_{t \in S} \bigwedge_{\psi \in \Psi} (\psi \leftrightarrow \psi(x//t)) \right).$$

We call T a *Skolem theory* with respect to X, Z, θ, if for every $x \in X$ and every finite $\Psi \subseteq \theta$, x, Ψ has a set of Skolem terms.

LEMMA 1.2. *Let T be a Skolem theory with respect to X, Z, θ.*

(i) *If $\varphi \in \Phi$ is quantifier-free, Ψ is the set of atomic subformulas of φ, $x \in X$, and S is a set of Skolem terms for x, Ψ, then*

(*)
$$T \models \exists\, x(\varphi) \leftrightarrow \bigvee_{t \in S} \varphi(x//t)$$

and

(**)
$$T \models \forall\, x(\varphi) \leftrightarrow \bigwedge_{t \in S} \varphi x//t).$$

(ii) *Suppose modified substitution $(x, t, t') \mapsto t(x//t')$, and the assignment of sets S of Skolem terms to pairs (x, Ψ) is recursive for the theory T. Then there is a recursive quantifier elimination procedure for Φ, T.*

PROOF. (i) (*) is obvious from the definitions; (**) follows from (*) by replacing $\forall\, x(\varphi)$ by $\neg \exists\, x \neg (\varphi)$. (ii) Notice that the formulas on the right-hand side of (*) and (**) are in Φ. So the following describes a recursive quantifier elimination procedure for Φ, T:

Input: φ
Output: φ'
$\varphi' := \varphi$;
while φ' contains a quantifier *do*
   begin find the first quantifier $\exists\, x$ or $\forall\, x$ in φ', whose scope ψ is quantifier-free; replace
     $\exists\, x(\psi)$ or $\forall\, x(\psi)$ in φ' by the corresponding right-hand side of (*) or (**),
     respectively.
   end
end.

The following technical lemma will be useful in sections 2 and 3 to avoid explicit case distinctions on the parameters of linear problems.

LEMMA 1.3. *Let $x \in X$, let Ψ, Ψ' be finite subsets of θ, and assume for every $\varphi \in \Psi$ there exists a finite set J_φ of pairs of formulas such that*

(i) *$x \notin X(\rho)$ and $\sigma \in \Psi'$ for all $(\rho, \sigma) \in J_\varphi$;*

(ii) *$T \models \bigvee_{(\rho, \sigma) \in J_\varphi} \rho$;*

(iii) *$T \models \bigwedge_{(\rho, \sigma) \in J_\varphi} \rho \to (\sigma \leftrightarrow \varphi)$.*

Then any set S of Skolem terms for x, Ψ' is also a set of Skolem terms for x, Ψ.

PROOF. Let all the variables of formulas in Ψ, Ψ' be interpreted in a model A of T. Then for any $\varphi \in \Psi$ there exists $(\rho_\varphi, \sigma_\varphi) \in J_\varphi$ such that $A \models \rho_\varphi$, and so $A \models \sigma_\varphi \leftrightarrow \varphi$. Furthermore, there exists $t \in S$ with

$$A \models \bigwedge_{\varphi \in \Psi} \sigma_\varphi \leftrightarrow \sigma_\varphi(x//t).$$

Since $x \notin X(\rho_\varphi)$, $A \models \rho_\varphi(x//t)$, and so $A \models \sigma_\varphi(x//t) \leftrightarrow \varphi(x//t)$ for any $\varphi \in \Psi$. Consequently,

$$A \models \bigwedge_{\varphi \in \Psi} \varphi \leftrightarrow \varphi(x//t).$$

Our next goal is to compute upper bounds on the complexity of the quantifier elimination procedure described in lemma 1.2. To this end, we assume that we have a notion of *rank* for terms $t \in Z$ with the following properties:

1.4 There is a positive constant c such that for all $x \in X$, $t, t' \in Z$,

    (i) $\text{rank}(t)$ is a positive integer;
    (ii) $\text{rank}(t) \leqslant \text{length}(t) \leqslant c \cdot \text{rank}(t) \cdot |X(t)|$;
    (iii) $\text{rank}(t(x//t')) \leqslant c \cdot \text{rank}(t) + \text{rank}(t')$.

For $\varphi \in \Phi$, we let $rank(\varphi) = \max(\text{rank}(t) : t \text{ occurs in } \varphi)$; $atom(\varphi)$ is the number of atomic subformulas of φ, and $quant(\varphi)$ is the number of quantifiers in φ. If $S \subseteq Z$, $\Psi \subseteq \Phi$ are finite, then $rank(S)$, $rank(\Psi)$ is the maximum of all ranks of terms (formulas) in S, Ψ, respectively.

LEMMA 1.5. *Let g, h be weakly monotonic functions from $\mathbb{N}$ in $\mathbb{N}$, and put $g_1(n) = g(n) \cdot n$, $h_1(n) = h(n) + cn$ with c as in 1.4. Let T be a Skolem theory with respect to X, Z, θ and assume that for $x \in X$, Ψ a finite subset of θ, a set S of Skolem terms for x, Ψ can be found with $|S| \leqslant g(|\Psi|)$, $\text{rank}(S) \leqslant h(\text{rank}(\Psi))$. Let $\varphi \mapsto \varphi'$ be the quantifier elimination procedure of 1.2(ii). Then*

    (i) $\text{atom}(\varphi') \leqslant g_1^{(\text{quant}(\varphi))}(\text{atom}(\varphi))$,
    (ii) $\text{rank}(\varphi') \leqslant h_1^{(\text{quant}(\varphi))}(\text{rank}(\varphi))$,
    (iii) $\text{length}(\varphi') \leqslant c' \cdot |X(\varphi)| \cdot g_1^{(\text{quant}(\varphi))}(\text{length}(\varphi)) \cdot h_1^{(\text{quant}(\varphi))}(\text{length}(\varphi))$ *for a positive constant c'.*

PROOF. Notice that

$$\text{atom}\left(\bigvee_{t \in S} \varphi(x//t)\right) = |S| \cdot \text{atom}(\varphi),$$

and by 1.4,

$$\text{rank}\left(\bigvee_{t \in S} \varphi(x//t)\right) = c \cdot \text{rank}(\varphi) + \text{rank}(S).$$

From this remark and the proof of 1.2, (i) and (ii) follow now by induction on $\text{quant}(\varphi)$. For (iii), let k be the maximal arity of a relation symbol in L. For $\vartheta \in \theta$, we have by 1.4,

$$\text{length}(\vartheta) \leqslant (k+2) \cdot \max(\text{length}(t) : t \text{ occurs in } \vartheta)$$
$$\leqslant (k+2) \cdot c \cdot \text{rank}(\vartheta) \cdot |X(\vartheta)|.$$

In 1.2(i), we may assume that $X(t) \subseteq X(\varphi)$ for all $t \in S$. So by (i) and (ii), we get

$$\text{length}(\varphi') \leqslant \text{atom}(\varphi') \cdot (k+2) \cdot c \cdot \text{rank}(\varphi') \cdot |X(\varphi')|$$
$$\leqslant c' \cdot g_1^{(\text{quant}(\varphi))}(\text{length}(\varphi)) \cdot h_1^{(\text{quant}(\varphi))}(\text{length}(\varphi)) \cdot |X(\varphi)|,$$

for $c' = (k+2)c$.

Specialising 1.5 to the case of a polynomial bound $g(n) = c_1 \cdot n^k$ and a linear bound $h(n) = c_1 \cdot n$ with $c_1 > 0$, we obtain:

COROLLARY 1.6. *Assume the hypothesis of 1.5 for g, h as specified above. Then*

    (i) $\text{atom}(\varphi') \leqslant (1 + \text{atom}(\varphi))**2**d*\text{quant}(\varphi)$,
    (ii) $\text{rank}(\varphi') \leqslant (2**d*\text{quant}(\varphi))*\text{rank}(\varphi)$,
    (iii) $\text{length}(\varphi') \leqslant 2**2**d*\text{length}(\varphi)$,

for some positive constant d.

THEOREM 1.7. *Let T be a Skolem theory with respect to X, Z, θ, and suppose that modified substitution $(x, t, t') \mapsto t(x//t')$ and the assignment of sets S of Skolem terms to pairs (x, Ψ)*

can be performed in polynomial time. Let $\varphi \mapsto \varphi'$ be the quantifier elimination procedure of 1.2. Then φ' can be constructed from φ in $\mathrm{TIME}_{qe}(\varphi) \leqslant 2^{**}2^{**}d^{*}\mathrm{length}(\varphi)$ for some positive constant d.

PROOF. By the hypothesis, the elimination of one quantifier according to 1.2(i) can be performed in polynomial time. So by induction on $\mathrm{quant}(\varphi)$, the construction of a quantifier-free equivalent φ' for a formula in 1.2 requires at most double exponential time.

COROLLARY 1.8. *Assume the hypothesis of 1.7, and suppose in addition that the question whether $T \models \varphi$ for a quantifier-free sentence $\varphi \in \Phi$ can be decided in time $\mathrm{TIME}_{dec}(\varphi)$ polynomial in $\mathrm{length}(\varphi)$. Then for an arbitrary sentence $\varphi \in \Phi$, $\mathrm{TIME}_{dec}(\varphi)$ $\leqslant 2^{**}2^{**}d^{*}\mathrm{length}(\varphi)$ for some positive constant d.*

PROOF. Apply 1.7 to pass from φ to φ', and decide φ'.

With respect to the space required to decide $T \models \varphi$ for $\varphi \in \Phi$, we get the following simple exponential bound.

COROLLARY 1.9. *Assume the hypothesis of 1.6, and suppose in addition that the question whether $T \models \varphi$ for a quantifier-free sentence $\varphi \in \Phi$ can be decided in space $\mathrm{SPACE}_{dec}(\varphi)$ polynomial in $\mathrm{length}(\varphi)$. Then for an arbitrary sentence $\varphi \in \Phi$, $\mathrm{SPACE}_{dec}(\varphi)$ $\leqslant 2^{**}d^{*}\mathrm{length}(\varphi)$ for a positive constant d.*

PROOF. It suffices to produce in a suitable systematic fashion the tuples of variable-free Skolem terms corresponding to the quantifiers in the given sentence φ, and to decide the validity in T of the sentences φ^{*} obtained from φ by substituting (in the modified sense) these tuples for the corresponding bound variables. Since by 1.6(ii),

$$\mathrm{rank}(t) \leqslant (2^{**}d' \cdot \mathrm{quant}(\varphi)) \cdot \mathrm{rank}(\varphi),$$

we get from 1.4,

$$\mathrm{length}(\varphi^{*}) \leqslant \mathrm{atom}(\varphi) \cdot c' \cdot (2^{**}d' \cdot \mathrm{length}(\varphi)) \cdot \mathrm{rank}(\varphi) \cdot |X(\varphi)|^{2}$$
$$\leqslant 2^{**}d'' \cdot \mathrm{length}(\varphi)$$

for certain constants d', c', d''. So these decisions together with a suitable record of their outcome can be made in space bounded by $2^{**}d \cdot \mathrm{length}(\varphi)$ for some constant d.

A tighter complexity bound for the decision problem for Φ, T under the hypothesis of 1.8 is given by the *Berman complexity class* $\bigcup_{c \in \mathbb{N}} STA(*, 2^{cn}, n)$. $STA(*, 2^{cn}, dn)$ may be described as the class of all sets accepted by an alternating Turing machine running in time 2^{cn} which may make only $d \cdot n$ alternations of universal and existential states, where n is the length of the input (see Berman, 1977, 1980; Kozen, 1980). For any $d \in \mathbb{N}$, this class is contained in EXPSPACE. Then the argument given for 1.9 shows (comp. Kozen, 1980, pp. 234, 235):

THEOREM 1.10. *Assume the hypothesis of 1.6 and 1.8. Then the decision problem for Φ, T is in the class $\bigcup_{c \in \mathbb{N}} STA(*, 2^{cn}, n)$.*

The framework outlined so far is sufficient for handling linear problems in fields, ordered fields and valued fields. For multi-ordered, multi-valued fields, we now introduce a method of quantifier elimination via *multiple substitution of Skolem terms*.

Let L_0 be an elementary language, let L_i $(1 \leqslant i \leqslant r)$ be extensions of L_0 by new relation symbols such that $L_i \cap L_j = L_0$ for $i \neq j$, and let $L = \cup L_i$. Let $X \subseteq V$, Z a set of L_0-terms, and θ_i $(0 \leqslant i \leqslant r)$ pairwise disjoint sets of atomic L_i-formulas such that each triple X, Z, θ_i satisfies 1.1.

Let $\bar{\theta}_i$ be the closure of θ_i under $\wedge$, and let Φ be the closure of $\bigcup_{0 \leqslant i \leqslant r} \theta_i$ under $\wedge, \vee, \exists x, \forall x$ with $x \in X$ (note the absence of the negation). Then we define the *multiple substitution* of an $(r+1)$-tuple $(t_0, t_1, \ldots, t_r)$ of terms $t_i \in Z$ for a variable $x \in X$ in a quantifier-free formula $\varphi \in \Phi$ as follows: $\varphi(x//t_0, t_1, \ldots, t_r)$ is obtained from φ by replacing any term t occurring in an atomic subformula ψ of φ in θ_i by $t(x//t_i)$ for $0 \leqslant i \leqslant r$. If T_i are theories in L_i and

$$T \supseteq \bigcup_{0 \leqslant i \leqslant r} T_i,$$

then we say T satisfies the *abstract approximation theorem* (AAT) for $\theta_1, \ldots, \theta_r$, if for all $\vartheta_i \in \bar{\theta}_i$,

$$T \models \bigwedge_{1 \leqslant i \leqslant r} \exists x \, \vartheta_i \to \exists x \left(\bigwedge_{1 \leqslant i \leqslant r} \vartheta_i \right).$$

LEMMA 1.11. *Let T_i be Skolem theories in L_i with respect to X, Z, θ_i for $0 \leqslant i \leqslant r$, and assume*

$$T_0 \models \forall x \, \vartheta \vee \forall x \, \forall y \, ((\vartheta \wedge \vartheta(x//y)) \to x = y)$$

for $x, y \in X$, $\vartheta \in \theta_0$. Let

$$T \supseteq \bigcup_{0 \leqslant i \leqslant r} T_i,$$

and assume T has only infinite models and satisfies AAT with respect to $\theta_1, \ldots, \theta_r$.

(i) *Let $\varphi \in \Phi$ be quantifier-free, let Ψ_i be the set of atomic subformulas of φ in θ_i, and let S_i be a set of Skolem terms in L_0 for x, Ψ_i. Then*

$$(*) \qquad T \models \exists x(\varphi) \leftrightarrow \bigvee_{t_0 \in S_0} \varphi(x//t_0) \vee \bigvee_{t_1 \in S_1} \cdots \bigvee_{t_r \in S_r} \bigwedge_{t_0 \in S_0} \varphi(x//t_0, t_1, \ldots, t_r).$$

(ii) *Assume in addition:*

(1) *There are quantifier-free formulas $v_=, v_{R(x)}$ for any atomic formula $R(x_1, \ldots, x_n)$ in L with $x_i \in X$, such that $T \models x_1 \neq x_2 \leftrightarrow v_=$, $T \models \neg R(\mathbf{x}) \leftrightarrow v_R$. Moreover, $v_=, v_{R(x)}$ contain no negation.*

(2) *Modified substitution $(x, t, t') \mapsto t(x//t')$ and the assignment of sets S_i of Skolem terms to pairs x, Ψ_i with $\Psi_i \subseteq \theta_i$ is recursive.*

Then there exists a recursive quantifier elimination procedure for Φ, T.

PROOF. (i) Let A be a model of T in which all the free variables of $\exists x(\varphi)$ are interpreted. "$\Rightarrow$": Suppose $A \models \varphi$ when x is interpreted by $a \in A$. If $a = t_0$ for some $t_0 \in S_0$, we are done. Otherwise, for every $\psi \in \Psi_0$, either $A \models \neg \psi$ or $A \models \forall x \psi$, and so for every $t_0 \in S_0$, $A \models \psi \to \psi(x//t_0)$. For $1 \leqslant i \leqslant r$, find $t_i \in S_i$ with

$$A \models \bigwedge_{\psi \in \Psi_i} \psi \leftrightarrow \psi(x//t_i).$$

Then

$$A \models \bigwedge_{t_0 \in S_0} \varphi(x//t_0, t_1, \ldots, t_r).$$

"$\Leftarrow$": If $A \models \varphi(x//t_0)$ for some $t_0 \in S_0$, we are done. Otherwise, there exist $t_i \in S_i$ $(1 \leqslant i \leqslant r)$ such that

$$A \models \bigwedge_{t_0 \in S_0} \varphi(x//t_0, t_1, \ldots, t_r).$$

By choice of S_0, there exists $t_0 \in S_0$ such that for every $\psi \in \Psi_0$, $A \models \neg \psi(x//t_0) \vee \forall\, x\psi$. Let Ψ'_i be the set of all $\psi \in \Psi_i$ with $A \models \psi(x//t_i)$, and let $\mu_i = \bigwedge \Psi'_i$ for $1 \leqslant i \leqslant r$. By the AAT there exists $a \in A$ such that, when x is interpreted by a,

$$A \models \bigwedge_{1 \leqslant i \leqslant r} \mu_i.$$

So $A \models \psi(x//t_i) \to \psi$ for all $\psi \in \Psi_i$, $0 \leqslant i \leqslant r$, and so $A \models \varphi$.

(ii) We define a recursive map $\varphi \mapsto \varphi_{\mathrm{neg}}$ assigning to every quantifier-free formula $\varphi \in \Phi$ a quantifier-free formula φ_{neg} with $T \models \neg \varphi \leftrightarrow \varphi_{\mathrm{neg}}$: φ_{neg} is obtained from φ by interchanging $\wedge$ and $\vee$ in φ and replacing any atomic subformula $R(t_1, \ldots, t_n)$ of φ by $v_R(x_1//t_1, \ldots, x_n//t_n)$, and any equation $t_1 = t_2$ in φ by $v_=(x_1//t_1, x_2//t_2)$. A quantifier elimination procedure for Φ, T can now be described as follows:

Input: φ; Output: φ'; $\varphi' :=$ a negation-free formula equivalent to φ;
while φ' contains a quantifier *do*
   begin find the first quantifier $\exists\, x$ or $\forall\, x$ in φ', whose scope ψ is quantifier-free;
   if this quantifier is existential
   then replace $\exists\, x(\psi)$ in φ' by the corresponding right-hand side of $(*)$
   else begin compute ψ_{neg}; compute the corresponding right-hand side ψ' of $(*)$ for
      $\exists\, x(\psi_{\mathrm{neg}})$; compute ψ'_{neg}; replace $\forall\, x(\psi)$ in φ' by ψ'_{neg} *end*
  end
end.

Upper bounds for the complexity of the quantifier elimination procedure $\varphi \mapsto \varphi'$ provided by 1.11 and a resulting decision procedure for Φ, T can now be computed similar as for 1.5–1.10. Two additional complications occur:

(1) Due to the double formation of formulas ψ_{neg} during the elimination of one universal quantifier, the number of atoms of the resulting formula is additionally increased by a constant factor determined by the maximum of all numbers $\mathrm{atom}(v_=)$, $\mathrm{atom}(v_{R(x)})$. In the final outcome, this increase is absorbed in the double exponential growth of $\mathrm{atom}(\varphi')$ given by 1.6.

(2) For each existential quantifier, $(*)$ introduces an additional alternation of a disjunction and a conjunction. This leads to a doubling of the last parameter in the Berman complexity class, yielding $STA(*, 2^{cn}, 2n)$ instead of $STA(*, 2^{cn}, n)$.

We leave the details of the verification to the reader and state only the results required for section 4.

THEOREM 1.12. *Let L, T be as in 1.11(ii). Assume in addition:*

(1) *Modified substitution $(x, t, t') \mapsto t(x//t')$ can be performed in polynomial time.*
(2) *For every pair x, Ψ_i $(\Psi_i \subseteq \theta_i)$ a set S_i of Skolem terms can be computed in polynomial time such that $\mathrm{rank}(S_i)$ is linear in $\mathrm{rank}(\Psi_i)$.*
(3) *For quantifier-free sentences $\varphi \in \Phi$, "$T \models \varphi$" can be decided in polynomial time.*

Then:

(i) *There is a quantifier elimination procedure $\varphi \mapsto \varphi'$ for Φ, T with*

$$\mathrm{length}(\varphi') \leqslant 2{**}2{**}d{*}\mathrm{length}(\varphi), \quad \mathrm{TIME}_{\mathrm{qe}}(\varphi) \leqslant 2{**}2{**}d{*}\mathrm{length}(\varphi)$$

for some positive constant d.

(ii) The decision problem for Φ, T is in the Berman complexity class $\bigcup_{c \in \mathbb{N}} STA(*, c^{cn}, 2n)$.

In particular, it can be solved in exponential space and double exponential time.

2. Linear Problems in Fields and Ordered Fields

We let $L_F = \{0, 1, +, -, \cdot, ^{-1}\}$ be the elementary language of fields, and $L_{OF} = L_F \cup \{<\}$ the language of ordered fields. We treat $(\ )^{-1}$ as a total operation with the convention that $0^{-1} = 0$. V is the set of variables for formulas in L_F and L_{OF}. We fix an infinite subset X of V and a linear order $\prec$ of X. The variables in X will be denoted by $x, x', x_1, x_2, \ldots$, and called *linear variables*. A term t of L_F is *linear* if t is of the form $a_0 + a_1 x_1 + \ldots + a_n x_n$ with $x_i \in X$, $x_i \prec x_j$ for $i < j$, and a_i terms containing no linear variable. The *rank* of t is then defined as

$$\operatorname{rank}(t) = \max(\operatorname{length}(a_i) : 0 \leqslant i \leqslant n).$$

Deviating from the usual definition of addition, subtraction, multiplication and substitution for terms in L_F, we define these operations on linear terms in such a way that linearity is preserved. Let

$$t = a_0 + a_1 x_1 + \ldots + a_n x_n, \qquad t' = a'_0 + a'_1 x'_1 + \ldots + a'_{n'} x'_{n'}$$

be linear terms, and let c be a term containing no linear variable. Then

$$-t = -a_0 + -a_1 x_1 + \ldots + -a_n x_n, \qquad t + t' = a''_0 + a''_1 x''_1 + \ldots + a''_m x''_m,$$

where

$$\{x''_1, \ldots, x''_m\} = \{x_1, \ldots, x_n\} \cup \{x'_1, \ldots, x'_{n'}\}$$

ordered according to $\prec$, and $a''_0 = (a_0 + a'_0)$,

$$a''_h = \begin{cases} a_i & \text{if } x''_h = x_i \notin \{x'_1, \ldots, x'_{n'}\}, \\ a'_i & \text{if } x''_h = x'_i \notin \{x_1, \ldots, x_n\}, \qquad \text{for } 1 \leqslant h \leqslant m. \\ (a_i + a'_j) & \text{if } x''_h = x_i = x'_j, \end{cases}$$

$$c \cdot t = t \cdot c = ca_0 + ca_1 x_1 + \ldots + ca_n x_n.$$

Modified substitution for linear terms is defined as follows:

$$t(x//t') = t \quad \text{if } x \notin \{x_1, \ldots, x_n\};$$

$$t(x_j//t') = b_0 + b_1 x''_1 + \ldots + b_m x''_m$$

with

$$\{x''_1, \ldots, x''_m\} = (\{x_1, \ldots, x_n\} \setminus \{x_j\}) \cup \{x'_1, \ldots, x'_{n'}\}$$

ordered according to $\prec$, $b_0 = a_0 + a_j a'_0$,

$$b_h = \begin{cases} a_i & \text{if } x''_h = x_i \notin \{x'_1, \ldots, x'_{n'}\}, \\ a_j a'_i & \text{if } x''_h = x'_i \notin \{x_1, \ldots, x_n\}, \qquad \text{for } 1 \leqslant h \leqslant m. \\ (a_i + a_j a'_k) & \text{if } x''_h = x_i = x'_k, \end{cases}$$

Then the following properties are obvious:

LEMMA 2.1.

    (i) $\operatorname{rank}(-t) = \operatorname{rank}(t) + 1$,
    (ii) $\operatorname{rank}(t + t') \leqslant \operatorname{rank}(t) + \operatorname{rank}(t') + 3$,
    (iii) $\operatorname{rank}(c \cdot t) \leqslant \operatorname{length}(c) + \operatorname{rank}(t) + 1$,
    (iv) $\operatorname{rank}(t(x//t') \leqslant 2\operatorname{rank}(t) + \operatorname{rank}(t') + 3$,

(v) *all these operations can be performed in polynomial time,*
(vi) $\text{rank}(t) \leqslant \text{length}(t) \leqslant (\text{rank}(t)+1)\cdot|X(t)|$, *and so 1.3. holds*

We call a formula φ in $L_F(L_{OF})$ *linear,* if

(i) every atomic subformula of φ is of the form $t = 0$ (or $t > 0$) for a linear term t;
(ii) every bound variable in φ is linear.

If φ is a quantifier-free linear formula, x is a linear variable and t' is a linear term, then we denote by $\varphi(x//t')$ the formula obtained from φ by replacing any term t in φ by $t(x//t')$. Then $\varphi(x//t')$ is again linear, and in any field (ordered field) F, $\varphi(x//t')$ is equivalent to the formula $\varphi(x/t')$ obtained from φ by substituting t' for x in the usual manner.

We now have the following result on linear elimination in fields of characteristic zero and ordered fields.

THEOREM 2.2. *Let T be the theory T_{FO} of fields of characteristic 0 in L_F, or the theory T_{OF} of ordered fields in L_{OF}. There is a quantifier elimination procedure $\varphi \mapsto \varphi'$ for linear formulas with respect to T such that*

$$\text{length}(\varphi') \leqslant 2{*}{*}2{*}{*}c{*}\text{length}(\varphi) \quad and \quad \text{TIME}_{qe}(\varphi) \leqslant 2{*}{*}2{*}{*}c{*}\text{length}(\varphi)$$

for some positive constant c.

In the light of the general results 1.2, 1.6, 1.7, where Z is now the set of linear terms, $\Phi(\theta)$ is the set of (atomic) linear formulas, it suffices to prove the following facts.

LEMMA 2.3. *There exists a positive constant c and a natural number k such that for any finite set Ψ of atomic linear formulas and any linear variable x, Ψ, x has a set S of linear Skolem terms with respect to T with the following properties:*

(i) $|S| \leqslant c \cdot |\Psi|^k$,
(ii) $\text{rank}(t) \leqslant c \cdot \text{rank}(\Psi)$ *for any $t \in S$,*
(iii) *S can be constructed from Ψ, x in polynomial time.*

PROOF. By permuting summands, we may write $\psi \in \Psi$ in the form $ax+b\,\bar{\underset{(>)}{=}}\,0$ where $X(a) = \emptyset$, b is linear, $x \notin X(b)$. So we may assume that Ψ is of the form $\{ax+b\,\bar{\underset{(>)}{=}}\,0 : (a, b) \in I\}$, with linear terms b, $x \notin X(b)$, $X(a) = \emptyset$.

CASE 1: *Fields of characteristic 0. Since*

$$a = 0 \rightarrow (ax+b = 0 \leftrightarrow b = 0),$$
$$a \neq 0 \rightarrow (ax+b = 0 \leftrightarrow x = -b/a),$$

it suffices by 1.4 to find a set of Skolem terms for $\{x = -b/a : (a, b) \in I\}$. We claim that

$$S = \{-b/a, -b/a+1 : (a, b) \in I\}$$

is such a set. This is a consequence of the following lemma.

LEMMA 2.4. *Let F be a field, C a finite subset of F and let $\text{char}(F) = 0$ or $\text{char}(F) > |C|$. Then $\{c+1 : c \in C\} \not\subseteq C$.*

PROOF. $\{c+1 : c \in C\} \subseteq C$ implies that for any $c \in C$ there exist $h < k \leqslant |C|$ with $c+h = c+k$. So $\text{char}(F)$ divides $k-h$, and so $0 < \text{char}(F) \leqslant |C|$.

The proof of case 1 is now completed by the observation that $|S| \leqslant 2|\Psi|$,

$$\mathrm{rank}(t) \leqslant 10 \max(\mathrm{rank}(ax+b): (a, b) \in I),$$

and that all terms in S are linear.

CASE 2: *Ordered fields. Since*

$$a \geqslant 0 \rightarrow (ax+b > 0 \leftrightarrow x \geqslant -b/a),$$

it suffices by 1.4 to find a set of Skolem terms for

$$\{x = -b/a, \ x \geqslant -b/a : (a, b) \in I\}.$$

We claim that

$$S = \{-b/a+1, \ -b/a-1 : (a, b) \in I\} \cup \{(-b'/a'-b/a)/2 : (a, b), (a', b') \in I\}$$

is such a set. This follows from the following observation due to Ferrante & Rackoff (1975):

LEMMA 2.5. *Let C be a finite subset of an ordered field (or a 2-divisible ordered abelian group) F, and let $d \in F$. Then there exists*

$$e \in \{c+1, c-1 : c \in C\} \cup \{(c+c')/2 : c, c' \in C\}$$

such that for all $c \in C$, $d \gtreqqless c$ iff $e \gtreqqless c$.

PROOF. Let $C = \{c_1, \ldots, c_n\}$ with $c_1 \leqslant c_2 \leqslant \ldots \leqslant c_n$. If $d = c_i$, pick $e = c_i$; if $d < c_1$, pick $e = c_1 - 1$; if $d > c_n$, pick $e = c_n + 1$; if $c_i < d < c_{i+1}$, pick $d = (c_i + c_{i+1})/2$.

To complete the proof of 2.3, we notice that all terms in S are linear, $|S| \leqslant 3|I|^2$,

$$\mathrm{rank}(t) \leqslant 10 \max(\mathrm{rank}(ax+b): (a, b) \in I)$$

for $t \in S$.

THEOREM 2.6. *The decision problem for linear sentences in the theory T_{FO} of fields of characteristic 0 and in theory T_{FO} of ordered fields is in the Berman complexity class $\cup STA(*, 2^{cn}, n)$. In particular, it can be solved in exponential space and double exponential time.*

PROOF. By 1.8–1.10, it suffices to verify that the validity of an equation $t = 0$ and an inequality $t > 0$ for variable-free t can be tested in the field $\mathbb{Q}$ of rationals in polynomial time. For this purpose, one evaluates t as a quotient m/n of integers in binary expansion, and observes that the number of digits required for m and n can be bounded by 2 length(t).

COROLLARY 2.7. *There exists a positive constant c such that any linear sentence φ that holds in some field of characteristic 0, also holds in any field of characteristic p with $p > 2**2**c*\mathrm{length}(\varphi)$.*

PROOF. Reduce φ equivalently in T_{FO} to a quantifier-free linear sentence φ'. Then by 2.2, length$(\varphi') \leqslant 2**2**c*\mathrm{length}(\varphi)$, and by 2.4, the reduction holds in all fields of characteristic $> \mathrm{length}(\varphi')$. By the proof of 2.6, the atomic subformulas $t = 0$ of φ' may be written in the form $m/n = 0$, where the number of binary digits required for m and n is bounded by

$$\text{length}(t) = \text{rank}(t) \leqslant \text{rank}(\varphi').$$

So by 1.6(ii),

$$m, n \leqslant 2^{**}\text{rank}(\varphi') \leqslant 2^{**}2^{**}d^{*}\text{length}(\varphi)$$

for some positive constant d. So the truth-value of $m/n = 0$ will be the same in all fields of characteristic 0 or characteristic $> 2^{**}2^{**}c'^{*}\text{length}(\varphi)$ for $c' = \max(c, d)$.

3. Linear Problems in Discretely Valued Fields

A *discretely valued field* (F, v, Γ) is a field F with a Krull valuation $v : F \to \Gamma \cup \{\infty\}$, where the value group Γ has a smallest positive element $1 = 1_\Gamma$. We treat discretely valued fields as one-sorted structures for the language $L_{DVF} = L_F \cup \{\pi, \text{div}\}$, where π is a field constant of value 1_Γ (i.e. a uniformising parameter) and div is a *strict linear divisibility* relation, i.e. a div b iff $va < vb$ (cf. Macintyre *et al.*, 1983).† Accordingly, we define a *linear formula* in L_{DVF} as a formula φ of L_{DVF} such that

(i) every atomic subformula of φ is of the form $t = 0$ or t div t' with linear terms t, t';
(ii) every bound variable in φ is linear.

LEMMA 3.1. *The following hold in any valued field.*

(i) $\qquad\qquad\qquad vd \geqslant 0 \to (vx + vd < v(x+c) \leftrightarrow v(cd) < v(x+c)),$

(ii) $\qquad\qquad\qquad vd < 0 \to (vx + vd < v(x+c) \leftrightarrow vx + vd < vc),$

(iii)

$$v(ax+b) < v(a'x+b') \leftrightarrow \begin{cases} vb < vb' & \text{if } a = a' = 0, \\ v(b/a') < v(x+b'/a') & \text{if } a = 0,\ a' \neq 0, \\ v(x+b/a) < v(b'/a) & \text{if } a \neq 0,\ a' = 0, \\ v(a/a'(b'/a' - b/a)) < v(x+(b'/a' - b/a)), & \text{if } a, a' \neq 0,\ va \geqslant va', \\ v(x+b/a) < v(a'/a(b'/a' - b/a)), & \text{if } a, a' \neq 0,\ va < va'. \end{cases}$$

PROOF. (i) By the strong triangle inequality, $vd \geqslant 0$ and $vx + vd < v(x+c)$ implies $vx = vc$, and so $v(cd) < v(x+c)$; conversely, this implies $vc < v(x+c)$, and so $vx = vc$, and so $vx + vd < v(x+c)$.

(ii) "$\to$": Assume $vx + vd \geqslant vc$. Then $vx > vc$, and so $v(x+c) = vc \leqslant vx + vd$, a contradiction. "$\leftarrow$": If $vx \geqslant vc$, then $vx + vd < vc = v(x+c)$; if $vx < vc$, then $vx + vd < vx = v(x+c)$.

(iii) The first three equivalences are obvious; for the last two, observe that for $a, a' \neq 0$, $v(ax+b) < v(a'x+b')$ is equivalent to

$$v(x+b/a) + v(a/a') < v((x+b/a) + (b'/a' - b/a)),$$

and apply (i) and (ii).

Next, let T_{DVFp} be the theory of discretely valued fields F with residue class field F_v of characteristic p (p zero or prime) in the language L_{DVF}.

LEMMA 3.2. *Let x be a linear variable, let I be a finite set of pairs (c, d) of linear terms, and put*

$$\Psi = \{x + c \text{ div } d, d \text{ div } x + c : (c, d) \in I\}.$$

---

† The reason for taking this relation rather than $va \leqslant vb$ as basic is technical: it guarantees that the relation is open in F^2 under the valuation topology. This will be useful in the next section.

Then

$$S = \{-c, d-c, \pi d-c, \pi(c'-c)-c, \pi^{-1}d-c, \pi^{-1}(c'-c)-c, c-c'-c'' :$$
$$(c, d), (c', d'), (c'', d'') \in I\}$$

is a set of linear Skolem terms for Ψ, x *with respect to* T_{DVFp}, *provided* $p = 0$ *or* $p > |I|$.

PROOF. Let F be a model of T_{DVFp}, $a \in F$, and let all c, d with $(c, d) \in I$ be interpreted in F. We are going to show that for some $a' \in S$,

$$F \models \psi(a) \text{ iff } F \models \psi(a') \text{ for all } \psi(x) \in \Psi. \tag{*}$$

Choose $(c_0, d_0) \in I$ such that $v(a+c_0) = \max(v(a+c): (c, d) \in I)$.

CASE 1. *For all* $(c, d) \in I$, $v(c-c_0) \neq v(a+c_0)$.

SUBCASE 1.1. There is $(c, d) \in I$ with $vd \leqslant v(a+c_0)$. Pick $(c_1, d_1) \in I$ with

$$vd_1 = \max(vd : (c, d) \in I, vd \leqslant v(a+c_0)).$$

SUBSUBCASE 1.1.1. There is no $(c, d) \in I$ with $vd_1 < v(c-c_0) \leqslant v(a+c_0)$. Then we may put $a' = d_1 - c_0$ if $vd = v(a+c_0)$, and $a' = \pi d_1 - c_0$, if $vd_1 < v(a+c_0)$. Then the triangle inequality guarantees that

$$v(a+c) \gtrless vd \text{ iff } v(a'+c) \gtrless vd \text{ for all } (c, d) \in I.$$

SUBSUBCASE 1.1.2. There exists $(c_2, d_2) \in I$ such that

$$vd_1 < v(c_2-c_0) = \max(v(c-c_0) : v(c-c_0) < v(a+c_0)).$$

Then we put $a' = \pi(c_2-c_0)-c_0$ and again use the triangle inequality.

SUBCASE 1.2. $v(a+c_0) < vd$ for all $(c, d) \in I$. Pick $(c_1, d_1) \in I$ with

$$vd_1 = \min(vd : (c, d) \in I).$$

SUBSUBCASE 1.2.1. There is no $(c, d) \in I$ with $v(a+c_0) \leqslant v(c-c_0) < vd_1$. Then we put $a' = \pi^{-1}d_1 - c_0$.

SUBSUBCASE 1.2.2. There exists $(c_2, d_2) \in I$ with

$$v(a+c_0) < v(c_2-c_0) = \min(v(c-c_0) : (c, d) \in I) < vd_1.$$

Then we put $a' = \pi^{-1}(c_2-c_0)-c_0$. Again, the triangle inequality proves (*) in both subsubcases.

CASE 2. *There exist* $(c_1, d_1) \in J \subseteq I$ *such that for all* $(c, d) \in J$,

$$v(a+c_0) = v(c-c_0).$$

SUBCASE 2.1. $a = -c_0$; then we put $a' = -c_0$.

SUBCASE 2.2. $a \neq -c_0$, and hence $c \neq c_0$ for all $(c, d) \in J$. Put

$$J' = \{c-c_0 : (c, d) \in J\}$$

and notice that $0 \notin J'$.

CLAIM. *There exists $(c_2, d_2) \in J$ such that with $e_2 = c_2 - c_0$, $e_1 = c_1 - c_0$,*

$$v(e_2 + e_1 - e) = v e_1$$

for all $e \in J'$.

PROOF of the claim: If the claim fails, there exists a finite sequence $e_1, e_2, \ldots, e_k$ of elements of J' and $1 \leqslant j < k \leqslant |J'| + 1$ such that $e_j = e_k$ and $v(e_i + e_1 - e_{i+1}) > v e_1$ for $1 \leqslant i \leqslant k$. Let $f_i = \mathrm{res}(e_i/e_1) \neq 0$ be the residues of e_i/e_1 in the field F_v. Then $f_i + 1 = f_{i+1}$ for $1 \leqslant i \leqslant k$ and $f_j = f_k$. So $f_j + (k - j)1 = f_j$, and so $(k - j)1 = 0$, and so $p = \mathrm{char}(F_v)$ divides $(k - j)$, and so $0 \neq p \leqslant |I|$, a contradiction.

We now put

$$a' = -(e_2 + e_1 + c_0) = -(c_2 + c_1 - c_0).$$

Then for every $(c, d) \in J$,

$$v(a' + c) = v((c - c_0) - e_2 - e_1) = v e_1 = v(a + c),$$

and for $(c, d) \notin J$, $v(a' + c) = v(a + c)$ by the triangle inequality.

This completes the proof of lemma 3.2.

We can now prove a counterpart to theorem 2.2.

THEOREM 3.3. *There is a quantifier elimination procedure $\varphi \mapsto \varphi'$ for linear formulas in L_{DVF} with respect to T_{DVFO} such that*

$$\mathrm{length}(\varphi') \leqslant 2{**}2{**}c{*}\mathrm{length}(\varphi), \quad \mathrm{TIME}_{qe}(\varphi) \leqslant 2{**}2{**}c{*}\mathrm{length}(\varphi)$$

for some positive constant c. Moreover, the equivalence $\varphi \leftrightarrow \varphi'$ holds in T_{DVFp} for any $p > 2{**}2{**}c{*}\mathrm{length}(\varphi)$.

PROOF. By 1.2, 1.6, 1.7, it suffices to verify the hypothesis of lemma 2.3 for the language L_{DVF} and the theory T_{DVFO}. Since $t = 0$ is equivalent to $\neg(t \text{ div } 0)$, we may assume that Ψ consists entirely of linear atomic formulas of the form $ax + b$ div $a'x + b'$, say

$$\Psi = \{ax + b \text{ div } a'x + b' : (a, b, a', b') \in I\}.$$

By 1.4 and 3.1(iii), it suffices to consider instead Ψ', x with

$$\Psi' = \{x + c \text{ div } d, d \text{ div } x + c : (c, c) \in I'\}$$

with

$$I' = \{(b'/a', b/a'), (b/a, b'/a), ((b'/a' - b/a), a/a'(b'/a' - b/a)),$$
$$(b/a, a'/a(b'/a' - b/a)) : (a, b, a', b') \in I\}.$$

By 3.2, S can now be taken as

$$\{-c, d-c, \pi d-c, \pi(c'-c)-c, \pi^{-1}d-c, \pi^{-1}(c'-c)-c, c-c'-c'' :$$
$$(c, d), (c', d'), (c'', d'') \in I'\}.$$

Notice that I' consists entirely of pairs of linear terms; so all the terms in S are linear. Furthermore,

$$|S| \leqslant 4|I'| + 2|I'|^2 + |I'|^3 \leqslant 7|I'|^3 \leqslant 28|I|^3 = 28|\Psi|^3; \text{ for } t \in S,$$

$$\mathrm{rank}(t) \leqslant 10 \max(\mathrm{rank}(c), \mathrm{rank}(d) : (c, d) \in I') \leqslant 100 \, \mathrm{rank}(\Psi).$$

S is obviously constructible in polynomial time. The last statement of the theorem follows from the fact that in the equivalence proof for φ and φ', lemma 3.2 is applied to sets I of pairs of terms with

$$|I| \leqslant \text{length}(\varphi') \leqslant 2^{**}2^{**}c^*\text{length}(\varphi).$$

THEOREM 3.4. *The decision problem for linear sentences in the theory T_{DVFO} is in the Berman complexity class $\cup STA(*, 2^{cn}, n)$. In particular, it can be solved in exponential space and double exponential time.*

PROOF. Any model F of T_{DVFO} contains the rational function field $\mathbb{Q}(\pi)$, and for $f(X) = \Sigma f_i X^i \in \mathbb{Q}[X]$, the valuation v induced by F on $\mathbb{Q}(\pi)$ is determined by

$$vf(\pi) = k1_\Gamma \text{ iff } f_0 = \ldots = f_{k-1} = 0 \neq f_k.$$

Since any variable-free term t in L_{DVF} can be rewritten as a rational function term $f(\pi)/g(\pi)$ with $\deg(f), \deg(g) \leqslant \text{length}(t)$, equations $t = 0$ and relations t div t' can be decided in time polynomial in the length of t and t'. By 1.8–1.10, this proves the theorem.

COROLLARY 3.5. *There is a positive constant c, such that for any linear sentence φ in L_{DVF}, φ has the same truth-value in any model of T_{DVFp} with $p = 0$ or $p > 2^{**}2^{**}c^*\text{length}(\varphi)$.*

PROOF. Similar to the proof of 2.7.

Next, we consider discretely valued fields with finite residue fields.

Let $L_{DVF\alpha}$ be L_{DVF} extended by a constant α. For k a positive integer, p prime, we let $T_{DVFp,k}$ be the theory of all models F of T_{DVFp} in the language $L_{DVF\alpha}$, whose residue field F_v has p^k elements and is obtained from the prime field $\mathbb{F}_p$ by adjunction of α. Furthermore, we assume that for each (p, k), the irreducible polynomial $f_\alpha(X)$ of α is specified with coefficients in $\{0, \ldots, p-1\}$. In particular for $k = 1$, we assume $\alpha = 1$.

THEOREM 3.6. *Let p, k be fixed. There is a quantifier elimination procedure $\varphi \mapsto \varphi'$ for linear formulas in $L_{DVF\alpha}$ with respect to $T_{DVFp,k}$ such that*

$$\text{length}(\varphi') \leqslant 2^{**}2^{**}c^*\text{length}(\varphi), \quad \text{TIME}_{qe}(\varphi) \leqslant 2^{**}2^{**}c^*\text{length}(\varphi)$$

for some positive constant c.

PROOF. For the proof, it suffices to replace the application of lemma 3.2 in the proof of 3.3 by the following lemma.

LEMMA 3.7. *Under the hypothesis of 3.2,*

$$S = \{-c, d-c, \pi d-c, \pi(c'-c)-c, \pi^{-1}d-c, \pi^{-1}(c'-c)-c, g(\alpha)(c'-c)-c :$$
$$(c, d), (c', d') \in I, g(X)$$

polynomial of degree $< k$ with coefficients in $\{0, \ldots, p-1\}\}$ is a set of linear Skolem terms for Ψ, x with respect to $T_{DVFp,k}$.

The *proof* is the same as for 3.2, except in subcase 2.2: There is $(c_1, d_1) \in J \subseteq I$ such that for all $(c, d) \in J$,

$$\infty > v(a+c_0) = v(c-c_0) = \max(v(a+c) : (c, d) \in I).$$

Then $v((a+c_0)/(c-c_0)) = 0$, and so there is a polynomial $g(X)$ of degree $<k$ with coefficients in $\{0, \ldots, p-1\}$ such that

$$v((a+c_0)/(c-c_0)-g(\alpha)) > 0,$$

and so

$$v(a+c_0-(c-c_0)g(\alpha)) > v(a+c_0).$$

So with $a' = (c-c_0)g(\alpha)-c_0$, we find that $v(a+c) \geq vd$ iff $v(a'+c) \geq vd$.

Concerning the decision of linear sentences in models of $T_{DVFp,k}$, the situation is somewhat more intricate than in the characteristic zero case (cf. Weispfenning, 1985).

THEOREM 3.8. *Let F be a model of $T_{DVFp,k}$ such that for all polynomials $f(X, Y)$, $g(X, Y) \in \mathbb{Z}[X, Y]$ the relation $f(\pi, \alpha)$ div $g(\pi, \alpha)$ can be decided in polynomial time. Then the decision problem for linear sentences in F is in the Berman complexity class $\cup STA(*, 2^{cn}, n)$, and hence solvable in exponential space and double exponential time.*

PROOF. It suffices to remark that any variable-free term t in L_{DVF} can be rewritten in polynomial time as a rational function term $f(\pi, \alpha)/g(\pi, \alpha)$ with integer coefficients. The result follows now from the hypothesis and 1.10.

The hypothesis of 3.8 is satisfied, e.g. in the following cases:

3.9 (i) $\text{char}(F) = 0$, $\alpha = 1$, $\pi = p$; in particular for the p-adic valuation on $\mathbb{Q}$.

(ii) $\text{char}(F) = p$ and the residue field F_v is embedded in F. Then $vf(\pi, \alpha)$ is determined as the index of the lowest non-vanishing coefficient of $f(\pi, \alpha)$, when regarded as polynomial in π; in particular this is the case for any rational function field or Laurent series field over a finite field.

4. Linear Problems in Multiordered, Multivalued Fields

In this section, we consider fields with a finite number of independent orderings and discrete valuations. We describe such fields as structures for a language L^* formed as follows: $L^* = L(\text{Ord}, \text{Val})$ is obtained from the language L_F of fields by adding a finite set Ord of binary relation symbols $<, <', <_1, \ldots$ for orders, and finite families $\{\text{div}_v : v \in \text{Val}\}$ of binary relation symbols for strict linear divisibilities and $\{\pi_v, \alpha_v : v \in \text{Val}\}$ of field constants. The case that Ord or Val is empty is admitted.

Let us call a theory T in L^* *distinguished* if for any model F of T:

(i) F is a field of characteristic zero;

(ii) every $< \in \text{Ord}$ is a field ordering of F;

(iii) for every $v \in \text{Val}$, div_v is the strict linear divisibility associated with a discrete valuation v of F with uniformising parameter π_v;

(iv) for every $v \in \text{Val}$, *either* $\text{char}(F_v) = 0$ and $\alpha_v = 1$ *or* $F_v = \mathbb{F}_p(\alpha_v)$ is finite and T specifies p and an irreducible polynomial f_{α_v} for α_v over F_p; moreover, this decision is made uniformly for all models of T (but may vary with $v \in \text{Val}$).

If F is a model of a distinguished theory T in L^*, we let $\tau_<, \tau_v$ be the topologies associated with the orders $< \in \text{Ord}$ and the valuations $v \in \text{Val}$. We call these topologies *independent* if for every family

$$\{U_< : \, < \in \text{Ord}\} \cup \{U_v : v \in \text{Val}\}$$

of non-empty subsets of F such that $U_<$ is $\tau_<$-open and U_v is τ_v-open, it follows that

$$\bigcap_{<\,\in\,\mathrm{Ord}} U_< \cap \bigcap_{v\,\in\,\mathrm{Val}} U_v$$

is non-empty. By Stone's approximation theorem for V-topologies on fields (see Macintyre *et al.*, 1983), this is the case iff the topologies $\{\tau_< : \, <\in\mathrm{Ord}\}\cup\{\tau_v : v\in\mathrm{Val}\}$ are pairwise different. The independence of these topologies can be expressed by the following (linear) sentence IND in L^*: Let

$$\mathrm{Ord} = \{<_1, \ldots, <_n\}, \quad \mathrm{Val} = \{v_1, \ldots, v_m\}.$$

Then IND is the sentence

$$\forall x_1 \ldots x_{n+m} \, \forall y_1 \ldots y_{n+m} \left(\bigwedge_{1\leqslant i\leqslant n} y_i > 0 \wedge \bigwedge_{n<i\leqslant n+m} y_i \neq 0 \right)$$

$$\rightarrow \exists z \left(\bigwedge_{1\leqslant i\leqslant n} x_i - y_i \underset{i}{<} z \underset{i}{<} x_i + y_i \wedge \bigwedge_{n<i\leqslant n+m} y_i \, \mathrm{div}_{v_{i-n}} z - x_i \right).$$

Next, we describe how distinguished theories T in L^* fit into the framework of multiple Skolem terms and the abstract approximation theorem presented in section 1 (last part): L_0 is the language of fields extended by the constants π_v, α_v for $v\in\mathrm{Val}$. If Ord and Val are as above, then $L_i = L_0\cup\{\underset{i}{<}\}$ for $1\leqslant i\leqslant n$, and $L_i = L_0\cup\{\mathrm{div}_{v_i}\}$ for $n<i\leqslant n+m$. X is an infinite subset of V and Z is the set of linear terms (with respect to X) in L_0. θ_0 is the set of all linear equations $t = 0$ in L_0; for $1\leqslant i\leqslant n$, θ_i is the set of all linear inequalities $t \underset{i}{>} 0$ in L_i, and for $n<i\leqslant n+m$, θ_i is the set of all strict divisibilities $t \, \mathrm{div}_{v_{i-n}} t'$ between linear terms in L_i. T_0 is the theory T_{FO} in L_0; for $1\leqslant i\leqslant n$, T_i is T_{OF} in L_i; for $n<i\leqslant n+m$, T_i is T_{DVFO} or $T_{DVFp,k}$ in L_i. To complete the picture, we now show:

LEMMA 4.1. *Let T be a distinguished theory in L^*. Then $T\models IND \leftrightarrow AAT$, where AAT is the abstract approximation theorem of section 1.*

PROOF. "$\rightarrow$": Let $F\models T$, $\vartheta_i(x)\in\bar\theta_i$, $x\in X$. Let the free variables of ϑ_i, except x, be interpreted in F and let $\vartheta_i^F = \{a\in F : F\models\vartheta_i(a)\}\neq\emptyset$. Observe that terms in $\vartheta_i(x)$ are interpreted as linear functions in $F[x]$, and hence are continuous (with respect to all order and valuation topologies) at any point in F. So (by the way div_{v_i} are defined) all φ_i^F are open in their respective topology τ_i. Hence, by INT their intersection is non-empty, which proves AAT.

"$\leftarrow$": Suppose $F\models T$ and A_i are non-empty subsets of F that are open in the topology τ_i induced by $\underset{i}{<}$ for v_{i-n} for $1\leqslant i\leqslant n+m$. Pick $a_i\in A_i$ and a basic τ_i-open neighbourhood U_i of a_i (i.e. an interval or a circle) with $U_i\subseteq A_i$. Then there exist formulas $\vartheta_i(x)\in\bar\theta_i$ and an interpretation of the free variables of ϑ_i except x in F, such that $U_i = \vartheta_i^F$. Since

$$F \models \bigwedge_{1\leqslant i\leqslant n+m} \exists x\vartheta_i(x),$$

AAT entails

$$F \models \exists x \left(\bigwedge_{1\leqslant i\leqslant n+m} \vartheta_i \right),$$

and so

$$\bigcap_{1\leqslant i\leqslant n+m} A_i \neq \emptyset.$$

We are now in a position to apply theorem 1.12.

THEOREM 4.2. *Let T be a distinguished theory in L^* satisfying IND. Suppose that for every $v \in \mathrm{Val}$, for which F_v is finite in any model F of T, the relations $f(\pi_v, \alpha_v) \operatorname{div}_v g(\pi_v, \alpha_v)$ with $f(X, Y), g(X, Y) \in \mathbb{Z}[X, Y]$ are decidable in polynomial time. Then the following hold:*

(i) *There is a quantifier elimination procedure $\varphi \mapsto \varphi'$ for linear formulas in L^* with respect to T such that $\operatorname{length}(\varphi') \leqslant 2^{**}2^{**}c^*\operatorname{length}(\varphi)$, $\mathrm{TIME}_{qe}(\varphi) \leqslant 2^{**}2^{**}c^*\operatorname{length}(\varphi)$ for some positive constant c.*

(ii) *The decision problem for linear sentences in L^* with respect to T is in the Berman complexity class $\cup STA(*, 2^{cn}, 2n)$, and hence solvable in exponential space and double exponential time.*

For the case $\mathrm{Ord} = \emptyset$, we have a characteristic transfer principle similar to 2.7 and 3.5:

THEOREM 4.3. *Let $L^* = L(\emptyset, \mathrm{Val})$. Then there exists a positive constant c such that for every linear sentence φ in L^*, φ has the same truth-value in all fields F with independent discrete valuations $v \in \mathrm{Val}$ such that $\operatorname{char}(F_v) = 0$ or $\operatorname{char}(F_v) > 2^{**}2^{**}c^*\operatorname{length}(\varphi)$.*

PROOF. 1.6(ii) remains valid for the quantifier elimination procedure of 4.2(i), when T specifies $\operatorname{char}(F_v) = 0$ for all $v \in \mathrm{Val}$. The argument is now as for 2.7 and 3.5.

5. Lower Complexity Bounds

In the previous sections, we have shown that for various classes of fields with additional structure:

5.1 Quantifier elimination for linear formulas can be performed in double exponential space and time.

5.2 Transfer of linear statements φ from fields of characteristic zero to fields of prime characteristic p works for p at least double exponential in $\operatorname{length}(\varphi)$.

5.3 The decision problem for linear sentences is in the Berman complexity class $\cup STA(*, 2^{cn}, n)$ or $\cup STA(*, 2^{cn}, 2n)$.

The purpose of this section is to show that these upper complexity bounds are tight in the following sense:

5.1' Any quantifier elimination procedure for linear formulas in any of the fields considered so far requires double exponential space on a set of linear formulas of unbounded length in L_F.

5.2' There is a sequence ρ_n of linear sentences of unbounded length in L_F and a positive constant c such that ρ_n holds in any field of characteristic zero, but fails in fields of prime characteristic $< 2^{**}2^{**}c^*\operatorname{length}(\varphi)$.

5.3' The decision problem for linear sentences in fields of characteristic zero is $\cup STA(*, 2^{cn}, n)$—hard under polynomial time reductions. (So, for linear problems in multiordered, multivalued fields there remains a potential gap between upper and lower complexity bounds.)

The last statement 5.3' was essentially proved by Berman (1977, 1980) by analysing Fischer & Rabin (1974). He shows that $\cup STA(*, 2^{cn}, n)$ is polynomial-time reducible to the decision problem for the theory of reals in the language $\{0, 1, +, -, <\}$. His proof— as well as that of Fischer & Rabin—makes no essential use of the order relation, and hence holds for the language $L_{G1} = \{0, 1, +, -\}$. In this language, the theory of reals can be completely axiomatised by the axioms for torsion-free, divisible, abelian groups and the axiom $0 \neq 1$. This yields the following version of Berman's result which covers 5.3′:

THEOREM 5.4. *Let G be a torsion-free, divisible, abelian group with distinguished element $1 \neq 0$. G may carry additional structure for a language L extending L_{G1}. Then the decision problem for elementary sentences in the L-theory of G is $\cup STA(*, 2^{cn}, n)$—hard under polynomial time reductions.*

The proof of 5.1′ and 5.2′ also employs a construction of Fischer & Rabin (1974, thm. 8, cor. 9) in a somewhat extended setting.

LEMMA 5.5 (Fischer–Rabin). *There is a positive constant c and a sequence $\mu_n(x)$ of L_{G1}-formulas with one free variable x such that:*

(i) *In any abelian group G, where 1 is an element of infinite order,*

$$\mu_n^G = \{a \in G : G \models \mu_n(a)\} = \{0, 1, 2, \ldots, 2**2**n - 1\}.$$

(ii) *In any abelian group G with distinguished element 1,*

$$\{0, 1, 2, \ldots, 2**2**n - 1\} \subseteq \mu_n^G.$$

(iii) $\text{length}(\mu_{n+1}) \leqslant c(n+1)$.

PROOF. Fischer & Rabin define a sequence of L_{G1}-formulas $M_n(x, y, z)$ such that in the field of reals—and in fact in any abelian group G with a distinguished element 1 of infinite order, $G \models M_n(a, b, c)$ iff $a \in \{0, 1, \ldots, 2**2**n - 1\}$ and $a \cdot b = c$ (when a is regarded as a natural number). Moreover, $\text{length}(M_{n+1}) \leqslant c(n+1)$ for a suitable constant c, and M_n can be equivalently expressed by a positive existential formula. So the formulas $\mu_n(x) = M_n(x, 0, 0)$ satisfy the lemma.

We prove 5.2′ in the following more general form:

THEOREM 5.6. *There is a sequence ρ_n of sentences of unbounded length in L_{G1} and a positive constant c such that ρ_n holds in any abelian group, where 1 is an element of infinite order, and ρ_n fails in any abelian group, where 1 has order $\leqslant 2**2**c*\text{length}(\rho_n)$ for $n > 0$.*

PROOF. Let ρ_n be the sentence $\forall x(\mu_n(x) \to x + 1 \neq 0)$. Then for some constant $c > 0$, 5.5(iii) shows that $\text{length}(\rho_{n+1}) \leqslant c(n+1)$. By 5.5(i), ρ_n holds if 1 has infinite order. If

$$m = \text{order}(1) \leqslant 2**2**c^{-1}\text{length}(\rho_{n+1}) \leqslant 2**2**(n+1),$$

then by 5.5(ii), $G \models \mu_{n+1}(m-1)$, and so $G \models \neg \rho_{n+1}$.

Finally, 5.1′ will be a consequence of the following very general, but somewhat technical result 5.7. As in section 4, we call a finite, non-empty set Top of topologies τ on a set G *independent*, if whenever $\mathscr{U}$ is a set of non-empty subsets of G, then $\cap \mathscr{U} \neq \emptyset$, provided there is an injective map $\mathscr{U} \to \text{Top}$, $U \mapsto \tau_U$ such that U is τ_U-open. For $A \subseteq G$, $\tau \in \text{Top}$, $\delta_\tau(A) = cl_\tau(A) \cap cl_\tau(G \setminus A)$ denotes the *τ-boundary* of A.

THEOREM 5.7. *Let G be an abelian group with distinguished element 1 of infinite order. G may carry additional structure for a language L extending L_{G1}. Let Top be a finite independent* *set of T_1-topologies on G, such that for any $\tau \in \mathrm{Top}$, no $a \in G$ is τ-isolated. Assume, furthermore, that there is a positive constant d and a map $\psi(x) \mapsto \tau_\psi$ assigning to any atomic L-formula $\psi(x)$ in one variable x a topology $\tau_\psi \in \mathrm{Top}$ such that $|\delta_\tau(\psi^G)| \leqslant d \cdot \mathrm{length}(\psi)$. Let c and $\{\mu_n(x)\}$ be as in lemma 5.5. Then for any sequence $\{\sigma_n(x)\}$ of quantifier-free L-formulas with $G \models \mu_n \leftrightarrow \sigma_n$, we have $d \cdot \mathrm{length}(\sigma_n) \geqslant 2{**}2{**}c^{-1}\mathrm{length}(\mu_n)$ for positive n.*

PROOF. By 5.5, we know that $\mu_n^G = \{0, 1, \ldots, 2{**}2{**}n - 1\}$, and that $\mathrm{length}(\mu_{n+1}) \leqslant c \cdot (n+1)$ for some positive constant c. So it suffices to prove the following claim:

CLAIM 5.8. *Let $\sigma(x)$ be a quantifier-free L-formula in one variable x, and let Ψ be the set of all atomic subformulas of σ. If σ^G is finite, then*

$$\sigma^G \subseteq \cup\{\delta_{\tau_\psi}(\psi^G) : \psi \in \Psi\}.$$

Assuming the claim, we argue as follows: If $G \models \mu_n \leftrightarrow \sigma$ for a quantifier-free σ, then

$$\{0, \ldots, 2{**}2{**}n - 1\} = \mu_n^G = \sigma^G \subseteq \cup\{\delta_{\tau_\psi}(\psi^G) : \psi \in \Psi\},$$

and so

$$2{**}2{**}c^{-1}\mathrm{length}(\mu_{n+1}) \leqslant 2{**}2{**}(n+1) \leqslant \Sigma\,\{|\delta_{\tau_\psi}(\psi^G)| : \psi \in \Psi\}$$
$$\leqslant d \cdot \Sigma(\mathrm{length}(\psi) : \psi \in \Psi) = d \cdot \mathrm{length}(\sigma).$$

PROOF of the claim. Assume for a contradiction that σ^G is finite

$$a \in \sigma^G \backslash \cup\{\delta_{\tau_\psi}(\psi^G) : \psi \in \Psi\}.$$

Then $a \in \mathrm{int}_{\tau_\psi}(\psi^G) \cup \mathrm{int}_{\tau_\psi}(\neg\psi^G)$ for every $\psi \in \Psi$. By the hypothesis on Top, we find τ_ψ-open sets U_ψ with $U_\psi \subseteq ((\neg)\psi)^G$, $U_\psi \cap \sigma^G = \{a\}$. Let $V_\psi = \cap\{U_{\psi'} : \tau_{\psi'} = \tau_\psi\}$. Then $V_\psi \cap \sigma^G = \{a\}$, $V_\psi \subseteq ((\neg)\psi)^G$ and $V_\psi \neq V_{\psi'}$ implies $\tau_\psi \neq \tau_{\psi'}$. Since a is not τ_ψ-isolated, we find for each V_ψ a non-empty τ_ψ-open subset V'_ψ with $a \notin V'_\psi$. Since Top is independent, there exists $b \in \cap\{V'_\psi\}$. But then $b \in \psi^G$ iff $a \in \psi^G$ for all $\psi \in \Psi$. Consequently, $b \in \sigma^G$, since $a \in \sigma^G$, and so

$$b \in \sigma^G \cap \bigcap\{V'_\psi\} \subseteq \sigma^G \cap \bigcap\{U_\psi \backslash \{a\}\} = \emptyset.$$

This completes the proof of theorem 5.7.

VERIFICATION of 5.1′ by means of theorem 5.7:

Notice to begin with at any L_F-term $t(x)$ in one variable can be rewritten as a rational function term $f(x)/g(x)$ with integer coefficients and $\deg f$, $\deg g \leqslant \mathrm{length}(t)$.

If G is a field of characteristic zero, we may take $\mathrm{Top} = \{\tau\}$, where τ is the cofinite topology on G, and $d = 2$. This gives us a new proof of the fact (see Heintz, 1983) that no algebraically closed field admits quantifier elimination in less than double exponential space, for the characteristic zero case.

If G is an ordered field, we take $\mathrm{Top} = \{\tau_<\}$, where $\tau_<$ is the order topology, and observe that $\delta_{\tau_<}(t(x) > 0)^G \subseteq (t(x) = 0)^G$. This shows in particular that no real closed field admits quantifier elimination in less than double exponential space.

If G is a valued field, regarded as structure for a sublanguage of $LDVF$, possibly with additional constants, then we take $\mathrm{Top} = \{\tau_v\}$ with the valuation topology τ_v, and observe that

$$\delta_{\tau_v}(t(x) \text{ div } t'(x))^G \subseteq (t(x) = 0)^G \cup (t'(x) = 0)^G.$$

So we may take $d = 4$. The same argument is valid for languages including in addition nth root predicates $W_n(x) \leftrightarrow \exists z(z^n = x)$, provided G satisfies Hensel's lemma. For Hensel's lemma guarantees that $\delta_{\tau_v}(W_n(t(x))^G \subseteq (t(x) = 0)^G$. So none of the following fields admits quantifier elimination in its natural language (see Macintyre *et al.*, 1983; Weispfenning, 1985) in less than double exponential space: Algebraically closed valued fields of characteristic zero, p-adic fields, fields of Laurent series $F((t))$ over a real closed or algebraically closed field F of characteristic zero.

Finally, if G is a field with finitely many orderings $< \in \mathrm{Ord}$ and finitely many valuations $v \in \mathrm{Val}$ inducing different topologies on G, we take

$$\mathrm{Top} = \{\tau_< : \, < \in \mathrm{Ord}\} \cup \{\tau_v : v \in \mathrm{Val}\}$$

and argue as before.

6. Linear Problems of Bounded Dimension or Bounded Quantifier Alternation

We have defined the concept of a linear problem very broadly; accordingly, the upper and lower bounds on the complexity of these problems have turned out to be quite high. So it is natural to look for more restricted classes of linear problems that are still comprehensive enough for applications but computationally less complex.

Recall from 1.5 and 1.6 that the number of quantifiers $\mathrm{quant}(\varphi)$ occurring in a linear formula φ is the most decisive parameter for the computational complexity of φ. Since $\mathrm{quant}(\varphi)$ measures the number of essential variables of the problem φ, we refer to $\mathrm{quant}(\varphi)$ as the *dimension* of φ. A first obvious restriction on linear problems is thus to bound the dimension q of the problems considered. This has the drastic effect that all problems considered so far become solvable in polynomial time. More precisely, we have the following result.

THEOREM 6.1. *Let T be any of the theories of fields considered in sections 2, 3, 4 and let q be a fixed non-negative integer. Then the following hold:*

(i) *There is a quantifier elimination procedure $\varphi \mapsto \varphi'$ for linear formulas of dimension $\leqslant q$ with respect to T such that*

$$\mathrm{length}(\varphi') \leqslant c*(\mathrm{length}(\varphi))k \quad and \quad \mathrm{TIME}_{qe}(\varphi) \leqslant c*(\mathrm{length}(\varphi))^k$$

for some constants $c, k \in \mathbb{N}$.
(ii) *The decision problem for linear sentences of dimension $\leqslant q$ in the theory T is solvable in polynomial time.*

PROOF. The upper bounds in sections 2–4 are all based on the inequalities in 1.6, in particular the very generous bound in 1.6(iii) on the increase of length of a formula during quantifier elimination. Using the tighter inequality 1.5(iii) together with the uniform bound $\mathrm{quant}(\varphi) \leqslant q$, we immediately obtain $\mathrm{length}(\varphi') \leqslant c*(\mathrm{length}(\varphi))^k$ for some constants $c, k \in \mathbb{N}$. This bound applies to the quantifier elimination by multiple substitutions of Skolem terms in 1.12 as well, and thus proves (i). (ii) follows from (i) and the fact that for all theories considered, quantifier-free linear sentences can be decided in polynomial time.

For some applications, e.g. in computational geometry, a uniform bound on the dimension may be felt to be too restrictive. Consider, e.g. the following problem from robotics:

Input: $n, k \in \mathbb{N}$, a (not necessarily convex) polyhedron P in $\mathbb{R}^n$ with distinguished point $p \in P$, a polyhedral environment U of P in $\mathbb{R}^n$, a finite set D of vectors (directions) in $\mathbb{R}^n$ and a goal vector $g \in \mathbb{R}^n$.

Question: Does there exist a sequence of k translations of P along directions in D moving P into a position, where p coincides with g without collision with U?

The problem can be expressed by a linear formula φ in L_{OF} of the form $\exists x_1 \ldots \exists x_k \forall y_1 \ldots \forall y_{n+1}\psi$, where ψ is quantifier-free. So it is not sensible to bound k and hence quant(φ) in advance; on the other hand, φ will always contain only two *blocks* of quantifiers $\exists x_1 \ldots \exists x_k$ and $\forall y_1 \ldots \forall y_{n+1}$. A similar observation can be made with other problems from computational geometry. (In fact, no human being is likely to comprehend a linear sentence with, say, 10 alternating blocks of quantifiers, unless quantifiers can be "hidden" in defined concepts, for which a "higher level" intuition is available.) This motivates the study of linear formulas with a bounded number of alternating quantifier blocks.

A linear formula φ is *prenex* if it is of the form $Q_1 x_1 \ldots Q_n x_n\psi$, where Q_i are quantifiers $\exists, \forall$ and ψ is quantifier-free. By collecting adjacent quantifiers of the same kind into blocks, we may write φ in the form $\mathbf{Q}_1\mathbf{x}_1 \ldots \mathbf{Q}_a\mathbf{x}_a\psi$, where each $\mathbf{Q}_i\mathbf{x}_i$ is a *block of quantifiers* $\exists x_{i_1} \ldots \exists x_{i_{b_i}}$ or $\forall x_{i_1} \ldots \forall x_{i_{b_i}}$. We let $\Phi_a(\Phi_{a,b})$ be the set of all linear sentences φ in the language considered such that φ is prenex with at most a blocks of quantifiers (each comprising at most b single quantifiers). For formulas in Φ_a, the upper complexity bounds derived in sections 2 and 3 can then be improved as follows.

THEOREM 6.2. *Let T be any of the theories considered in sections 2 and 3, and let $0 < a, b \in \mathbb{N}$.*

(i) *There is a quantifier elimination procedure $\varphi \mapsto \varphi'$ for formulas $\varphi \in \Phi_{a,b}$ with respect to T such that*

$$\text{length}(\varphi') \leqslant \text{length}(\varphi) **(a*(c*b)**a)$$

and

$$\text{TIME}_{qe}(\varphi) \leqslant \text{length}(\varphi) **(a*(c*b)**a)$$

for some constant $c \in \mathbb{N}$.

(ii) *Linear sentences $\varphi \in \Phi_{a,b}$ can be decided in T in*

$$\text{TIME}_{dec}(\varphi) \leqslant \text{length}(\varphi) **(a*(c*b)**a)$$

for some constant $c \in \mathbb{N}$.

PROOF. Since quantifier-free linear sentences can be decided in T in polynomial time, (ii) is an immediate consequence of (i). To prove (i), we use an observation made by Reddy & Loveland (1978) for Presburger arithmetic: In all cases considered, the elimination of an existential quantifier is achieved via the equivalence

$$\exists x\varphi \leftrightarrow \bigvee_{t \in S} \varphi(x//t)$$

established in 1.2, where $S = S(x, \varphi)$ is a set of Skolem terms for x and the set Atom(φ) of atomic subformulas of φ. Let now y be a second linear variable,

$$\varphi_t = \varphi(x//t), \qquad \varphi' = \bigvee_{t \in S} \varphi_t.$$

Then we may replace the equivalence used so far

$$\exists y \exists x\varphi \leftrightarrow \exists y\varphi \leftrightarrow \bigvee_{t' \in S'} \varphi'(y//t'),$$

where $S' = S(y, \varphi')$, by the much more economical equivalence

$$\exists\, y\, \exists\, x\varphi \leftrightarrow \exists\, y\varphi' \leftrightarrow \bigvee_{t \in S} \exists\, y\varphi_t \leftrightarrow \bigvee_{t \in S}\; \bigvee_{t' \in S_t} \varphi_t(y//t'),$$

where $S_t = S(y, \varphi_t)$. The same applies to an arbitrary block of existential or universal quantifiers.

Suppose now $\varphi \in \Phi_{a,b}$ and φ' is obtained from φ by quantifier elimination in T employing the modified equivalence described above. Using the notation of 1.5, we then get the following bounds. Let

$$g_2^{(0)}(n, b) = n, \quad g_2^{(m+1)}(n, b) = (g(g_2^{(m)}(n, b)))^b \cdot g_2^{(m)}(n, b).$$

Then

$$\text{atom}(\varphi') \leqslant g_2^{(a)}(\text{atom}(\varphi), b)$$

and

$$\text{length}(\varphi') \leqslant c' \cdot |X(\varphi)| \cdot g_2^{(a)}(\text{length}(\varphi), b) \cdot h_1^{(a \cdot b)}(\text{length}(\varphi)).$$

Specialising g and h as in 1.6, this yields

$$\text{atom}(\varphi') \leqslant \text{atom}(\varphi)^{(kb+1)^a} \quad \text{and} \quad \text{length}(\varphi') \leqslant \text{length}(\varphi)^{a(cb)^a}$$

for some constant $c \in \mathbb{N}$.

REMARK. This proof does not work for the theories T of multivalued, multiordered fields considered in section 4, since here an existential quantifier is eliminated using a disjunction of conjunctions in place of a simple disjunction.

In Fürer (1982, theorem 5), is proved a lower bound for the decision of sentences of bounded quantifier alternation concerning real addition. His proof is valid for all the theories considered in sections 2–4:

THEOREM 6.3 (Fürer). *Let T be any of the theories considered in sections 2–4, and let $a \in \mathbb{N}$. There exists a positive constant c such that $\varphi \in \Phi_a$ cannot be decided in T in*

$$\text{NTIME}_{\text{dec}}(\varphi) \leqslant (\text{length}(\varphi)/a)^{**}\lfloor c \cdot a \rfloor.$$

A lower space bound for quantifier elimination on Φ_a can be obtained by the following variant of 5.5, which results from Fürer's modification of the Fischer–Rabin trick (Fürer, 1982, theorem 1].

LEMMA 6.3. *For all $0 < n$, $a \in \mathbb{N}$ there is a L_{G1}-formula $\mu_{n,a}(x)$ in Φ_{2a-1} such that:*

(i) *In any abelian group G, where 1 is an element of infinite order,*

$$\mu_{n,a}^G = \{0, 1, 2, \ldots, 2^{**}n^{**}a\}.$$

(ii) *$\text{length}(\mu_{n,a}) \leqslant c(a \cdot n \cdot \log n + 1)$ for some positive constant c.*

From 6.3(ii) we get $n \geqslant c'^{*}(\text{length}(\mu_{n,a})/a)^{**}(1 - \delta)$ for some $c' > 0$ and arbitrary $\delta > 0$. As a consequence, theorem 5.7 holds with μ_n replaced by $\mu_{n,a}$ and the lower bound $2^{**}2^{**}c^{-1}\text{length}(\mu_n)$ by $2^{**}(c'(\text{length}(\mu_{n,a})/a))^{**}(a(1-\delta))$. This yields the following variant of 5.1':

THEOREM 6.4. *Let $0 < a \in \mathbb{N}$ and let T be any of the theories considered in sections 2–4. Then there is a set M of formulas of unbounded length in Φ_{2a-1} and a positive constant c such* *that any quantifier elimination procedure* $\varphi \mapsto \varphi'$ *on M requires space*

$$\text{length}(\varphi') \geqslant 2**(c*\text{length}(\varphi)/a)**(a*(1-\delta))$$

for arbitrary $\delta > 0$.

It is well known (see von zur Gathen & Sieveking, 1976) that *existential* linear sentences can be decided in the theory T_{OF} of ordered fields in non-deterministic polynomial time. A use of the approximation theorem shows that this fact holds for multiordered fields as well. Is the corresponding statement true for the theories of valued fields considered in section 3?

References

Ben-Or, M., Kozen, D., Reif, J. (1986). The complexity of elementary algebra and geometry. *ACM Symp. on Computing*, 457–464.

Berman, L. (1977). Precise bounds for Presburger arithmetic and the reals with addition (preliminary report). *Proc. 18 IEEE Symp. FOCS*, pp. 95–99.

Berman, L. (1980). The complexity of logical theories. *Theor. Comp. Sci.* **11**, 71–77.

Brown, S. S. (1978). Bounds on transfer principles for algebraically closed and complete, discretely valued fields. *AMS Memoirs* **204**.

Collins, G. E. (1983). Quantifier elimination for real closed fields, a guide to the literature. In: (Buchberger, B., Collins, G. E., Loos, R., eds) *Computer Algebra,* 2nd edn. Berlin: Springer.

Cooper, D. C. (1972). Theorem-proving in arithmetic without multiplication. In: (B. Meltzer & D. Michie, eds.) *Machine Intelligence* 7, 91–100. Univ. of Edinburgh Press.

Ferrante, J., Rackoff, Ch. (1975). A decision procedure for the first order theory of real addition with order. *SIAM J. Comp.* **4**, 69–77.

Ferrante, J., Rackoff, Ch. (1979). The computational complexity of logical theories. *Springer Lec. Notes Math.* **718**.

Fischer, M. J., Rabin, M. O. (1974). Super-exponential complexity of Presburger arithmetic. *SIAM–AMS Proc.* **7**, 27–41.

Fürer, M. (1982). The complexity of Presburger arithmetic with bounded quantifier alternation depth. *Theor. Comp. Sci.* **18**, 105–111.

Heintz, J. (1983). Definability and fast quantifier elimination in algebraically closed fields. *Theor. Comp. Sci.* **24**, 239–277.

Kozen, D. (1980). Complexity of Boolean algebras. *Theor. Comp. Sci.* **10**, 221–247.

Macintyre, A., McKenna, K., van den Dries, L. (1983). Elimination of quantifiers in algebraic structures. *Adv. Math.* **47**, 74–87.

Point, F. (1983). *Quantifier elimination for projectable L-groups and linear elimination for rings.* Thèse, Mons, Belgium.

Prestel, A., Ziegler, M. (1978). Model theoretic methods in the theory of topological fields. *J. reine u. ang. Math.* **299/300**, 318–341.

Reddy, C. R., Loveland, D. W. (1978). Presburger arithmetic with bounded quantifier alternation. *Proc. 10 ACM Symp. Th. of Comp.*, pp. 320–325.

van den Dries. L. (1981). Quantifier elimination for linear formulas over ordered and valued fields. *Bull. Sos. Math. Belg.* **33**, ser. B, 19–32.

von zur Gathen, J., Sieveking, M. (1976). Weitere zum Erfüllungsproblem polynomial äquivalente kombinatorische Aufgaben. In: Komplexität von Entscheidungsproblemen. (Specker, E., Strassen, V., eds) *Springer Lec. Notes Comp. Sci.* 43.

Weispfenning, V. (1984). Aspects of quantifier elimination in algebra. In: *Universal Algebra and its Links.* Proc. 25. Arbeitstagung über Allgemeine Algebra, Darmstadt, 1983. Berlin: Heldermann.

Weispfenning, V. (1985). Quantifier elimination and decision procedures for valued fields. Proc. Logic Coll. '83, Aachen. Part I: Models and sets. (Müller, G. H., Richter, M. M., eds). *Springer Lec. Notes Math.* **1103**, 419–472.

Weispfenning, V. (1986). The complexity of elementary problems in archimedean ordered groups. Proc. EUROCAL '85, vol. 2. (Caviness, B. F., ed.) *Springer Lec. Notes Comp. Sci.* **204**, 87–88.

J. Symbolic Computation (1988) **5**, 29–35

Real Quantifier Elimination is Doubly Exponential

JAMES H. DAVENPORT AND JOOS HEINTZ

*School of Mathematical Sciences, University of Bath,
Claverton Down, Bath BA2 7AY, England and*

*F.B. Mathematik, J.W. Goethe Universität, D-6000 Frankfurt/Main F.R.G.
and Universidad Nacional de La Plata, La Plata, Argentina.*

(Received 7 March 1987)

We show that quantifier elimination over real closed fields can require doubly exponential space (and hence time). This is done by explicitly constructing a sequence of expressions whose length is linear in the number of quantifiers, but whose quantifier-free expression has length doubly exponential in the number of quantifiers. The results can be applied to cylindrical algebraic decomposition, showing that this can be doubly exponential. The double exponents of our lower bounds are about one fifth of the double exponents of the best-known upper bounds.

1. Introduction.

Tarski [1951] was the first to show that quantifier elimination was possible for the first-order theory of real-closed fields. His method was totally impractical, and was substantially modified and improved by Collins [1975], who gave a doubly-exponential algorithm for *cylindrical algebraic decomposition*, a process which divided $\mathbf{R}^n$ into a series of regions, on each of which the members of a given family of polynomials were identically zero, identically positive or identically negative. Here, and throughout, the expression *doubly exponential* refers to the dependence of the running time on n, the number of dimensions. The running time is polynomially dependent on all the other measures of size involved: the number of polynomials, their degree and the length of their coefficients.

It is the purpose of this paper to show that this doubly-exponential behaviour is intrinsic to the problem. This result has already been proved by Weispfenning [1985] by completely different methods, but we feel that our method is of independent interest. The method proceeds by adapting a construction of Heintz [1983] which was used to show that complex-valued quantifier elimination was doubly exponential. In the next section we construct the formulae, and explain the method, which is essentially a technique for writing $x_1^{2^{2^{k+1}}} = x_2$ in a form whose length, when considered as a dense polynomial, is only linear in k. The following section proves that any quantifier-free expression for this requires at least 2^{2^k} symbols if written densely. It should be noted that our formulation is different from that of Ben-Or *et al.* [1986], who take satisfiability as their criterion, i.e. they are considering formulae with no free variables. It seems that there is still much to be understood on the relationship between their view and ours.

This can be applied to cylindrical algebraic decomposition, since the corresponding decomposition of space requires at least one region for each zero of this polynomial. The result for cylindrical decomposition is independent of any considerations of density or sparsity, since we can show that the number of regions that have to be described is doubly exponential.

2. The Formulae.

We consider a sequence of formulae $\phi_0, \phi_1, \ldots, \phi_k$, each with four free variables. The even-numbered formulae will have x_{1R}, x_{1I}, x_{2R} and x_{2I} as free variables, while the odd-numbered formulae will have z_{1R}, z_{1I}, z_{2R} and z_{2I} as free variables. The basic formula ϕ_0 is given as

$$\left(x_{1R}^4 - 6x_{1R}^2 x_{1I}^2 + x_{1I}^4 = x_{2R}\right) \wedge \left(4x_{1R}^3 x_{1I} - 4x_{1R} x_{1I}^3 = x_{2I}\right).$$

While this formula may appear mysterious, it is simply an expression of the real and imaginary parts of the equality

$$(x_{1R} + ix_{1I})^4 = x_{2R} + ix_{2I}.$$

Hence we can regard ϕ_0 as being the *complex* equation

$$x_1^4 = x_2. \tag{1}$$

The rule for computing ϕ_{j+1} from ϕ_j depends on the parity of j: here we give the rule for even j, while that for odd j is obtained by interchanging the roles of x and z.

$$\phi_{j+1}(z_{1R}, z_{1I}, z_{2R}, z_{2I}) = \exists y_R \exists y_I \forall x_{1R} \forall x_{1I} \forall x_{2R} \forall x_{2I}$$
$$(((x_{1R} = z_{1R} \wedge x_{1I} = z_{1I} \wedge x_{2R} = y_R \wedge x_{2I} = y_I)$$
$$\vee (x_{1R} = y_R \wedge x_{1I} = y_I \wedge x_{2R} = z_{2R} \wedge x_{2I} = z_{2I}))$$
$$\Rightarrow \phi_j(x_{1R}, x_{1I}, x_{2R}, x_{2I})).$$

In its complex form, this equation reduces to

$$\phi_{j+1}(z_1, z_2) = \exists y \forall x_1 \forall x_2 \left(((x_1 = z_1 \wedge x_2 = y) \vee (x_1 = y \wedge x_2 = z_2)) \Rightarrow \phi_j(x_1, x_2)\right).$$

This formula is logically equivalent to the formula

$$\exists y \forall x_1 \forall x_2 \left(((x_1 = z_1 \wedge x_2 = y) \Rightarrow \phi_j(x_1, x_2)) \wedge ((x_1 = y \wedge x_2 = z_2) \Rightarrow \phi_j(x_1, x_2))\right).$$

In turn, we can simplify this, since the implications are trivially true unless x_1 and x_2 have the values given in the hypotheses of the implications. Hence $\phi_{j+1}(z_1, z_2)$ is logically equivalent to

$$\exists y \left(\phi_j(z_1, y) \wedge \phi_j(y, z_2)\right). \tag{2}$$

While we have only made these transformations on the complex version of ϕ_{j+1}, the same transformations could clearly be made on the real version.

Proposition 1. *The complex version of ϕ_j is equivalent to the equation*

$$x_1^{2^{2^{j+1}}} = x_2$$

(or the same equation in terms of z_1 and z_2 if j is odd).

The proof is a trivial induction on equations (1) and (2).

Corollary. ϕ_j is equivalent to the following logical formula, where $\mathcal{R}$ and $\mathcal{I}$ stand for the real part and the imaginary part respectively:

$$\mathcal{R}\left((x_{1R} + ix_{1I})^{2^{2^{j+1}}}\right) = x_{2R} \wedge \mathcal{I}\left((x_{1R} + ix_{1I})^{2^{2^{j+1}}}\right) = x_{2I}$$

(or the same equations written in terms of z_{1R}, z_{1I}, z_{2R} and z_{2I} if j is odd).

Although these equations are phrased in terms of complex variables, they do in fact have a purely real expression, as

$$\left(\sum_{l=0}^{2^{2^{j+1}-1}} (-1)^l \binom{2^{2^{j+1}}}{2l} x_{1R}^{2^{2^{j+1}}-2l} x_{1I}^{2l}\right) = x_{2R}$$

$$\wedge \left(\sum_{l=0}^{2^{2^{j+1}-1}-1} (-1)^l \binom{2^{2^{j+1}}}{2l+1} x_{1R}^{2^{2^{j+1}}-2l-1} x_{1I}^{2l+1}\right) = x_{2I}.$$

Proposition 2. $\phi_k(x_{1R}, x_{1I}, 1, 0)$ (or $\phi_k(z_{1R}, z_{1I}, 1, 0)$ if k is odd) defines a semi-algebraic set in $\mathbf{R}^2$ consisting of $2^{2^{k+1}}$ isolated points.

Each point of the set defined by ϕ_k corresponds to a $2^{2^{k+1}}$-th complex root of unity.

It is worth noting that the alphabet required to define ϕ_k is independent of k, and hence the length of ϕ_k is linear in k — $44k + 37$ symbols the way we have written it.

3. The size of a quantifier-free expression.

In the complex case, Heintz [1983] was able to argue that the only way to define a subset of $\mathbf{C}^1$ consisting of $2^{2^{k+1}}$ points is via a polynomial of that degree (or a set of polynomials of which that is the sum of the degrees). This is not so obvious in the real case, as a quantifier-free expression may well contain inequalities as well as equalities, and several levels of logical conjunctions and disjunctions, such as $A \wedge (B \vee C \vee (D \wedge E))$.

If $p(x, y)$ and $q(x, y)$ are two polynomials in $\mathbf{R}[x, y]$ we define $D(p)$ to be the set of isolated points of $p = 0$ and $D(p, q)$ to be the set of isolated points of the intersection of the curves $p = 0$ and $q = 0$ which are not isolated points of the curves considered separately. We note that $D(p_1 p_2) \subset D(p_1) \cup D(p_2)$ and that $D(p_1 p_2, q) \subset D(p_1, q) \cup D(p_2, q)$.

Proposition 3. Given any quantifier-free polynomial expression in two real variables x and y, involving polynomials $p_1(x, y)$, ..., $p_n(x, y)$, the set of isolated points of the subset of $\mathbf{R}^2$ defined by the formula is a subset of

$$\left(\bigcup_{i=1}^{m} D(q_i)\right) \cup \left(\bigcup_{i=1}^{m} \bigcup_{j=i+1}^{m} D(q_i, q_j)\right)$$

where the q_i are the irreducible factors of the p_i (in fact, it suffices to ensure that the q_i are without multiple factors, and relatively prime).

Proof. We can re-write the expression in terms of the q_i, using the facts that $fg = 0$ is equivalent to $f = 0 \vee g = 0$, and $fg > 0$ is equivalent to $(f > 0 \wedge g >$

$0) \vee (f < 0 \wedge g < 0)$. After applying the distributive laws to our expression, we can assume that each constituent of the expression is given by the conjunction of a number of elementary formulae of the form $q_i > 0$, $q_i = 0$ or their negations. Furthermore, the negation of $q_i > 0$ can be written as $q_i = 0 \vee q_i < 0$, and this term can be expanded with the other disjunctions, so that we need not consider such expressions. Hence each constituent of the expression is of the form

$$a_1 = 0 \wedge \ldots \wedge a_k = 0 \wedge b_1 > 0 \wedge \ldots \wedge b_l > 0 \wedge c_1 \neq 0 \wedge \ldots \wedge c_m \neq 0,$$

where the a_i, b_i and c_i are some of the q_i or their negatives. It is, of course, perfectly concievable that some of k, l or m could be zero. However, if

$$a_1 = 0 \wedge \cdots \wedge a_k = 0$$

does not define an isolated point (and possibly some other connected components), then the expression as a whole can not define an isolated point, since the inequalities are open conditions, while isolated points are closed. Hence the isolated points defined by the original expression are defined by the combination of certain equalities in the transformed expression (which may have come from inequalities in the original: for example

$$x^2 + y^2 - 1 \leq 0 \wedge y \geq 1$$

defines an isolated point $x = 0 \wedge y = 1$, but this will be transformed into the two equalities, as well as into three other sets that are actually void). In particular $k > 0$ for an isolated point. There are then two possibilities: either $a_1 = 0$ defines this isolated point, or it defines a curve which passes through this point. In the former case, the point is an element of $D(a_1)$, which may well contain other components as well. In the latter case, it would be natural to assume that the point was an element of $D(a_1, a_2)$, but this is not necessarily the case, since $a_2 = 0$ may define the same curve. However, there has to exist an i such that

$$a_1 = 0 \wedge \ldots \wedge a_{i-1} = 0$$

defines a curve passing through the point in question, but

$$a_1 = 0 \wedge \ldots \wedge a_i = 0$$

defines the point precisely. Since a_1 and a_i are relatively prime, $\{(x,y) : a_1(x,y) = a_i(x,y) = 0\}$ is a variety of dimension zero, and so consists only of isolated points. Thus the point belongs to $D(a_1, a_i)$ (as well as possibly to several other such sets).

Proposition 4. *If p_i is a polynomial of total degree d_i, the total number of isolated points is at most*

$$\left(\sum_{i=1}^{n} d_i \right)^2.$$

Proof. After the previous proposition, it suffices to show that

$$\left| \left(\bigcup_{i=1}^{m} D(q_i) \right) \cup \left(\bigcup_{i=1}^{m} \bigcup_{j=i+1}^{m} D(q_i, q_j) \right) \right| < \left(\sum_{i=1}^{n} d_i \right)^2.$$

If q_i is a polynomial of degree e_i, then we have that $\sum_{i=1}^{n} d_i \geq \sum_{i=1}^{m} e_i$. But $D(q_i)$ is the set of isolated real points of one equation, and any such point has to be a multiple point (since in the neighbourhood of the point we must have complex y values corresponding to real x values, and these must come in conjugate pairs). Hence the x-coordinate of any such point is a root of the discriminant of q_i. Furthermore, the number of y values of multiple points with a given x value is at most half the multiplicity of this root of the discriminant. Since the discriminant of a polynomial of total degree d has degree at most $2d^2$, we deduce that $|D(q_i)| \leq e_i^2$. An alternative approach proceeds via the Bezout inequality (Theorem 1 of Heintz [1983]). The multiple point is a common root of q_i and dq_i/dy, and hence belongs to a set of degree bounded by the products of the total degrees of q_i and dq_i/dy, i.e. $e_i(e_i - 1) \leq e_i^2$.

Similarly $D(q_i, q_j)$ is the set of isolated points of two equations, and any such point has an x coordinate which is a root of the resultant of q_i and q_j (which is non-zero by the irreducibility of the q_i). Furthermore, the number of common points with this x coordinate is at most the multiplicity of this root of the resultant. Since the degree of a resultant is at most twice the product of the degrees, we deduce that $|D(q_i, q_j)| \leq 2e_i e_j$. Again, it would be possible to proceed via Bezout's Inequality and an argument on total degree. Hence

$$\left| \left(\bigcup_{i=1}^{m} D(q_i) \right) \cup \left(\bigcup_{i=1}^{m} \bigcup_{j=i+1}^{m} D(q_i, q_j) \right) \right| \leq \left(\sum_{i=1}^{m} |D(q_i)| \right) + \left(\sum_{i=1}^{m} \sum_{j=i+1}^{m} |D(q_i, q_j)| \right)$$

$$\leq \left(\sum_{i=1}^{m} e_i^2 \right) + \left(\sum_{i=1}^{m} \sum_{j=i+1}^{m} 2e_i e_j \right)$$

$$= \left(\sum_{i=1}^{m} e_i \right)^2 .$$

Remarking that, in a dense representation of polynomials, a polynomial in two variables of total degree n requires at least n symbols to write it, we can combine propositions 2 and 4 to deduce

Theorem 1. *There exist formulae, of length linear in k, containing $6k$ quantifiers and two free variables, such that the real quantifier-free expressions corresponding to them require at least 2^{2^k} symbols to write down.*

We started ϕ_0 as a quartic equation in order to obtain an extra factor of two which would cancel with the square-root. If we had made ϕ_0 into a quadratic, the current result would have been $2^{2^{k-1}}$. Since every data item must be written, we deduce the following result.

Corollary. *The time required to eliminate n quantifiers over $\mathbf{R}$ is doubly exponential in n.*

4. Applications to Cylindrical Algebraic Decomposition.

It is well-known that quantifier elimination is an easy consequence of a cylindrical algebraic decomposition, and hence it would be possible to deduce that cylindrical algebraic decomposition requires doubly exponential time. In fact, we can proceed directly, since the cylindrical algebraic decomposition of $\mathbf{R}^{6k+2}$ induced by ϕ_k contains at least $2^{2^{k+1}}$ 0-dimensional regions — one corresponding to each root of unity. Hence we have proved

Theorem 2. *The time required to decompose $\mathbf{R}^{6k+2}$ cylindrically according to $8k+2$ polynomials of degree at most 4 is at least $2^{2^{k+1}}$.*

In terms of the dimensionality of the space, this becomes $2^{2^{(n+4)/6}}$. The exponent $(n+4)/6$ in this lower bound should be compared with Collins' [1975] upper bound of $2n+8$, since improved by McCallum [1985b] (see also McCallum [1985a]) to $n+\log n+7$, and by Davenport [1985] to $n + \log n + 5$.

It is possible to do slightly better than the previous results would indicate. For example, if we replace the induction rule for the ϕ_j (in its complex form) by

$$\phi_{j+1}(z_1, z_2) = \exists y \exists w \forall x_1 \forall x_2$$
$$(((x_1 = z_1 \wedge x_2 = y) \vee (x_1 = y \wedge x_2 = w) \vee (x_1 = w \wedge x_2 = z_2)) \Rightarrow \phi_j(x_1, x_2)),$$

we get a formula logically equivalent to

$$\exists y \exists w \left(\phi_j(z_1, y) \wedge \phi_j(y, w) \wedge \phi_j(w, z_2) \right),$$

and Proposition 2 would contain the number 4^{3^k} instead of $2^{2^{k+1}}$ (at the cost of needing two more symbols w_R and w_I and two more quantifiers at each step). Adding another two variables and quantifiers gives us $4^{4^k} = 2^{2^{2k+1}}$. Since we are now in dimension $10k + 2$, this gives us the following variant of Theorem 2:

Theorem 2'. *The time required to decompose $\mathbf{R}^{10k+2}$ cylindrically according to $16k+2$ polynomials of degree at most 4 is at least $2^{2^{2k+1}}$.*

In terms of the dimensionality of the space, this becomes $2^{2^{(n+3)/5}}$, which seems to be about the limit of this method.

5. Conclusions

There seems little to add to Theorems one and two. Clearly, there is still quite a gap between the exponents of these lower bounds and the best-known upper bounds, and it would be interesting to know how to narrow this. It would also be interesting to know whether quantifier elimination is still doubly-exponential if the number of bound variables is constant. Weispfenning [1985] has shown that, for *linear* formulae, bounding the number of quantifiers makes the problem polynomial, and bounding the number of alternations makes it simply exponential.

Note added in proof. Lars Langemyr (Stockholm) and the first author have applied McCallum's method to a special case, viz.

$$\exists c \forall b \forall a(((a = d \wedge b = c) \vee (a = c \wedge b = 1)) \Rightarrow a^2 = b),$$

which reduces to the formula $d^4 = 1$, whose *real* solution consists of $d = 1$ and $d = -1$. The equations (four linear and one quadratic) induce a cellular algebraic decomposition of $\mathbf{R}^4$ into 3837 cells. The computation took 2 hours (7202.8 seconds) on an otherwise unloaded discless SUN 3/160 (with 8 Mb real memory). SAC-2 needed a 4 Mb heap to compute the decomposition and the formulae for the cells. Further details of this calculation are to appear in the SIGSAM Bulletin.

The authors wish to thank the Centre de Calcul de l'Esplanade, Université Louis Pasteur, Strasbourg, where this work was done, and in particular Maurice Mignotte, who introduced the authors to each other. The first author was visiting the Centre de Mathématiques de l'École Polytechnique, Palaiseau, France at the time.

References

Ben-Or, M., Kozen, D. & Reif, J. (1986). The Complexity of Elementary Algebra and Geometry. J. Comput. Syst. Sci. **32** 251–264.

Collins, G. E. (1975). Quantifier Elimination for Real Closed Fields by Cylindrical Algebraic Decomposition. Proc. 2nd. GI Conf. Automata Theory & Formal Languages (Springer Lecture Notes in Computer Science 33) pp. 134–183. MR 55 #771.

Davenport, J. H. (1985). Computer Algebra for Cylindrical Algebraic Decomposition. TRITA–NA–8511, NADA, KTH, Stockholm, Sept. 1985.

Heintz, J. (1983). Definability and Fast Quantifier Elimination in Algebraically Closed Fields. Theor. Comp. Sci. **26** (1983) pp. 239–277. MR 85a:68061.

McCallum, S. (1985a). An Improved Projection Operation for Cylindrical Algebraic Decomposition. Proc. EUROCAL 85, Vol. 2 (Springer Lecture Notes in Computer Science 204) pp. 277–278.

McCallum, S. (1985b). An Improved Projection Operation for Cylindrical Algebraic Decomposition. Computer Science Tech. Report 548, Univ. Wisconsin at Madison, Feb., 1985.

Tarski, A. (1951). A Decision Method for Elementary Algebra and Geometry. 2nd. ed., Univ. Cal. Press, Berkeley, 1951. MR 10 #499.

Weispfenning, V. (1985). The Complexity of Linear Problems in Fields. Manuscript, 1985 (revised March 1986).

J. Symbolic Computation (1988) **5**, 37–64

Solving Systems of Polynomial Inequalities in Subexponential Time

D. Yu. GRIGOR'EV AND N. N. VOROBJOV (Jr)

*Leningrad Department of Mathematical Steklov Institute
of the Academy of Sciences of the USSR,
Fontanka embankment, 27, Leningrad, 191011, USSR and
Leningrad State University, Universitetskaya embankment, 7/9,
Leningrad, 199164, USSR*

(*Received* 20 *March* 1985)

Let the polynomials $f_1, \ldots, f_k \in \mathbb{Z}[X_1, \ldots, X_n]$ have degrees $\deg (f_i) < d$ and absolute value of any coefficient of f_i less than or equal to 2^M for all $1 \le i \le k$. We describe an algorithm which recognises the existence of a real solution of the system of inequalities $f_1 \ge 0, \ldots, f_k \ge 0$. In the case of a positive answer the algorithm constructs a certain finite set of solutions (which is, in fact, a representative set for the family of components of connectivity of the set of all real solutions of the system). The algorithm runs in time polynomial in $M(kd)^{n^2}$. The previously known upper time bound for this problem was $(Mkd)^{2^{O(n)}}$.

Introduction

The problem of finding real solutions of systems of polynomial inequalities is of known significance for symbolic computation. For the first time the decidability of this problem was proven by Tarski (1951). However, the time-bound of the algorithm from Tarski (1951) is non-elementary (in particular, the time-bound cannot be estimated by any tower of exponents). Later, exponential-time algorithms were devised for this problem (Collins, 1975; Wüthrich, 1976). In fact, Tarski (1951), Collins (1975) and Wüthrich (1976) consider a more general problem, namely, quantifier elimination in the first order theory of real closed fields.

In the present paper we describe a subexponential-time algorithm for finding real solutions of systems of polynomial inequalities (see also Vorobjov & Grigor'ev, 1985). This algorithm essentially involves the subexponential-time algorithm for solving systems of polynomial equations over an algebraically closed field (Chistov & Grigor'ev, 1983a,b; Chistov, 1984; Grigor'ev, 1984; see also Chistov & Grigor'ev, 1984; Grigor'ev, 1987). Before the papers by Chistov & Grigor'ev (1983a, b) only exponential-time algorithms were known for the latter problem (see e.g. Collins, 1975; Wüthrich, 1976; Heintz, 1983). On the other hand, it is clear that the problem of solving systems of inequalities over real numbers is more general than the problem of solving systems of equations, for example, over the field of complex numbers.

Let $f_1, \ldots, f_k \in \mathbb{Z}[X_1, \ldots, X_n]$ be input polynomials. An algorithm described in the present paper finds a certain set of solutions in $\mathbb{R}^n$ (or indicates their absence) of a system of inequalities

$$f_1 > 0, \ldots, f_m > 0, f_{m+1} \ge 0, \ldots, f_k \ge 0. \tag{1}$$

0747–7171/88/010037+28 $03.00/0

A rational function $g \in \mathbb{Q}(Y_1, \ldots, Y_s)$ can be represented as $g = g_1/g_2$ where the polynomials $g_1, g_2 \in \mathbb{Z}[Y_1, \ldots, Y_s]$ are relatively prime. Denote by $l(g)$ the maximum bit lengths of the (integer) coefficients of the polynomials g_1, g_2. Throughout this paper we suppose that the following inequalities are valid:

$$\deg_{X_1, \ldots, X_n}(f_i) < d, \; l(f_i) \leqslant M, \quad 1 \leqslant i \leqslant k. \tag{2}$$

We estimate the size of system (1) by the value $\mathscr{L} = kMd^n$ (cf. Chistov & Grigor'ev, 1983a, b; Chistov, 1984; Grigor'ev, 1984; also Chistov & Grigor'ev, 1984).

The running time of the algorithms for solving systems of inequalities from Collins (1975) and Wüthrich (1976) is bounded by $(Mkd)^{2^{O(n)}}$.

The notation $h_1 \leqslant \mathscr{P}(h_2, \ldots, h_t)$ for functions $h_1, \ldots, h_t > 0$ means that for suitable natural numbers q, p an inequality $h_1 \leqslant p(h_2 \ldots h_t)^q$ is true.

A subset in $\mathbb{R}^n$ is called semi-algebraic if it consists of all points in $\mathbb{R}^n$ satisfying an appropriate quantifier-free formula Π with the atomic subformulas of the form $(g_j \geqslant 0)$ with $g_j \in \mathbb{R}[X_1, \ldots, X_n]$. We denote this subset by $\{\Pi\} \subset \mathbb{R}^n$.

Let

$$\mathscr{V} = \{(f_1 > 0) \& \ldots \& (f_m > 0) \& (f_{m+1} \geqslant 0) \& \ldots \& (f_k \geqslant 0)\} \subset \mathbb{R}^n$$

be a semi-algebraic set consisting of all solutions of system (1). The set $\mathscr{V}$ is decomposable (uniquely) in the disjoint union of its components of connectivity $\mathscr{V}_i$, i.e.

$$\mathscr{V} = \bigcup_i \mathscr{V}_i.$$

Moreover, every set $\mathscr{V}_i$ is also semi-algebraic (see e.g. Collins, 1975; Wüthrich, 1976). A finite set $\mathscr{T} \subset \mathscr{V}$ is called a representative set for $\mathscr{V}$ (or in other words, for the system (1)) iff for each index i the intersection $\mathscr{V}_i \cap \mathscr{T} \neq \phi$. Denote by $\tilde{\mathbb{Q}} \subset \mathbb{R}$ the field of all real algebraic numbers. The main result of the present paper is the following theorem (see also Vorobjov & Grigor'ev, 1985; Grigor'ev, 1987).

THEOREM. *There is an algorithm which, for any system of inequalities of the kind (1), satisfying (2), produces some representative set $\mathscr{T} \subset \mathscr{V} \cap \tilde{\mathbb{Q}}^n$ with a number of points not exceeding $\mathscr{P}((kd)^{n^2})$. The running time of the algorithm is less than $\mathscr{P}(M, (kd)^{n^2}) \leqslant \mathscr{P}(\mathscr{L}^{\log^2 \mathscr{L}})$ (i.e. the time-bound is subexponential in $\mathscr{L}$). For every point $(\chi_1, \ldots, \chi_n) \in \mathscr{T}$ the algorithm constructs a corresponding polynomial $\Phi \in \mathbb{Q}[Z]$, that is irreducible over $\mathbb{Q}$, and the expressions*

$$\chi_i = \chi_i(\omega) = \sum_j \beta_j^{(i)} \omega^j \in \mathbb{Q}[\omega],$$

where $\beta_j^{(i)} \in \mathbb{Q}$, $1 \leqslant i \leqslant n$, $0 \leqslant j < \deg(\Phi)$ and $\omega \in \tilde{\mathbb{Q}}$, $\Phi(\omega) = 0$. Besides that, the algorithm produces a pair of rational numbers $b_1, b_2 \in \mathbb{Q}$ such that inside the interval $(b_1, b_2) \subset \mathbb{R}$ there is a unique real root $\omega \in (b_1, b_2)$ of the polynomial Φ. In addition, the equality

$$\omega = \sum_{1 \leqslant j \leqslant n} \lambda_i \chi_i(\omega)$$

is fulfilled for certain natural numbers $1 \leqslant \lambda_i \leqslant \deg(\Phi)$, $1 \leqslant i \leqslant n$. Finally, the polynomials and expressions constructed satisfy the following bounds:

$$\deg(\Phi) \leqslant \mathscr{P}((kd)^n); \quad l(\Phi), l(\chi_i(\omega)), l(b_1), l(b_2) \leqslant M\mathscr{P}((kd)^n).$$

REMARK. Based on the description of the points constructed in the theorem and using, e.g. Heindel (1971), one can find, for any rational $0 < \delta \leqslant 1$, rational δ-approximations to the points from the set $\mathscr{T}$ within time $\mathscr{P}(\log(1/\delta), M, (kd)^{n^2})$.

For the proof of the theorem we need the algorithms from Chistov & Grigor'ev (1982, 1983a, b), Chistov (1984), Grigor'ev (1984), also Chistov & Grigor'ev (1984) on polynomial factoring and on solving systems of algebraic equations. Now we formulate exactly these results. Taking into account that only fields of characteristic zero are considered in the present paper, and in order to avoid complex formulas due to inseparable fields extensions, we restrict ourselves here to the zero characteristic case.

Thus, consider a ground field $F = \mathbb{Q}(T_1, \ldots, T_e)[\eta]$, where the elements $T_1, \ldots, T_e$ are algebraically independent over $\mathbb{Q}$, the element η is algebraic over the field $\mathbb{Q}(T_1, \ldots, T_e)$. Denote by

$$\varphi = \sum_{0 \leqslant i \leqslant \deg_Z(\varphi)} (\varphi_i^{(1)}/\varphi^{(2)}) Z^i \in \mathbb{Q}(T_1, \ldots, T_e)[Z]$$

its minimal polynomial over $\mathbb{Q}(T_1, \ldots, T_e)$ with leading coefficient $lc_z(\varphi) = 1$, where $\varphi_i^{(1)}, \varphi^{(2)} \in \mathbb{Z}[T_1, \ldots, T_e]$ and the degree $\deg(\varphi^{(2)})$ is the least possible. Any polynomial $f \in F[X_1, \ldots, X_n]$ can be uniquely represented in a form

$$f = \sum_{0 \leqslant i < \deg_z(\varphi);\, i_1, \ldots, i_n} (a_{i, i_1, \ldots, i_n}/b) \eta^i X_1^{i_1} \ldots X_n^{i_n},$$

where $a_{i, i_1, \ldots, i_n}, b \in \mathbb{Z}[T_1, \ldots, T_e]$ and the degree $\deg(b)$ is the least possible. Define

$$\deg_{T_j}(f) = \max_{i_1, \ldots, i_n} \{\deg_{T_j}(a_{i, i_1, \ldots, i_n}), \deg_{T_j}(b)\}.$$

Let

$$\deg_{X_m}(f) < \tau, \quad \deg_{T_j}(f) < \tau_2, \quad \deg_{T_j}(\varphi) < \tau_1, \quad \deg_z(\varphi) < \tau_1,$$

$$l(f) \leqslant M_2, \quad l(\varphi) \leqslant M_1 \quad \text{for all} \quad 1 \leqslant m \leqslant n, \quad 1 \leqslant j \leqslant e.$$

As the size $L_1(f)$ of the polynomial f we consider in proposition 1 the value $\tau^{n+e} \tau_2^e \tau_1 M_2$ and analogously $L_1(\varphi) = \tau_1^{e+1} M_1$.

PROPOSITION 1 (Chistov & Grigor'ev, 1982, 1984; Chistov, 1984; Grigor'ev, 1984). *One can factor a polynomial f over F within time polynomial in the sizes $L_1(f)$, $L_1(\varphi)$. Furthermore, for any divisor $f_1 | f$ where a polynomial $f_1 \in F[X_1, \ldots, X_n]$ has a certain coefficient equal to 1, the following bounds are true:*

$$\deg_{T_j}(f_1) \leqslant \tau_2 \mathscr{P}(\tau, \tau_1), \quad l(f_1) \leqslant (M_1 + M_2 + e\tau_2 + n) \mathscr{P}(\tau, \tau_1).$$

For the cases when the field F is finite or F is a finite extension of $\mathbb{Q}$, other polynomial-time algorithms for factoring are described in Lenstra (1984).

Now we proceed to the problem of solving systems of algebraic equations. Let the input system $f_1 = \ldots = f_k = 0$ be given, where the polynomials $f_1, \ldots, f_k \in F[X_1, \ldots, X_n]$. Let

$$\deg_{X_1, \ldots, X_n}(f_i) < d, \quad \deg_{T_1, \ldots, T_e, Z}(\varphi) < d_1,$$

$$\deg_{T_1, \ldots, T_e}(f_i) < d_2, \quad l(f_i) \leqslant M_2 \quad \text{for all} \quad 1 \leqslant i \leqslant k.$$

As the size L of the system is proposition 2 we consider the value $kd^n d_1 d_2^e M_2 + d_1^{e+1} M_1$.

The variety $\mathscr{W} \subset \bar{F}^n$ of all roots (defined over the algebraic closure $\bar{F}$ of the field F) of the system $f_1 = \ldots = f_k = 0$ is decomposable as the union of its components

$$\mathscr{W} = \bigcup_\alpha W_\alpha$$

defined and irreducible over the field F. The algorithm from proposition 2 finds the components W_α and outputs every W_α in the two following manners: by its general point (see below) and, on the other hand, by a certain system of algebraic equations such that W_α coincides with the variety of all roots of this system.

Let $W \subset \bar{F}^n$ be a closed variety of dimension $\dim W = n - m$ defined and irreducible over F. Denote by $t_1, \ldots, t_{n-m}$ some algebraically independent elements over F. A general point of the variety W can be given by the following field isomorphism:

$$F(t_1, \ldots, t_{n-m})[\theta] \simeq F(X_1, \ldots, X_n) = F(W), \qquad (*)$$

where the element θ is algebraic over the field $F(t_1, \ldots, t_{n-m})$. Denote by $\Phi(Z) \in F(t_1, \ldots, t_{n-m})[Z]$ its minimal polynomial over $F(t_1, \ldots, t_{n-m})$ with leading coefficient $lc_z(\Phi) = 1$. The elements $X_1, \ldots, X_n$ are considered as the rational (coordinate) functions on the variety W. Under the isomorphism $(*)$ $t_i \to X_{j_i}$ for suitable $1 \leqslant j_1 < \ldots < j_{n-m} \leqslant n$, where $1 \leqslant i \leqslant n - m$. Besides, θ is the image under isomorphism $(*)$ of an appropriate linear function $\sum_{1 \leqslant i \leqslant n} c_i X_i$, where c_i are integers. The algorithm from proposition 2 represents the isomorphism $(*)$ by the integers $c_1, \ldots, c_n$ and apart from that by the images of the coordinate functions $X_1, \ldots, X_n$ in the field $F(t_1, \ldots, t_{n-m})[\theta]$. Sometimes in the formulation of proposition 2 we identify a rational function with its image under the isomorphism.

PROPOSITION 2 (Chistov & Grigor'ev, 1983a, b, 1984; Chistov, 1984; Grigor'ev, 1984). *An algorithm can be designed which produces a general point of every component W_α and constructs a certain family of polynomials $\psi_\alpha^{(1)}, \ldots, \psi_\alpha^{(N)} \in F[X_1, \ldots, X_n]$ such that W_α coincides with the variety of all roots of the system $\psi_\alpha^{(1)} = \ldots = \psi_\alpha^{(N)} = 0$. Denote by $n - m = \dim W_\alpha$, $\theta_\alpha = \theta$, $\Phi_\alpha = \Phi$ (see $(*)$). Then $\deg_z(\Phi_\alpha) \leqslant \deg(W_\alpha) \leqslant d^m$, for all j, s, the degrees*

and
$$\deg_{T_1, \ldots, T_e, t_1, \ldots, t_{n-m}}(\Phi_\alpha), \deg_{T_1, \ldots, T_e, t_1, \ldots, t_{n-m}}(X_j), \deg_{T_1, \ldots, T_e}(\psi_\alpha^{(s)}) \leqslant d_2 \mathscr{P}(d^m, d_1),$$

$$\deg_{X_1, \ldots, X_n}(\psi_\alpha^{(s)}) \leqslant d^{2m}.$$

The number of equations $N \leqslant m^2 d^{4m}$. Furthermore,

and
$$l(\Phi_\alpha), l(X_j) \leqslant (M_1 + M_2 + (n + e)d_2)\mathscr{P}(d^m, d_1)$$

$$l(\psi_\alpha^{(s)}) \leqslant (M_1 + M_2 + ed_2)\mathscr{P}(d^n, d_1).$$

Finally, the total running time of the algorithm can be bounded by $\mathscr{P}(M_1, M_2, (d^n d_1 d_2)^{n+e}, k)$. Obviously, the latter value does not exceed $\mathscr{P}(L^{\log L})$, in other words, is subexponential in the size.

The contents of the paper are briefly as follows. In section 1 a device is introduced for justifying the calculations with infinitesimals which are involved below in sections 2, 3. Some properties of semi-algebraic sets over ordered extensions by infinitesimals of the field $\tilde{\mathbb{Q}}$ are ascertained.

In section 2 an algorithm is suggested that produces a representative set for the variety of all real roots of a given polynomial. For this purpose an infinitesimal "perturbation" of the initial polynomial is considered, so that the variety of all the real roots of the "perturbed" polynomial turns out to be a smooth hypersurface. The algorithm finds on this hypersurface points with some fixed directions of the gradient, solving an appropriate system of algebraic equations over an algebraically closed field with the help of proposition 2.

In section 3 we prove at first some bounds on real algebraic solutions of the system (1).

After that, for the given system (1), the algorithm yields a relevant polynomial and applies the construction from section 2 in order to produce a representative set for the variety of real roots of this polynomial. Then among the points produced the algorithm picks out all the points satisfying system (1). This completes the proof of the theorem.

In section 4 an outline of the whole algorithm is given, omitting some details covered in sections 2, 3.

1. Calculations with Infinitesimals

Let K, in the course of this section, denote an arbitrary real closed field (see e.g. Lang, 1965) and an element $\varepsilon > 0$ infinitesimal relatively to the elements of the field K, i.e. for any positive element $0 < \alpha \in K$ in the ordered field $K(\varepsilon)$ the inequalities $0 < \varepsilon < \alpha$ are valid. Obviously, the element ε is transcendental over K. For an ordered field K_1 we denote by $\tilde{K}_1 \supset K_1$ its unique (up to isomorphism) real closure, preserving the order on K_1 (see e.g. Lang, 1965).

Let us remind some well-known statements about real closed fields. A Puiseux series (or in other words power-fractional series) over the field K is a series of the kind

$$a = \sum_{i \geqslant 0} \alpha_i \varepsilon^{v_i/\mu},$$

where $0 \neq \alpha_i \in K$ for all $i \geqslant 0$, the integers $v_0 < v_1 < \ldots$ increase and the natural number $\mu \geqslant 1$. The field $K((\varepsilon^{1/\infty}))$ consisting of all Puiseux series (with added zero) is real closed, and hence $K((\varepsilon^{1/\infty})) \supset \widetilde{K(\varepsilon)} \supset K(\varepsilon)$. Besides, the field $K[\sqrt{-1}]((\varepsilon^{1/\infty})) = \bar{K}((\varepsilon^{1/\infty}))$ is algebraically closed (here and further a bar over a field denotes its algebraic closure).

If $v_0 < 0$, then the element $a \in K((\varepsilon^{1/\infty}))$ is infinitely large; if $v_0 > 0$, then a is infinitesimal (relatively to the elements of the field K). A vector $(a_1, \ldots, a_n) \in (K((\varepsilon^{1/\infty})))^n$ is called K-finite if each coordinate a_i $(1 \leqslant i \leqslant n)$ is not infinitely large relatively to the elements of K. For any K-finite element $a \in K((\varepsilon^{1/\infty}))$ its standard part $st(a) \in K$ is definable, namely $st(a) = \alpha_0$ in the case $v_0 = 0$ and $st(a) = 0$ if $v_0 > 0$. Analogously one can define the standard part of a Puiseux series from the field $\bar{K}((\varepsilon^{1/\infty}))$. For any K-finite vector $(a_1, \ldots, a_n) \in (K((\varepsilon^{1/\infty})))^n$ its standard part is defined by an equality

$$st(a_1, \ldots, a_n) = (st(a_1), \ldots, st(a_n)).$$

For a set $W \subset (K((\varepsilon^{1/\infty})))^n$ consisting of only K-finite vectors we define

$$st(W) = \{st(w) : w \in W\}.$$

It is well known (Tarski, 1951) that all real closed fields are elementary equivalent and that any extension between real closed fields is elementary. This means that if K_1, K_2 are real closed fields where $K_1 \subset K_2$ and Π is any closed formula (without free variables) of the first order theory of the field K_1, then the truth values of Π in the fields K_1 and K_2 coincide. We refer below to this statement as to the "transfer principle". Sometimes the first order theory of the real closed fields is called Tarski algebra.

Now we shall demonstrate, how the transfer principle can work and show (a known fact) that any semi-algebraic set over a real closed field K can be represented uniquely as a union of its components of connectivity, each in its turn being a semi-algebraic set. Consider a semi-algebraic set $W = \{\Pi\} \subset K^n$, determined by a quantifier-free formula Π of Tarski algebra with the atomic subformulas of the kind $(f \geqslant 0)$, where the polynomials $f \in K[X_1, \ldots, X_n]$. By the format of the formula Π we mean the sum of the number of its variables, the number of atomic subformulas and the degrees of the polynomials f.

In the case of the field $K = \mathbb{R}$ the set W is uniquely representable as the union of its components of connectivity

$$W = \bigcup_i W_i,$$

where every W_i is in its turn a semi-algebraic set (and connected in the euclidean topology). From the papers by Collins (1975) and Wüthrich (1976) one can deduce the existence of a function $\mathfrak{R}$ such that if the format of the formula Π is less than $\mathcal{N}$, then the number of the components W_i is less than $\mathfrak{R}(\mathcal{N})$ and, moreover, one can find quantifier-free formulas Π_i of Tarski algebra, each of format less than $\mathfrak{R}(\mathcal{N})$, such that $W_i = \{\Pi_i\}$. Indeed, the algorithms from Collins (1975) and Wüthrich (1976) allow to produce a cylindrical algebraic decomposition of a semi-algebraic set and as a corollary to produce the decomposition in the components of connectivity. For a given format $\mathcal{N}$ of an initial formula (with symbolic coefficients) each of the two algorithms can be represented as a rooted tree (directed outward the root) having vertices either with the out-degree one or out-degree three. To the root corresponds the initial formula, to any vertex of the tree with out-degree one corresponds an arithmetic operation, finally, to any vertex with out-degree three corresponds a polynomial. The computation for an arbitrary initial formula, with the specified coefficients substituted instead of the symbolic ones, proceeds along a suitable path of the tree starting from the root, performing the corresponding arithmetic operation in a vertex with out-degree one, and branching in a vertex with out-degree three according to the sign of the corresponding polynomial. This representation as a tree provides the desired function $\mathfrak{R}$.

Thus, for a given $\mathcal{N}$, one can obtain a formula $\Omega_{\mathcal{N}}$ of Tarski algebra (for the case of the field $K = \mathbb{R}$), expressing the existence of a decomposition of any semi-algebraic set $W = \{\Pi\}$ with the format of Π less than $\mathcal{N}$, into its components of connectivity

$$W = \bigcup_i \{\Pi_i\}$$

such that the format of every Π_i, and the number of them, are less than $\mathfrak{R}(\mathcal{N})$. Moreover, the formula $\Omega_{\mathcal{N}}$ states that for each pair of indices $i \neq j$ the components $\{\Pi_i\}$ and $\{\Pi_j\}$ are "separated", i.e. the following formula of Tarski algebra is valid:

$$\forall\,((a_1, \ldots, a_n) \in \{\Pi_i\})\ \exists\, z > 0 (\forall\, (b_1, \ldots, b_n) \in \{\Pi_j\})(\sum_{1 \leqslant l \leqslant n} (a_l - b_l)^2 \geqslant z).$$

Besides, the formula $\Omega_{\mathcal{N}}$ claims the "connectedness" of every $\{\Pi_i\}$, this means that there do not exist two "separated" semi-algebraic subsets of $\{\Pi_i\}$, each determined by a quantifier-free formula of Tarski algebra with format less than $\mathfrak{R}(\mathfrak{R}(\mathcal{N}))$.

Apart from that, for given $\mathcal{N}, \mathcal{M}$ one can prove (for the case of the field $K = \mathbb{R}$) a formula $\Omega_{\mathcal{N},\mathcal{M}}$ of Tarski algebra expressing the following. If $\{\Pi\}$ (where the format of Π is less than $\mathcal{N}$) can be represented as a union of more than one and less than $\mathcal{M}$ pairwise "separated" semi-algebraic sets, each being determined by a quantifier-free formula of Tarski algebra of format less than $\mathcal{M}$, then $\{\Pi\}$ can be represented as a union of more than one and less than $\mathfrak{R}(\mathcal{N})$ pairwise "separated" semi-algebraic "connected" sets, each being determined by a quantifier-free formula of Tarski algebra of format less than $\mathfrak{R}(\mathcal{N})$.

Applying the transfer principle to all the formulas $\Omega_{\mathcal{N}}, \Omega_{\mathcal{N},\mathcal{M}}$, one concludes that any semi-algebraic set (over a real closed field K) can be uniquely represented as a union of its pairwise "separated" "components of connectivity", moreover, each component is semi-algebraic and is "connected", i.e. cannot be represented as a union of a finite number of pairwise "separated" semi-algebraic sets. Below we utilise the terms "connected semi-algebraic set" and "components of connectivity of a semi-algebraic set" without quotation marks since the notion of connectedness in any topology will not be considered.

Denote by

$$\mathscr{D}_w(R) = \{(X_1 - w_1)^2 + \ldots + (X_n - w_n)^2 \leqslant R^2\}$$

the closed ball of radius $R \geqslant 0$ with the centre in the point $w = (w_1, \ldots, w_n)$.

LEMMA 1.

(a) *Any semi-algebraic set $W \subset (\widetilde{K(\varepsilon)})^n$ consisting only of K-finite points lies in a ball $\mathscr{D}_0(R)$ for a certain radius $R \in K$.*

(b) *Let $V \subset K^n$, $W \subset (\widetilde{K(\varepsilon)})^n$ be semi-algebraic sets and let, apart from that, W consist only of K-finite points and $st(W) = V$. Let*

$$V = \bigcup_m V_m, \quad W = \bigcup_l W_l$$

be the decompositions of the sets V, W respectively, into their components of connectivity. Then, for every index m, there exist such indices $l_1, \ldots, l_s$ that $st(W_{l_1} \cup \ldots \cup W_{l_s}) = V_m$. Moreover, for each index l there is a unique index m such that $st(W_l) \subset V_m$.

PROOF. (a) Assume that, on the contrary, for some index $1 \leqslant i \leqslant n$ and any $\alpha \in K$, there is a point $w = (w_1, \ldots, w_n) \in W$ such that the absolute value $|w_i| > \alpha$. Consider the projection $\pi : (\widetilde{K(\varepsilon)})^n \to \widetilde{K(\varepsilon)}$ on the coordinate X_i. Then $\pi(W) \subset \widetilde{K(\varepsilon)}$ is a semi-algebraic set (Tarski, 1951). Therefore, $\pi(W)$ coincides with the union of a finite number of intervals (maybe endless in one or both sides). Consider the extreme right interval (analogously one can consider the extreme left interval). It cannot be of the form $\{X_i > a\}$ or $\{X_i \geqslant a\}$, otherwise the set W would contain points with infinitely large coordinate X_i. Thus, the extreme right interval has one of the four following forms: $\{a \leqslant X_i \leqslant b\}$, or $\{a < X_i \leqslant b\}$, or $\{a \leqslant X_i < b\}$, or $\{a < X_i < b\}$. If the element b is infinitely large (relatively to the field K), then there exists an infinitely large element $c \in \widetilde{K(\varepsilon)}$ such that $a < c < b$ in the case when $a < b$. Otherwise, put $c = a = b$, which again leads to a contradiction. Hence, the element b is K-finite. The extreme left interval satisfies the analogous property. So, we arrive at a contradiction with the assumption at the beginning of the proof.

(b) First of all we shall show that for any semi-algebraic set $V \subset K^n$ and its component of connectivity V_1, one can construct an open (in the topology with the base consisting of all open balls) semi-algebraic set $U \subset K^n$ such that $V_1 \subset U$ and, besides that, for every point $v \in V \setminus V_1$ there exists $\tau > 0$, for which the intersection $U \cap \mathscr{D}_v(\tau) = \phi$ (in other words, $V \cap \bar{U} = V_1$, where $\bar{U}$ denotes the topological closure of U). Namely, as U one can take the set of all points $u \in K^n$, satisfying the following requirement. There exist $0 \leqslant \tau^{(u)} \in K$, $0 < \tau_0^{(u)} \in K$ and a point $v_1 \in V_1$ such that the distance $\|v_1 - u\| \leqslant \tau^{(u)}$ and for each point $v \in V \setminus V_1$ the distance $\|v - u\| \geqslant \tau^{(u)} + \tau_0^{(u)}$.

The latter requirement can be expressed by a formula of Tarski algebra, therefore the set $U \subset K^n$ is semi-algebraic (Tarski, 1951). The set $V_1 \subset U$ since, for any point $v_1 \in V_1$, one can put $\tau^{(v_1)} = 0$ and take $\tau_0^{(v_1)}$ from the definition of "separation" (see above). Now let us show that U is open. Let a point $u \in U$, let us prove the inclusion $\mathscr{D}_u(\tau_0^{(u)}/3) \subset U$. For every point $u_1 \in \mathscr{D}_u(\tau_0^{(u)}/3)$ put $\tau^{(u_1)} = \tau^{(u)} + \tau_0^{(u)}/3$, $\tau_0^{(u_1)} = \tau_0^{(u)}/3$. Then these $\tau^{(u_1)}$, $\tau_0^{(u_1)}$ are the required ones. Indeed, there is such a point $v_1 \in V_1$ that $\|u - v_1\| \leqslant \tau^{(u)}$, henceforth,

$$\|u_1 - v_1\| \leqslant \|u_1 - u\| + \|u - v_1\| \leqslant \tau^{(u_1)}.$$

Apart from that, for each point $v \in V \setminus V_1$, the following inequalities are valid:

$$\|v - u_1\| \geqslant \|v - u\| - \|u - u_1\| \geqslant \tau^{(u)} + \tau_0^{(u)} - \tau_0^{(u)}/3 = \tau^{(u_1)} + \tau_0^{(u_1)}.$$

Finally, let a point $v \in V \setminus V_1$. According to the definition of the "separation", there exists $0 < \tau_1 \in K$ such that the intersection $\mathscr{D}_v(\tau_1) \cap V_1 = \phi$. Let us check that $\mathscr{D}_v(\tau_1/2) \cap U = \phi$. Suppose the contrary, let a certain point $u_2 \in \mathscr{D}_v(\tau_1/2) \cap U$. Then, by virtue of the requirement formulated above,

$$\tau^{(u_2)} + \tau_0^{(u_2)} \leqslant \|u_2 - v\| \leqslant \tau_1/2,$$

but on the other hand,

$$\mathscr{D}_{u_2}(\tau_1/2) \cap V_1 \subset \mathscr{D}_v(\tau_1) \cap V_1 = \phi,$$

therefore, $\tau^{(u_2)} > \tau_1/2$. The obtained contradiction completes the proof of the properties of the constructed set U.

In order to prove (b) it is sufficient to show that for any component of connectivity W_l there is a unique component V_m containing the set $st(W_l)$. Assume that, on the contrary, there exist for definiteness some points

$$v_1^{(0)} \in V_1 \cap st(W_l), \quad v_2^{(0)} \in V_2 \cap st(W_l).$$

Consider points $w_1^{(0)}, w_2^{(0)} \in W_l$ such that $st(w_1^{(0)}) = v_1^{(0)}$, $st(w_2^{(0)}) = v_2^{(0)}$. The semi-algebraic set constructed above U is $\{\Pi\} \subset K^n$ for a relevant quantifier-free formula Π of Tarski algebra. Introduce a semi-algebraic set $U^{(\varepsilon)} = \{\Pi\} \subset (\widetilde{K(\varepsilon)})^n$ determined by the same formula Π. Let a point $v_1 \in V_1$. There exists $0 < \tau_1 \in K$ such that $\mathscr{D}_{v_1}(\tau_1) \subset U$. In other words, the following statement is true:

$$\forall x \, (\|x - v_1\| \leqslant \tau_1 \Rightarrow \Pi(x)).$$

The latter statement can be expressed by a formula of Tarski algebra. Hence, this formula is true also over the field $\widetilde{K(\varepsilon)}$ by the transfer principle. This entails the inclusion $\mathscr{D}_{v_1}(\tau_1) \subset U^{(\varepsilon)}$. Let a point $v_2 \in V \setminus V_1$. Then for a suitable $0 < \tau_2 \in K$ the intersection $\mathscr{D}_{v_2}(\tau_2) \cap U = \phi$. Reasoning analogously as above, one can conclude that $\mathscr{D}_{v_2}(\tau_2) \cap U^{(\varepsilon)} = \phi$. Thus, if a point $w_1 \in W_l \cap U^{(\varepsilon)}$, then $st(w_1) \in (V \cap U) = V_1$, and if a point $w_2 \in W_l \setminus U^{(\varepsilon)}$, then $st(w_2) \in V \setminus U = V \setminus V_1$, taking into account that the standard parts $st(w_1), st(w_2)$ are definable and the distances $\|w_1 - st(w_1)\|$, $\|w_2 - st(w_2)\|$ are infinitesimals.

Let us check that the semi-algebraic set $W_l \cap U^{(\varepsilon)}$ is separated from its complement $W_l \setminus U^{(\varepsilon)}$ in the set W_l. According to what we proved above, the points $w_1^{(0)} \in W_l \cap U^{(\varepsilon)}$, $w_2^{(0)} \in W_l \setminus U^{(\varepsilon)}$. Therefore both sets $W_l \cap U^{(\varepsilon)}$ and $W_l \setminus U^{(\varepsilon)}$ are non-empty. Let a point $w_1 \in W_l \cap U^{(\varepsilon)}$. Then for a suitable $0 < \tau_3 \in K$ the inclusion $\mathscr{D}_{st(w_1)}(\tau_3) \subset U$ is true. This implies $\mathscr{D}_{w_1}(\tau_3/2) \cap (W_l \setminus U^{(\varepsilon)}) = \phi$. Similarly, one can consider a point $w_2 \in W_l \setminus U^{(\varepsilon)}$. This contradicts to the connectivity of W_l and concludes the proof of the lemma.

Apparently, the set $st(W) \subset K^n$ is semi-algebraic for any semi-algebraic set $W \subset (\widetilde{K(\varepsilon)})^n$ (consisting only of not infinitely large points relatively to the field K), but we shall not need this statement further and shall not dwell on its proof.

Define the boundary ∂W of a set $W \subset K^n$ as the set of points $w \in K^n$ such that for each $\tau > 0$ both sets $\mathscr{D}_w(\tau) \cap W \neq \phi$, $\mathscr{D}_w(\tau) \setminus W \neq \phi$. In the following lemma the polynomials $f_1, \ldots, f_k \in K[X_1, \ldots, X_n]$ and the element $0 < R \in K$. Introduce the semi-algebraic sets

$$V = \{(f_1 \geqslant 0) \& \ldots \& (f_k \geqslant 0)\} \subset K^n,$$

$$V^{(\varepsilon)} = \{(f_1 + \varepsilon > 0) \& \ldots \& (f_k + \varepsilon > 0)\} \subset (\widetilde{K(\varepsilon)})^n$$

and the polynomial

$$g = (f_1 + \varepsilon) \ldots (f_k + \varepsilon) - \varepsilon^k \in K[\varepsilon][X_1, \ldots, X_n].$$

LEMMA 2.

(a) $V \cap \mathscr{D}_0(R) = st(V^{(\varepsilon)} \cap \mathscr{D}_0(R))$

$\qquad = st(V^{(\varepsilon)} \cap \{g \geqslant 0\} \cap \mathscr{D}_0(R)) \subset (V^{(\varepsilon)} \cap \{g \geqslant 0\} \cap \mathscr{D}_0(R));$

(b) *The boundary*

$$\partial(V^{(\varepsilon)} \cap \{g \geqslant 0\}) \subset (V^{(\varepsilon)} \cap \{g \geqslant 0\})$$

and, for every point $u \in \partial(V^{(\varepsilon)} \cap \{g \geqslant 0\})$, *the equality* $g(u) = 0$ *is fulfilled.*

(c) *On any component of connectivity* $U_1 \subset (\widetilde{K(\varepsilon)})^n$ *of the semi-algebraic set* $\{g \geqslant 0\}$ *each polynomial* $f_i + \varepsilon$, $1 \leqslant i \leqslant k$ *has a constant sign.*

(d) *Let a point* $x \in \partial V \subset V \subset K^n$. *Then there exists a point*

$$z \in (V^{(\varepsilon)} \cap \{g = 0\}) \subset (\widetilde{K(\varepsilon)})^n$$

such that the distance $\|z - x\|$ *is infinitesimal relatively to the field* K.

PROOF. (a) Let $w \in V^{(\varepsilon)} \cap \mathscr{D}_0(R)$. Then the elements $f_i(w) - f_i(st(w))$, $1 \leqslant i \leqslant k$ and $\|w - st(w)\|$ are infinitesimals, therefore, $st(w) \in V \cap \mathscr{D}_0(R)$. Now let $v \in V \cap \mathscr{D}_0(R)$. Then $f_i(v) + \varepsilon \geqslant \varepsilon > 0$ and $g(v) \geqslant 0$. This entails $v \in V^{(\varepsilon)} \cap \{g \geqslant 0\} \cap \mathscr{D}_0(R)$ and, taking into account the equality $v = st(v)$, one deduces that

$$v \in st(V^{(\varepsilon)} \cap \{g \geqslant 0\} \cap \mathscr{D}_0(R)).$$

(b) Let $w \in \partial(V^{(\varepsilon)} \cap \{g \geqslant 0\})$. Assume that either $f_i(w) + \varepsilon < \beta < 0$ for a certain $1 \leqslant i \leqslant k$ or, respectively, $g(w) < \beta < 0$, where $\beta \in \widetilde{K(\varepsilon)}$. There exists $0 < \tau \in \widetilde{K(\varepsilon)}$ such that for any point $w_1 \in \mathscr{D}_w(\tau)$ the inequalities $|f_i(w_1) - f_i(w)| \leqslant -\beta/2$, $1 \leqslant i \leqslant k$ and $|g(w_1) - g(w)| \leqslant -\beta/2$ are valid. On the other hand, according to the definition of the boundary, one can find a point $w_2 \in (V^{(\varepsilon)} \cap \{g \geqslant 0\} \cap \mathscr{D}_w(\tau))$. This leads to a contradiction to the assumption. Thus, we have obtained the inequalities $f_i(w) + \varepsilon \geqslant 0$, $1 \leqslant i \leqslant k$ and $g(w) \geqslant 0$. If $f_i(w) + \varepsilon = 0$ for a certain i, then $g(w) = -\varepsilon^k < 0$. Hence, $w \in (V^{(\varepsilon)} \cap \{g \geqslant 0\})$, i.e.

$$\partial(V^{(\varepsilon)} \cap \{g \geqslant 0\}) \subset (V^{(\varepsilon)} \cap \{g \geqslant 0\}).$$

Reasoning analogously one can also prove that in every point v on the boundary of an arbitrary semi-algebraic set

$$\{(g^{(1)} > 0) \& \ldots \& (g^{(s)} > 0) \& (g^{(s+1)} \geqslant 0) \& \ldots \& (g^{(m)} \geqslant 0)\}$$

the inequalities $g^{(i)}(v) \geqslant 0$ for all $1 \leqslant i \leqslant m$ are true and an equality $g^{(i_0)}(v) = 0$ is fulfilled for a certain $1 \leqslant i_0 \leqslant m$. This implies the equality $g(w) = 0$ for any point $w \in \partial(V^{(\varepsilon)} \cap \{g \geqslant 0\})$ by that proved above.

(c) Suppose that a polynomial $f_i + \varepsilon$ changes its sign on U_1 for some $1 \leqslant i \leqslant k$. Then there exists such a point $x \in U_1$ that $(f_i + \varepsilon)(x) = 0$ by virtue of the connectivity of U_1. This leads to a contradiction since $g(x) = -\varepsilon^k < 0$.

(d) For each fixed m, the following formula of Tarski algebra

$$\forall x \; \forall h(((\deg(h) \leqslant m) \& (h(x) > 0))$$

$$\Rightarrow ((\forall y(h(y) > 0)) \vee \exists \tau \exists z \forall z_1((\|x - z\| = \tau) \& (h(z) = 0) \& ((\|x - z_1\| < z)$$

$$\Rightarrow (h(z_1) > 0)))))$$

is true where $h \in K[X_1, \ldots, X_n]$ denotes a polynomial with degree $\deg h \leqslant m$. One can prove it, first, for the case $K = \mathbb{R}$ and then use the transfer principle. Apply the formula to the polynomial $h = g$ and the point $x \in \partial V$. Note that $f_i(x) \geqslant 0$ for all $1 \leqslant i \leqslant k$ (see the proof of item (b) of the present lemma). Therefore $g(x) \geqslant 0$. If $g(x) = 0$ we can take $z = x$ because of item (a) of the present lemma. So we assume that $g(x) > 0$.

If $\forall\, y(g(y) > 0)$ is fulfilled, then $\forall\, y((f_i + \varepsilon)(y) > 0)$ is valid for all $1 \leqslant i \leqslant k$ by virtue of item (c) of the present lemma. But, on the other hand, there exists a point $y_0 \in K^n$ and an index $1 \leqslant i_0 \leqslant k$ satisfying the inequality $0 > f_{i_0}(y_0) \in K$, i.e. $y_0 \notin V$ taking into account that $x \in \partial V$. Hence, $f_{i_0}(y_0) + \varepsilon < 0$ and we obtain the contradiction.

Thus, one can find a point $z \in (\widetilde{K(\varepsilon)})^n$ such that $g(z) = 0$ and for any point $z_1 \in \mathscr{D}_x(\tau)$, where $\tau = \|z - x\|$, the inequality $g(z_1) \geqslant 0$ is true. Then item (c) of the present lemma implies the inclusion $\mathscr{D}_x(\tau) \subset V^{(\varepsilon)} \cap \{g \geqslant 0\}$. So, if τ is infinitesimal, then z is the desired point. Otherwise, let $0 < \tau_1 < \tau$ and $\tau_1 \in K$. Then item (a) of the present lemma implies that for each point $y \in \mathscr{D}_x(\tau_1) \cap K^n$, the inclusion $y \in V$ is valid. In other words, $\mathscr{D}_x(\tau_1) \subset V$. This contradicts with the condition $x \in \partial V$ and completes the proof of the lemma.

LEMMA 3. *Let* $0 \not\equiv f \in K[X_1, \ldots, X_n]$ *and assume that, for any point* $x \in (\widetilde{K(\varepsilon)})^n$, *the inequality* $f(x) \geqslant 0$ *is fulfilled. Assume that* $f(a) = 0$ *for some* $a \in K^n$. *Then one can find a point* $\alpha \in (\widetilde{K(\varepsilon)})^n$ *such that the distance* $\|a - \alpha\|$ *is infinitesimal, the equality* $f(\alpha) = \varepsilon$ *is correct, and apart from that,* $\|\alpha\| < \|a\|$ *provided that* $a \neq 0$.

PROOF. One can find a point $b \in K^n$ satisfying the requirement $0 < \beta = f(b) \in K$, and besides that, if $a \neq 0$, then $\|b\| < \|a\|$. We can reduce the whole proof to the case of one variable $(n = 1)$ introducing a polynomial

$$h(Z) = f(a + Z(b - a)) \in K[Z].$$

Evidently $h(0) = 0, h(1) = \beta$. There exists $z \in \widetilde{K(\varepsilon)}$ such that $0 < z < 1$ and $h(z) = \varepsilon$ (obviously, the number of elements z satisfying the latter conditions does not exceed $\deg(h)$). Denote by z_0 the least among these elements z. Then, for any $0 < z_1 < z_0$, where $z_1 \in \widetilde{K(\varepsilon)}$, the inequalities $0 \leqslant h(z_1) < \varepsilon$ are true. The element z_0 is infinitesimal, since otherwise there exists an element $z_1 \in K$, for which the inequalities $0 < z_1 < z_0$ and $0 < h(z_1) \in K$ are valid. This leads to the contradiction. One can put $\alpha = a + z_0(b - a)$, with $\|\alpha\| \leqslant (1 - z_0)\|a\| + z_0\|b\| < \|a\|$, provided that $a \neq 0$. The lemma is proved.

Now we proceed to considering the critical values of a polynomial $f \in K[X_1, \ldots, X_n]$. An element $z \in K$ is called a critical value of the polynomial f if the system of equations

$$f - z = \frac{\partial f}{\partial X_1} = \ldots = \frac{\partial f}{\partial X_n} = 0 \tag{3}$$

has a solution in the space K^n.

Let Π_1 be any quantifier-free formula of the first order theory of the field $\mathbb{R}$ and $W = \{\Pi_1\} \subset \mathbb{R}^n$ be the semi-algebraic set. If $\pi : \mathbb{R}^n \to \mathbb{R}^m$ is a linear projection, then Tarski's theorem (Tarski, 1951) states that $\pi(W) \subset \mathbb{R}^m$ is also a semi-algebraic set. Moreover, one can construct a quantifier-free formula Π_2, defined over the same field as the formula Π_1, such that $\pi(W) = \{\Pi_2\}$ and the format of Π_2 is bounded by a function depending only on the format of Π_1 (see e.g. Collins, 1975; Wüthrich, 1976). Therefore, the statement of Tarski theorem is also correct for an arbitrary real closed field K.

Observe that the set of all critical values of a polynomial f coincides with the projection of the semi-algebraic set consisting of all roots of system (3) in the space K^{n+1} with coordinates $Z, X_1, \ldots, X_n$ on the line K^1 defined by coordinate Z. Therefore, the set of all critical values is semi-algebraic over the field generated by the coefficients of the polynomial f according to Tarski's theorem. From this statement and Sard's theorem (asserting that in the case $K = \mathbb{R}$, the set of critical values has the measure null, see e.g. Milnor (1965), one infers that, in the case $K = \mathbb{R}$, there is a finite number of critical values z (see (3)).

The result of Milnor (1964) implies that the number of components of connectivity of the semialgebraic set consisting of all roots of system (3) in the space $\mathbb{R}^{n+1}$ is not greater than $\mathscr{P}((\deg(f))^n)$. Hence, the number of critical values also does not exceed $\mathscr{P}((\deg(f))^n)$. For polynomials f of a given degree $\deg(f)$ the latter statement can be expressed by a formula of Tarski algebra. Therefore, the same bound on the number of critical values is true also for an arbitrary real closed field K. In particular, all critical values of a polynomial are algebraic over the field generated by the coefficients of the polynomial.

LEMMA 4.

    (a) If $f \in K[X_1, \ldots, X_n]$, then an element ε infinitesimal relatively to the field K is not a critical value of polynomial f (over the field $\widetilde{K(\varepsilon)}$).

    (b) Let $f \in K[X_1, \ldots, X_n]$ and $z \in K$. If z is not a critical value of polynomial f then, for any vector $0 \neq u \in K^n$ and every component of connectivity W of the variety $\{f = z\} \subset K^n$ such that $W \subset \mathscr{D}_0(R)$ for a certain $R \in K$, one can find a point $w \in W$ such that the gradient $((\partial f/\partial X_1)(w), \ldots, (\partial f/\partial X_n)(w)) \neq 0$ in this point is collinear to vector u.

PROOF. (a) Item (a) follows from the fact that all the critical values are algebraic over the field K.

    (b) In the case $K = \mathbb{R}$, the statement is proved, e.g. in Thorpe (1979). For an arbitrary real closed field K make use of the transfer principle.

2. Finding Real Roots of a Polynomial

Let a polynomial $g_1 \in \mathbb{Z}[\varepsilon_1][X_1, \ldots, X_{n-1}]$ be given where $\varepsilon_1 > 0$ is infinitesimal relatively to the field $\tilde{\mathbb{Q}}$. We assume the fulfilment of the following inequalities $\deg_{\varepsilon_1, X_1, \ldots, X_{n-1}}(g_1) < d$ and $l(g_1) \leqslant M$ (cf. (2) in the introduction). Define the field $F_1 = \tilde{\mathbb{Q}}(\varepsilon_1)$. In the course of this section we fix a natural number R (it will be specified in the next section).

In the present section we look for roots of the equation $g_1 = 0$ in a ball $\mathscr{D}_0(R) \subset F_1^{n-1}$. The general case of finding the solutions of a system of inequalities will be reduced to this one in section 3. We introduce a new variable X_n and the polynomial $g = g_1^2 + (X_1^2 + \ldots + X_n^2 - R^2)^2$, which yields the semi-algebraic set $V_0 = \{g = 0\} \subset F_1^n$. Evidently, $\pi(V_0) = \{g_1 = 0\} \cap \mathscr{D}_0(R)$ where the projection π is defined by the formula $\pi(X_1, \ldots, X_n) = (X_1, \ldots, X_{n-1})$. The algorithm described in the present section, produces a certain representative set $\mathscr{S}' \subset V_0$ for the semi-algebraic set V_0 (or determines that $V_0 = \phi$) and, incidentally, a representative set $\mathscr{S} = \pi(\mathscr{S}')$ for the set $\pi(V_0)$.

Let an element $\varepsilon > 0$ be infinitesimal relatively to the field F_1. Define $F = \widetilde{F_1(\varepsilon)}$. Then ε is not the critical value of the polynomial g according to lemma 4(a). Introduce the semi-algebraic set $V_\varepsilon = \{g = \varepsilon\} \subset F^n$, observe that $V_\varepsilon \subset \mathscr{D}_0(R + \varepsilon^{1/4}) \subset \mathscr{D}_0(R + 1)$. In the present section the term "standard part" st concerns the situation when the field $K = F_1$ (see section 1), i.e. for an element $a \in F$ its standard part $st(a) \in F_1$, provided that $st(a)$ is definable.

Denote by $N = (4d)^n$ and introduce the family $\Gamma \subset \mathbb{Z}^{n-1}$ consisting of N^{n-1} integer vectors $\Gamma = \{\gamma = (\gamma_2, \ldots, \gamma_n)\}$ where each γ_i runs independently over all values $1, \ldots, N$.

For every index $1 \leqslant i \leqslant n$ and a vector $\gamma = (\gamma_2, \ldots, \gamma_n) \in \Gamma$, consider the following system of equations where

$$\Delta = \sum_{1 \leqslant j \leqslant n} (\partial g / \partial X_j)^2 :$$

$$g - \varepsilon = \left(\frac{\partial g}{\partial X_2}\right)^2 - \frac{\gamma_2}{Nn} \Delta = \ldots = \left(\frac{\partial g}{\partial X_i}\right)^2 - \frac{\gamma_i}{Nn} \Delta = 0. \tag{4i}$$

Below, in the course of the following lemma, we consider $g \in \mathbb{Q}[\varepsilon_1][X_1, \ldots, X_n]$ as an arbitrary polynomial, and let $\bar{W}_{\gamma_2, \ldots, \gamma_i} \subset \bar{F}^n$ denote the variety of all points defined over the algebraic closure $\bar{F} = F[\sqrt{-1}]$ satisfying the system (4i) for the chosen polynomial g.

LEMMA 5. *There exist integers* $1 \leqslant \gamma_2, \ldots, \gamma_i \leqslant N$ *such that for each* $1 \leqslant i \leqslant n$ *any absolutely irreducible component* $\bar{U}_m^{(i)} \subset \bar{F}^n$ *of the variety* $\bar{W}_{\gamma_2, \ldots, \gamma_i}$, *containing at least one point from the space* F^n, *has a dimension* $\dim_{\bar{F}}(\bar{U}_m^{(i)}) = n - i$.

The proof of the lemma and the construction of the $\gamma_2, \ldots, \gamma_i$ will be conducted by induction on i. The base of induction for $i = 1$ is clear. Let us prove the lemma for arbitrary $i > 1$ supposing that for smaller numbers the lemma is already proved and the corresponding $1 \leqslant \gamma_2, \ldots, \gamma_{i-1} \leqslant N$ are constructed. Let an absolutely irreducible component $\bar{U}_l^{(i-1)}$ of the variety $\bar{W}_{\gamma_2, \ldots, \gamma_{i-1}}$ contain at least one point from F^n. Then for at most one $1 \leqslant \varkappa \leqslant N$ the polynomial $(\partial g / \partial X_i)^2 - \varkappa \Delta / Nn$ vanishes identically on $\bar{U}_l^{(i-1)}$. Otherwise, Δ would vanish on $\bar{U}_l^{(i-1)}$ and also would vanish all the partial derivatives $(\partial g / \partial X_j)$, $1 \leqslant j \leqslant n$. In particular, the partial derivatives would vanish at every point of the non-empty set $\bar{U}_l^{(i-1)} \cap F^n \subset \{g = \varepsilon\}$. We get a contradiction since ε is not a critical point of the polynomial g by virtue of lemma 4(a).

According to Bezout's inequality (see Shafarevich, 1974; Heintz, 1983) the number of components of the kind $\bar{U}_l^{(i-1)}$ is less than N. Therefore, for a suitable $1 \leqslant \varkappa \leqslant N$, the polynomial $(\partial g / \partial X_i)^2 - \varkappa \Delta / Nn$ does not vanish identically on any component of the kind $\bar{U}_l^{(i-1)}$ such that $\bar{U}_l^{(i-1)} \cap F^n \neq \phi$. Put $\gamma_i = \varkappa$. Then the dimension of each absolutely irreducible component of the variety

$$\{(\partial g / \partial X_i)^2 - \gamma_i \Delta / Nn = 0\} \cap \bar{U}_l^{(i-1)} \subset \bar{F}^n$$

equals to $(n - i)$, provided that $\bar{U}_l^{(i-1)} \cap F^n \neq \phi$, according to the inductive hypothesis and to the theorem on the dimension of intersection (Shafarevich, 1974). The lemma is proved.

Lemmas 4(a), 4(b), (5) entail (taking into account that the set V_ε is situated in the ball $\mathscr{D}_0(R+1)$) the following

COROLLARY. *There exists a vector* $\gamma^{(1)} = (\gamma_2, \ldots, \gamma_n) \in \Gamma$ *such that every solution from* F^n *of the system* (4n) *is an isolated point in the variety* $\bar{V}^{(\varepsilon)} = \bar{W}_{\gamma_2, \ldots, \gamma_n} \subset \bar{F}^n$ *of all the points satisfying the system* (4n). *Moreover, any component of connectivity of the semi-algebraic set* $V_\varepsilon = \{g = \varepsilon\} \subset F^n$ *contains a certain solution of the system* (4n) *for each vector* $\gamma^{(1)} \in \Gamma$.

Now we proceed to producing a representative set for the semi-algebraic set V_ε and later on for the semi-algebraic set V_0.

The algorithm looks over all the vectors $\gamma \in \Gamma$. Let us fix a certain vector

$\gamma = (\gamma_2, \ldots, \gamma_n) \in \Gamma$. Applying the algorithm from proposition 2 (see also theorem 2 from Chistov & Grigor'ev, 1984), to the system (4n) one can decompose the variety

$$\bar{V}^{(\varepsilon)} = \bigcup_j \bar{V}_j^{(\varepsilon)}$$

on the components $\bar{V}_j^{(\varepsilon)}$ defined and irreducible over the field $\mathbb{Q}(\varepsilon_1, \varepsilon)$.

Select among them all the null-dimensional components. Thus, let dim $\bar{V}_j^{(\varepsilon)} = 0$ for some j. The algorithm from proposition 2 represents the component $\bar{V}_j^{(\varepsilon)}$ in the following form. A polynomial $\Phi \in \mathbb{Q}[\varepsilon_1, \varepsilon][Z]$, irreducible over $\mathbb{Q}$, is constructed such that for every point $(\xi_1, \ldots, \xi_n) \in \bar{V}_j^{(\varepsilon)}$ the field

$$\mathbb{Q}(\varepsilon_1, \varepsilon)(\xi_1, \ldots, \xi_n) = \mathbb{Q}(\varepsilon_1, \varepsilon)[\theta] \simeq \mathbb{Q}(\varepsilon_1, \varepsilon)[Z]/(\Phi),$$

where $\Phi(\theta) = 0$ and the primitive element

$$\theta = \sum_{1 \leqslant j \leqslant n} \lambda_i \xi_i$$

for appropriate natural numbers $1 \leqslant \lambda_i \leqslant \deg_z(\Phi) \leqslant N$. Apart from that, the algorithm explicitly finds expressions

$$\xi_i = \xi_i(\theta) = \sum_{0 \leqslant e < \deg \Phi} \beta_i^{(e)} \theta^e$$

for the relevant $\beta_i^{(e)} \in \mathbb{Q}(\varepsilon_1, \varepsilon)$, $1 \leqslant i \leqslant n$, $0 \leqslant e < \deg_z(\Phi)$. All points of the component $\bar{V}_j^{(\varepsilon)}$ are conjugate over the field $\mathbb{Q}(\varepsilon_1, \varepsilon)$ and they correspond bijectively to roots (from the field $\bar{F}$) of the polynomial Φ. Finally, the algorithm yields irreducible (over $\mathbb{Q}$) polynomials $\Phi_1, \ldots, \Phi_n \in \mathbb{Q}[\varepsilon_1, \varepsilon][Z]$ such that $\Phi_i(\xi_i) = 0$, $1 \leqslant i \leqslant n$.

The following bounds are valid: the degrees

$$\deg_{\varepsilon_1, \varepsilon, Z}(\Phi), \ \deg_{\varepsilon_1, \varepsilon, Z}(\Phi_i), \ \deg_{\varepsilon_1, \varepsilon}(\xi_i(\theta)) \leqslant \mathscr{P}(d^n)$$

and the lengths of coefficients

$$l(\Phi), \ l(\Phi_i), \ l(\xi_i(\theta)) \leqslant (M + \log R)\mathscr{P}(d^n)$$

for the component $\bar{V}_j^{(\varepsilon)}$ and any index $1 \leqslant i \leqslant n$ by virtue of proposition 2.

Observe that a family of points of the kind $(\xi_1, \ldots, \xi_n) \in \bar{V}_j^{(\varepsilon)} \cap F^n$ for all possible vectors $\gamma \in \Gamma$ and null-dimensional components $\bar{V}_j^{(\varepsilon)}$ forms a representative set for the semi-algebraic set V_ε according to the corollary of lemma 5 (see above).

Now we shall turn to producing a representative set $\mathscr{S}'$ for the semi-algebraic set V_0.

If $(\xi_1, \ldots, \xi_n) \in \bar{V}_j^{(\varepsilon)} \cap F^n$, then taking into account the inequality $\xi_1^2 + \ldots + \xi_n^2 < R^2 + 1$ one concludes that the standard part $st(\xi_1, \ldots, \xi_n) \in F_1^n$ is defined. On the other hand, $0 = st(\phi_i(\xi_i)) = \psi_i(st(\xi_i))$, where we denote by $\psi_i = st(\phi_i) \in \mathbb{Q}[\varepsilon_1][Z]$ the polynomial obtained from Φ_i by coefficient-wise taking of a standard part. Notice that the polynomials ψ_i satisfy the same bounds as the polynomials Φ_i(see above).

In the sequel, we shall need the following auxiliary

LEMMA 6. *Let*

$$\theta' = \sum_{1 \leqslant j \leqslant n} \lambda_i' \xi_i \in \mathbb{Q}(\varepsilon_1, \varepsilon)(\xi_1, \ldots, \xi_n),$$

where the natural numbers $1 \leqslant \lambda_i' \leqslant \deg_z(\Phi)$. *Let* $\Phi^{(1)} \in \mathbb{Z}[\varepsilon_1, \varepsilon][Z]$ *be an irreducible (over* $\mathbb{Z}$) *polynomial such that* $\Phi^{(1)}(\theta') = 0$. *Then the following bounds are fulfilled:*

and
$$\deg_z(\Phi^{(1)}) \leqslant \deg_z(\Phi); \quad \deg_{\varepsilon_1, \varepsilon}(\Phi^{(1)}) \leqslant \mathscr{P}(d^n)$$

$$l(\Phi^{(1)}) \leqslant (M + \log R)\mathscr{P}(d^n).$$

PROOF. Represent $\theta' = q(\theta)$ for a suitable polynomial $q(Z) \in \mathbb{Q}(\varepsilon_1, \varepsilon)[Z]$, making use of the expressions $\xi_i(\theta)$, then the following bounds are true:

$$\deg_{\varepsilon_1, \varepsilon}(q) \leqslant \mathscr{P}(d^n) \quad \text{and} \quad l(q) \leqslant (M + \log R)\mathscr{P}(d^n).$$

From the latter bounds the following ones for the powers q^e, $1 \leqslant e \leqslant \deg_z(\Phi)$ can be deduced:

$$\deg_{\varepsilon_1, \varepsilon}(q^e) \leqslant \mathscr{P}(d^n, \deg_z(\Phi)) \leqslant \mathscr{P}(d^n) \quad \text{and} \quad l(q^e) \leqslant (M + \log R)\mathscr{P}(d^n).$$

For the remainders $\mathrm{rem}(q^e, \Phi)$ of dividing polynomial q^e by polynomial Φ over the field $\mathbb{Q}(\varepsilon_1, \varepsilon)$ the same bounds are correct. The following inequality is obvious:

$$\deg_z(\Phi^{(1)}) = [\mathbb{Q}(\varepsilon_1, \varepsilon)[\theta'] : \mathbb{Q}(\varepsilon_1, \varepsilon)] \leqslant [\mathbb{Q}(\varepsilon_1, \varepsilon)(\xi_1, \ldots, \xi_n) : \mathbb{Q}(\varepsilon_1, \varepsilon)] = \deg_z(\Phi)$$

(here and further $[H_1 : H_2]$ for a finite extension $H_2 \subset H_1$ of a field denotes its degree). The equality

$$\sum_{0 \leqslant e \leqslant \deg(\Phi(1))} \alpha_e \, \mathrm{rem}(q^e, \Phi) = 0,$$

where $\alpha_e \in \mathbb{Z}[\varepsilon_1, \varepsilon]$ are the coefficients of the polynomial $\Phi^{(1)}$, gives a system of homogeneous linear equations in the indeterminates α_e having a one-dimensional space of solutions. Involving Cramer's rule one can infer the bounds

$$\deg_{\varepsilon_1, \varepsilon}(\alpha_e) \leqslant \mathscr{P}(d^n) \quad \text{and} \quad l(\alpha_e) \leqslant (M + \log R)\mathscr{P}(d^n),$$

which completes the proof of the lemma.

Consider an arbitrary element of the sort

$$\tau = \sum_{1 \leqslant i \leqslant n} \lambda_i' st(\xi_i) \in \bar{F}_1,$$

provided that $st(\xi_i) \in F_1$, $1 \leqslant i \leqslant n$ are defined, where the natural numbers $1 \leqslant \lambda_i' \leqslant \deg_z(\Phi)$, $1 \leqslant i \leqslant n$. Then according to lemma 6, for an irreducible (over $\mathbb{Z}$) polynomial $\Phi^{(1)}(Z) \in \mathbb{Z}[\varepsilon_1, \varepsilon][Z]$ such that

$$\Phi^{(1)}\left(\sum_{1 \leqslant i \leqslant n} \lambda_i' \xi_i\right) = 0,$$

the bounds

$$\deg_Z(\Phi^{(1)}) \leqslant \deg_Z(\Phi); \quad \deg_{\varepsilon_1, \varepsilon}(\Phi^{(1)}) \leqslant \mathscr{P}(d^n) \quad \text{and} \quad l(\Phi^{(1)}) \leqslant (M + \log R)\mathscr{P}(d^n)$$

are valid. Evidently $(st(\Phi^{(1)}))(\tau) = 0$ where the polynomial $st(\Phi^{(1)}) \in \mathbb{Z}[\varepsilon_1][Z]$. Therefore, for an irreducible (over $\mathbb{Z}$) polynomial $\psi^{(1)}(Z) \in \mathbb{Z}[\varepsilon_1][Z]$ such that $\psi^{(1)}(\tau) = 0$, the following bounds are fulfilled:

$$\deg_Z(\Psi^{(1)}) \leqslant \deg_Z(\Phi^{(1)}); \quad \deg_{\varepsilon_1}(\psi^{(1)}) \leqslant p_1(d^n) \quad \text{and} \quad l(\psi^{(1)}) \leqslant (M + \log R).$$

$p_2(d^n)$ for certain polynomials p_1, p_2 by virtue of proposition 1 (cf. also Mignotte, 1974), taking into account that the polynomial $\psi^{(1)}$ divides $st(\Phi^{(1)})$.

Now let

$$\tau^{(2)} = \sum_{1 \leqslant i \leqslant n} \lambda_i^{(2)} st(\xi_i), \quad \tau^{(3)} = \sum_{1 \leqslant i \leqslant n} \lambda_i^{(3)} st(\xi_i),$$

where the natural numbers $1 \leqslant \lambda_i^{(2)}, \lambda_i^{(3)} \leqslant \deg_Z(\Phi)$, $1 \leqslant i \leqslant n$. Assume that

$$\tau^{(3)} = q(\tau^{(2)}) \in \mathbb{Q}(\varepsilon_1)[\tau^{(2)}]$$

for a polynomial $q(Z) \in \mathbb{Q}(\varepsilon_1)[Z]$ such that

$$\deg_Z(q) < [\mathbb{Q}(\varepsilon_1)[\tau^{(2)}] : \mathbb{Q}(\varepsilon_1)].$$

Let us show that $\deg_{\varepsilon_1}(q) \leqslant \mathscr{P}(d^n)$ and $l(q) \leqslant (M + \log R)\mathscr{P}(d^n)$. Indeed, suppose that $\psi^{(2)}(\tau^{(2)}) = 0$, $\psi^{(3)}(\tau^{(3)}) = 0$ for some irreducible (over $\mathbb{Z}$) polynomials $\psi^{(2)}, \psi^{(3)} \in \mathbb{Z}[\varepsilon_1][Z]$, then

$$\deg_{\varepsilon_1}(\psi^{(2)}), \deg_{\varepsilon_1}(\psi^{(3)}) \leqslant \mathscr{P}(d^n) \quad \text{and} \quad l(\psi^{(2)}), l(\psi^{(3)}) \leqslant (M + \log R)\mathscr{P}(d^n)$$

in view of lemma 6 and what was proved just after it. Polynomial $\psi^{(3)}$ has a root $\tau^{(3)} = q(\tau^{(2)})$ in the field $\mathbb{Q}(\varepsilon_1)[\tau^{(2)}] = \mathbb{Q}(\varepsilon_1)[Z]/(\psi^{(2)})$. Hence, $\deg_{\varepsilon_1}(q) \leqslant p_3(d^n)$ and $l(q) \leqslant (M + \log R)p_4(d^n)$ for the relevant polynomials p_3, p_4 according to proposition 1, since a polynomial $(Z - \tau^{(3)})$ divides the polynomial $\psi^{(3)}$ over the field $\mathbb{Q}(\varepsilon_1)[Z]/(\psi^{(2)})$.

Now we pass to describing a procedure which produces a family of some classes of vectors $(\zeta_1, \ldots, \zeta_n) \in \bar{F}_1^n$ conjugate over the field $\mathbb{Q}(\varepsilon_1)$ such that $\psi_i(\zeta_i) = 0$, $1 \leqslant i \leqslant n$ (let us recall that $\psi_i = st(\Phi_i)$) and, furthermore, the produced family contains vectors of the kind $(st(\xi_1), \ldots, st(\xi_n)) \in \bar{F}_1^n$ for all points $(\xi_1, \ldots, \xi_n) \in \bar{V}_j^{(\varepsilon)} \subset \bar{F}^n$ of the null-dimensional irreducible component $\bar{V}_j^{(\varepsilon)}$ of the variety $\bar{V}^{(\varepsilon)}$ constructed earlier, provided that $(st(\xi_1), \ldots, st(\xi_n))$ is definable.

In the first step, the procedure factorises the polynomial

$$\psi_1 = \prod_j \psi_{1,j}^{c_j}$$

over the field $\mathbb{Q}(\varepsilon_1)$ by proposition 1 (see also Lenstra, 1984). Here, the multipliers $\psi_{1j} \in \mathbb{Q}(\varepsilon_1)[Z]$. For each multiplier ψ_{1j} we introduce the field

$$\Xi_1^{(j)} = \mathbb{Q}(\varepsilon_1)[Z]/(\psi_{1j}) \supset \mathbb{Q}(\varepsilon_1),$$

which is a finite extension of the field $\mathbb{Q}(\varepsilon_1)$. Let $\psi_{1j}(\theta_1^{(j)}) = 0$ for a certain $\theta_1^{(j)} \in \Xi_1^{(j)}$, then $\Xi_1^{(j)} \simeq \mathbb{Q}(\varepsilon_1)[\theta_1^{(j)}]$. Put $\psi_1^{(j)} = \psi_{1j}$.

Assume that by recursion on i the following objects are already constructed by the procedure: a family of fields $\{\Xi_i^{(m)}\}_m$, for every field $\Xi_i^{(m)}$ an irreducible (over the field $\mathbb{Q}(\varepsilon_1)$) polynomial $\psi_i^{(m)} \in \mathbb{Q}(\varepsilon_1)[Z]$ such that

$$\Xi_i^{(m)} = \mathbb{Q}(\varepsilon_1)[\theta_i^{(m)}] \simeq \mathbb{Q}(\varepsilon_1)[Z]/(\psi_i^{(m)}),$$

where $\psi_i^{(m)}(\theta_i^{(m)}) = 0$, and a primitive element

$$\theta_i^{(m)} = \sum_{1 \leqslant e \leqslant i} \mu_e \zeta_e \in \Xi_i^{(m)},$$

with $\psi_e(\zeta_e) = 0$ and the natural numbers

$$1 \leqslant \mu_e \leqslant [\Xi_e^{(m_e)} : \mathbb{Q}(\varepsilon_1)] \leqslant \deg_Z(\Phi)$$

for suitable indices m_e, $1 \leqslant e \leqslant i$. Apart from that the procedure yields the expressions

$$\zeta_e = \sum_{0 \leqslant s < \deg_Z(\psi_i^{(m)})} \rho_{i,e,s}^{(m)}(\theta_i^{(m)})^s$$

for appropriate $\rho_{i,e,s}^{(m)} \in \mathbb{Q}(\varepsilon_1)$. Each root of the polynomial $\psi_i^{(m)}$ is of the same form as $\theta_i^{(m)}$, in addition the roots generate the different fields conjugate (over the field $\mathbb{Q}(\varepsilon_1)$) to the field $\mathbb{Q}(\varepsilon_1)[\theta_i^{(m)}]$.

For the realisation of the $(i+1)$th step of the procedure we fix a field $\Xi_i^{(m)}$ and factorise the polynomial

$$\psi_{i+1} = \prod_j \psi_{i+1,j}$$

over the field $\Xi_i^{(m)}$ by proposition 1, where $\psi_{i+1,j} \in \Xi_i^{(m)}[Z]$. For every factor $\psi_{i+1,t}$ consider the field $\Xi_{i+1}^{(t)} = \Xi_i^{(m)}[Z]/(\psi_{i+1,t})$ (for brevity we omit in its notation the dependence on m), provided that the product of degrees

$$[\Xi_i^{(m)} : \mathbb{Q}(\varepsilon_1)] \deg_Z(\psi_{i+1,t}) \leqslant \deg_Z(\Phi),$$

otherwise the procedure does not construct the field $\Xi_{i+1}^{(t)}$. Let $\psi_{i+1,t}(\zeta_{i+1}) = 0$ for a certain $\zeta_{i+1} \in \Xi_{i+1}^{(t)}$. Among the elements of the sort $\theta_i^{(m)} + \mu\zeta_{i+1}$ for all natural numbers

$$1 \leqslant \mu \leqslant [\Xi_i^{(m)} : \mathbb{Q}(\varepsilon_1)] \deg_Z(\psi_{i+1,t}) = [\Xi_{i+1}^{(t)} : \mathbb{Q}(\varepsilon_1)] \leqslant \deg_Z(\Phi),$$

one can find a primitive element of the field $\Xi_{i+1}^{(t)}$ over the field $\mathbb{Q}(\varepsilon_1)$ (see e.g. Lang, 1965).

The procedure finds for each element of the sort $\theta_i^{(m)} + \mu\zeta_{i+1}$ its minimal polynomial over the field $\mathbb{Q}(\varepsilon_1)$. For this goal, for any fixed $1 \leqslant s \leqslant [\Xi_{i+1}^{(t)} : \mathbb{Q}(\varepsilon_1)]$, the procedure checks whether there exist elements $\alpha_0, \ldots, \alpha_s \in \mathbb{Q}(\varepsilon_1)$ such that

$$\sum_{0 \leqslant e \leqslant s} \alpha_e (\theta_i^{(m)} + \mu\zeta_{i+1})^e = 0,$$

solving the system of linear equations over the field $\mathbb{Q}(\varepsilon_1)$ which arises from the latter equality. Namely, in order to obtain the system remove the parenthesis in the binoms $(\theta_i^{(m)} + \mu\zeta_{i+1})^e$ and compute the polynomials

$$\mathrm{rem}\,(Z^{e_1}, \psi_i^{(m)}(Z)) \in \mathbb{Q}(\varepsilon_1)[Z], \quad \mathrm{rem}\,(Z^{e_2}, \psi_{i+1,t}(Z)) \in \Xi_i^{(m)}[Z]$$

for all

$$0 \leqslant e, e_1, e_2 < [\Xi_{i+1}^{(t)} : \mathbb{Q}(\varepsilon_1)].$$

As a result, we obtain expressions of the powers $(\theta_i^{(m)} + \mu\zeta_{i+1})^e$ via the basis

$$(\theta_i^{(m)})^{d_1}(\zeta_{i+1})^{d_2}, \quad 0 \leqslant d_1 < \deg_Z(\psi_i^{(m)}), \quad 0 \leqslant d_2 < \deg_Z(\psi_{i+1,t})$$

of the field $\Xi_{i+1}^{(t)}$ over the field $\mathbb{Q}(\varepsilon_1)$. The desired system of linear equations is obtained by setting the coefficients at the monomials $(\theta_i^{(m)})^{d_1}(\zeta_{i+1})^{d_2}$ in the equality

$$\sum_{0 \leqslant e \leqslant s} \alpha_e (\theta_i^{(m)} + \mu\zeta_{i+1})^e = 0$$

equal to zero. The element $\theta_{i+1}^{(t)} = \theta_i^{(m)} + \mu_{i+1}\zeta_{i+1}$, for which the degree of its minimal polynomial $\psi_{i+1}^{(t)}$ is maximal, is primitive.

If at least one of the two inequalities

$$\deg_{\varepsilon_1}(\psi_{i+1}^{(t)}) \leqslant p_1(d^n) \quad \text{or} \quad l(\psi_{i+1}^{(t)}) \leqslant (M + \log R)p_2(d^n)$$

is not satisfied (cf. above), then the procedure does not construct the field $\Xi_{i+1}^{(t)}$.

The above procedure has produced the transforming matrix from the basis $(\theta_i^{(m)})^{d_1}(\zeta_{i+1})^{d_2}$ to the basis $(\theta_{i+1}^{(t)})^e$. Inverting this matrix, the procedure yields the expression of the former basis in terms of the latter one, in particular the expressions

$$\zeta_{i+1} = \sum_{0 \leqslant s < \deg_Z(\psi_{i+1}^{(t)})} \rho_{i+1,i+1,s}^{(t)}(\theta_{i+1}^{(t)})^s \quad \text{and} \quad (\theta_i^{(m)})^l = \sum_{0 \leqslant s < \deg_Z(\psi_{i+1}^{(t)})} \sigma_{i,s,l}^{(t)}(\theta_{i+1}^{(t)})^s.$$

Substituting the latter expressions in the equalities obtained in the previous step of the recursion, we compute the expressions

$$\zeta_e = \sum_{0 \leqslant s < \deg_Z(\psi_{i+1}^{(t)})} \rho_{i+1,e,s}^{(t)}(\theta_{i+1}^{(t)})^s, \quad 1 \leqslant \rho \leqslant i.$$

If at least one of the following inequalities

$$\deg_{\varepsilon_1}(\zeta_e(\theta_{i+1}^{(t)})) \leqslant p_3(d^n) \quad \text{or} \quad l(\zeta_e(\theta_{i+1}^{(t)})) \leqslant (M + \log R)p_4(d^n)$$

is not fulfilled (cf. above), then the procedure does not construct the field $\Xi_{i+1}^{(t)}$.

Thus, the procedure constructs the field $\Xi_{i+1}^{(t)}$ iff all the above inequalities are valid. This completes the recursive construction of the family of fields $\{\Xi_i^{(m)}\}_{m;1 \leqslant j \leqslant n}$. Observe that for every point $(\xi_1, \ldots, \xi_n) \in \bar{V}_j^{(\varepsilon)} \subset \bar{F}^n$ from the null-dimensional irreducible component $\bar{V}_j^{(\varepsilon)}$ of the variety $\bar{V}^{(\varepsilon)}$, to vector $(st(\xi_1), \ldots, st(\xi_n)) \in \bar{F}_1^n$ (provided that it is definable), corresponds a certain field among the fields

$$\Xi_n^{(m)} = \mathbb{Q}(\varepsilon_1)(st(\xi_1), \ldots, st(\xi_n))$$

constructed. In addition, the vector

$$(st(\xi_1), \ldots, st(\xi_n)) = (\zeta_1, \ldots, \zeta_n)$$

in view of the above-mentioned properties of the polynomials p_1, p_2, p_3, p_4.

Thereby, the following lemma is actually proved.

LEMMA 7. *Let a polynomial* $\Phi \in H[\varepsilon][Z]$ *be irreducible over an ordered field* H, *with* $\varepsilon > 0$ *infinitesimal relatively to* H. *Assume that the expressions*

$$\xi_i = \xi_i(\theta) = \sum_{0 \leqslant j < \deg_Z(\phi)} \beta_i^{(j)} \theta^j,$$

where $\beta_i^{(j)} \in H(\varepsilon)$, $1 \leqslant i \leqslant n$, $0 \leqslant j < \deg_Z(\Phi)$, *yield a class of conjugate points over the field* $H(\varepsilon)$ *of the kind* $(\xi_1(\theta), \ldots, \xi_n(\theta)) \in \overline{H(\varepsilon)}^n$, *where* $\phi(\theta) = 0$. *In addition*

$$\theta = \sum_{1 \leqslant i \leqslant n} \lambda_i \xi_i$$

for certain natural numbers $1 \leqslant \lambda_i \leqslant \deg_Z(\Phi)$, *i.e. the field*

$$H(\varepsilon)(\xi_1, \ldots, \xi_n) = H(\varepsilon)[\theta] \simeq H(\varepsilon)[Z]/(\phi).$$

Then one can produce a set $\mathcal{R} \subset \bar{H}^n$, *containing the set of standard parts* $(st(\xi_1), \ldots, st(\xi_n)) \in \bar{H}^n$ *of all those points from the above-mentioned class of conjugates such that the standard part is definable. Besides that,* $\mathcal{R}$ *is a union of classes of conjugate points over the field* H $(\zeta_1, \ldots, \zeta_n) \in \bar{H}^n$. *Here each class is represented analogously as the class of points* $(\xi_1(\theta), \ldots, \xi_n(\theta))$ *above.*

Note that the proof of lemma 7 was conducted for the field $H = \mathbb{Q}(\varepsilon_1)$, but one can easily carry it over to arbitrary ordered effectively given fields. We shall not make the meaning of the latter claim more precise, since in the present paper we apply lemma 8 only to the case of the field $H = \mathbb{Q}(\varepsilon_1)$ (in this section) and to the case of the field $H = \mathbb{Q}$ (in section 3). Estimations on the parameters of the points belonging to $\mathcal{R}$ were, in fact, given earlier in the case of the field $H = \mathbb{Q}(\varepsilon_1)$ (see the properties of polynomials p_1, p_2, p_3, p_4), in the case of the field $H = \mathbb{Q}$, the estimations on the bit lengths of the coefficients are the same; a time-bound on producing $\mathcal{R}$ will be obtained below in the case $H = \mathbb{Q}(\varepsilon_1)$ (cf. lemma 8), in the case $H = \mathbb{Q}$ the time-bound is the same.

Now we shall continue to describe the algorithm that produces a representative set for the semi-algebraic set V_0. Fix a certain class of conjugate vectors

$$(\zeta_1, \ldots, \zeta_n) \in \mathcal{R} \cap (\Xi_n^{(m)})^n$$

over the field $\mathbb{Q}(\varepsilon_1)$, which was constructed above (see lemma 7). Denote by $\eta = \theta_n^{(m)} \in \Xi_n^{(m)}$ the primitive element and by $\psi = \psi_n^{(m)} \in \mathbb{Q}(\varepsilon_1)[Z]$ its minimal polynomial over the field $\mathbb{Q}(\varepsilon_1)$. The algorithm tests, whether an equality $g(\zeta_1, \ldots, \zeta_n) = 0$ is fulfilled (observe that the fulfilment of this equality is independent of the choice of the particular vector in the conjugate class). Namely, involving expressions $\zeta_i(\eta) \in \mathbb{Q}(\varepsilon_1)[\eta]$ (see the proof of lemma 7) one can obtain a representation

$$g(\zeta_1(\eta), \ldots, \zeta_n(\eta)) = h(\eta)$$

for a certain polynomial $h(Z) \in \mathbb{Q}(\varepsilon_1)[Z]$. Then $g(\zeta_1, \ldots, \zeta_n) = 0$ iff ψ divides the polynomial h.

Suppose that $g(\zeta_1, \ldots, \zeta_n) = 0$. Determine whether ψ has at least one root in the field F_1, with the help of Sturm sequence (Lang, 1965; see also Heindel, 1971). If ψ has a root $\eta_0 \in F_1$, then the corresponding point

$$(\zeta_1(\eta_0), \ldots, \zeta_n(\eta_0)) \in V_0 = \{g = 0\} \subset F_1^n,$$

and the representative set $\mathscr{S}'$, which is produced by the algorithm described, consists of the points of the kind

$$(\zeta_1(\eta_1), \ldots, \zeta_n(\eta_1)) \in \mathscr{R} \cap F_1^n$$

for all roots $\eta_1 \in F_1$ of the polynomial ψ, all possible classes of conjugate points $(\zeta_1, \ldots, \zeta_n) \in F_1^n$, where $g(\zeta_1, \ldots, \zeta_n) = 0$; $(\xi_1, \ldots, \xi_n) \in F^n$, and lastly all vectors $\gamma \in \Gamma$.

Now we shall prove that $\mathscr{S}'$ is really a representative set for V_0. Lemma 3 from section 1 entails that $V_0 = st(V_\varepsilon)$. Let us consider an arbitrary component of connectivity V_1 of the set V_0 and make sure that the intersection $V_1 \cap \mathscr{S}' \neq \phi$. By virtue of lemma 1(b) from section 1 there exist components of connectivity $W_1, \ldots, W_e$ of the set V_ε such that $V_1 = st(W_1 \cup \ldots \cup W_e)$. According to the corollary of lemma 5, one can find a vector $\gamma \in \Gamma$, such that for any component of connectivity W_s of the set V_ε, there exists a null-dimensional irreducible component $\bar{V}_j^{(\varepsilon)}$ of the variety $\bar{V}^{(\varepsilon)} \subset \bar{F}^n$ of all roots of the system (4n), corresponding to the vector $\gamma \in \Gamma$ under consideration, for which $W_s \cap \bar{V}_j^{(\varepsilon)} \neq \phi$. This means that

$$\left(\bigcup_{\gamma \in \Gamma; \, \dim(\bar{V}_j^{(\varepsilon)}) = 0} \bar{V}_j^{(\varepsilon)} \right) \cap F^n$$

is a representative set for the set V_ε (see above). Fix some $1 \leqslant s \leqslant e$ and let a point $(\xi_1, \ldots, \xi_n) \in W_s \cap \bar{V}_j^{(\varepsilon)}$. Hence, $st(\xi_1, \ldots, \xi_n) \in V_1$. Starting with the class $\bar{V}_j^{(\varepsilon)}$ of conjugate points (see the proof of lemma 7), the above procedure constructs a polynomial $\psi \in \mathbb{Q}(\varepsilon_1)[Z]$, expressions $\zeta_i(\eta) \in \mathbb{Q}(\varepsilon_1)[\eta]$ and a set $\mathscr{R} \subset \bar{F}_1^n$ such that for an appropriate root

$$\eta_0 = \sum_{1 \leqslant i \leqslant n} \lambda_i^{(4)} \zeta_i(\eta_0) \in F_1$$

of the polynomial ψ, where the natural numbers $1 \leqslant \lambda_i^{(4)} \leqslant \deg_Z(\psi)$ for $1 \leqslant i \leqslant n$, the equality

$$st(\xi_1, \ldots, \xi_n) = (\zeta_1(\eta_0), \ldots, \zeta_n(\eta_0)) \in \mathscr{R}$$

holds. (Note that $st(\xi_1, \ldots, \xi_n)$ is definable since $(\xi_1, \ldots, \xi_n) \in V_\varepsilon \subset \mathscr{D}_0(R+1)$.) Thus, the point $st(\xi_1, \ldots, \xi_n) \in \mathscr{S}' \cap V_1$, which completes the proof of the fact that $\mathscr{S}'$ is a representative set for the set V_0.

Now we proceed to the time analysis of the algorithm suggested in the present section.

At the beginning of its work the algorithm scans all vectors $\gamma \in \Gamma = \{1, \ldots, (4d)^n\}^{n-1}$. For this time $\mathscr{P}(d^{n^2})$ is sufficient. After that the algorithm from proposition 2 is applied to the system (4n) (producing a representative set for the set V_ε), its running time can be bounded by $\mathscr{P}(M, \log R, d^{n^2})$. Then, recursively, the above procedure constructs a family of fields $\Xi_i^{(m)}$ and expressions $\zeta_e(\theta_i^{(m)})$ (see the proof of lemma 7). The field $\Xi_{i+1}^{(t)} = \Xi_i^{(m)}[Z]/(\psi_{i+1,t})$, where the polynomial $\psi_{i+1,t} \in \Xi_i^{(m)}[Z]$ is a certain factor of the polynomial $\psi_{i+1} = st(\Phi_{i+1})$ irreducible over $\Xi_i^{(m)}$. According to proposition 1, lemma 6 and the properties of the polynomials p_1, p_2, one can construct the polynomial $\psi_{i+1,t}$ within time $\mathscr{P}(M, \log R, d^n)$, furthermore, $\deg_{\varepsilon_1}(\psi_{i+1,t}) \leqslant \mathscr{P}(d^n)$ and

$$l(\psi_{i+1,t}) \leqslant (M + \log R)\mathscr{P}(d^n).$$

Next, the procedure searches for a primitive element $\theta_{i+1}^{(t)}$, solving for every element

$$\theta_i^{(m)} + \mu\zeta_{i+1}, \quad 1 \leqslant \mu \leqslant [\Xi_{i+1}^{(t)} : \mathbb{Q}(\varepsilon_1)] \leqslant \deg_Z(\Phi)$$

the respective system of linear equations by means of which the procedure finds the minimal polynomial of the element $\theta_i^{(m)} + \mu\zeta_{i+1}$ over the field $\mathbb{Q}(\varepsilon_1)$. The size of the system and, incidentally, the time for its solution can be estimated by $\mathscr{P}(M, \log R, d^n)$ in view of the properties of the polynomials p_1, p_2 and the bounds on $\deg_{\varepsilon_1}(\psi_{i+1,t})$, $l(\psi_{i+1,t})$ (see above). Therefore, the procedure constructs the expressions $\zeta_{i+1}(\theta_{i+1}^{(t)})$, $(\theta_i^{(m)})^l(\theta_{i+1}^{(t)})$ also within time $\mathscr{P}(M, \log R, d^n)$ and by virtue of the properties of the polynomials p_3, p_4 all expressions $\zeta_e(\theta_i^{(m)})$ can be constructed within the same time-bound. Thus, the whole computing time of the procedure constructing the set $\mathscr{R}$ mentioned in lemma 7 does not exceed $\mathscr{P}(M, \log R, d^{n^2})$ taking into account that the total number of fields of the kind $\Xi_n^{(m)}$ is less or equal to the product $\deg_Z(\psi_1) \ldots \deg_Z(\psi_n) \leqslant \mathscr{P}(d^{n^2})$.

After that, for each class $(\zeta_1, \ldots, \zeta_n) \in \mathscr{R}$ of conjugate points the algorithm checks, whether $g(\zeta_1, \ldots, \zeta_n) = 0$ is true, making use of the constructed expressions $\zeta_i(\eta)$, $1 \leqslant i \leqslant n$, where $\eta = \theta_n^{(m)}$ for an appropriate m. For this time $\mathscr{P}(M, \log R, d^n)$ is sufficient. Then the algorithm determines, whether the polynomial $\psi = \psi_n^{(m)}$ has a root in the field F_1, using Sturm sequence (see e.g. Lang, 1965). This requires also not more than $\mathscr{P}(M, \log R, d^n)$ time according, e.g. to Heindel (1971).

Summarising the results of this section we arrive at the following

LEMMA 8. *One can design an algorithm which, for a given polynomial $g_1 \in \mathbb{Z}[\varepsilon_1][X_1, \ldots, X_{n-1}]$ and for a natural number $R \geqslant 1$, produces a representative set $\mathscr{S} \subset F_1^{n-1}$ for the intersection $\{g_1 = 0\} \cap \mathscr{D}_0(R)$. The number of points in $\mathscr{S}$ does not exceed $\mathscr{P}(d^{n^2})$. Here, the algorithm outputs the set $\mathscr{S}$ as a union of classes $(\zeta_1, \ldots, \zeta_{n-1})$ of points conjugate over the field $\mathbb{Q}(\varepsilon_1)$. For every class the algorithm constructs a certain polynomial $\psi \in \mathbb{Z}[\varepsilon_1][Z]$ irreducible over $\mathbb{Q}(\varepsilon_1)$ and expressions*

$$\zeta_i(\eta) = \sum_j \rho_i^{(j)} \eta^j,$$

where

$$\eta \in F_1, \quad \rho_i^{(j)} \in \mathbb{Q}(\varepsilon_1), \quad 1 \leqslant i \leqslant n-1, \quad 0 \leqslant j < \deg_Z(\psi),$$

in addition $\psi(\eta) = 0$ and

$$\eta = \sum_{1 \leqslant i \leqslant n-1} \lambda_i' \zeta_i(\eta)$$

for suitable natural numbers $1 \leqslant \lambda_i' \leqslant \deg_Z(\psi)$, $1 \leqslant i \leqslant n-1$. The set $\mathscr{S}$ coincides with a family of points of the sort $(\zeta_1(\eta), \ldots, \zeta_{n-1}(\eta)) \in F_1^{n-1}$, where η runs over all roots $\eta \in F_1$ of the polynomial ψ and $(\zeta_1, \ldots, \zeta_{n-1})$ runs over all classes of conjugate points. Besides that, the following bounds are valid:

$$\deg_{\varepsilon_1, Z}(\psi), \deg_{\varepsilon_1}(\zeta_i(\eta)) \leqslant \mathscr{P}(d^n) \quad and \quad l(\psi), l(\zeta_i(\eta)) \leqslant (M + \log R)\mathscr{P}(d^n).$$

The algorithm works within time $\mathscr{P}(M, \log R, d^{n^2})$.

3. Finding Solutions of a System of Polynomial Inequalities

We turn now to considering the general case and complete the proof of the theorem (see the introduction).

Let the input system of polynomial inequalities $f_1 > 0, \ldots, f_m > 0, f_{m+1} \geqslant 0, \ldots, f_k \geqslant 0$ be given (cf. (1)), where the polynomials $f_1, \ldots, f_k \in \mathbb{Z}[X_2, \ldots, X_n]$ satisfy the bounds

$\deg_{X_2,\ldots,X_n}(f_i) < d$; $l(f_i) \leqslant M$, $1 \leqslant i \leqslant k$ (see (2)). Introduce a new variable X_1 and a polynomial $f_{k+1} = X_1 f_1, \ldots f_m - 1$. Denote by π_1 a linear projection defined by the formula $\pi_1(X_1, \ldots, X_n) = (X_2, \ldots, X_n)$. Consider the semi-algebraic set

$$V = \{(f_1 \geqslant 0) \& \ldots \& (f_m \geqslant 0) \& (f_{m+1} \geqslant 0) \& \ldots \& (f_k \geqslant 0) \& (f_{k+1} \geqslant 0)\} \subset \tilde{\mathbb{Q}}^n.$$

Then

$$\mathscr{V} = \pi_1(V) = \{(f_1 > 0) \& \ldots \& (f_m > 0) \& (f_{m+1} \geqslant 0) \& \ldots \& (f_k \geqslant 0)\} \subset \tilde{\mathbb{Q}}^{n-1}$$

and it is sufficient to produce a representative set $\mathscr{T}'$ for the set V, in this case $\mathscr{T} = \pi_1(\mathscr{T}')$ is the representative set for the set $\mathscr{V} = \pi_1(V)$ as desired in the theorem. Below, the standard part st concerns the situation $K = \tilde{\mathbb{Q}}$ (see section 1), i.e. for an element $a \in F_1$ its standard part $st(a) \in \tilde{\mathbb{Q}}$, provided that $st(a)$ is definable.

The following bound on real roots of a polynomial was originally proved in Vorobjov (1984). Here we expose a shorter proof.

LEMMA 9. *Let a polynomial $h \in \mathbb{Z}[X_1, \ldots, X_n]$ satisfy the inequalities $\deg_{X_1,\ldots,X_n}(h) < d$ and $l(h) \leqslant M$. Then any component of connectivity U_0 of the semi-algebraic set $\{h = 0\} \subset \tilde{\mathbb{Q}}^n$ has non-empty intersection with the ball $\mathscr{D}_0(R)$, where the natural number $R \leqslant \exp(M\mathscr{P}(d^n))$.*

Let us conduct the proof by induction on n. The base for $n = 1$ is well known (see e.g. Lang, 1965). Consider an arbitrary n. If U_0 has a non-empty intersection with one of the coordinate hyperplanes $\{X_i = 0\}$, where $1 \leqslant i \leqslant n$, then the statement of the lemma follows from the inductive hypothesis. So we can assume the opposite. In this case U_0 is situated in some cone of the kind

$$\{ \underset{1 \leqslant i \leqslant n}{\&} (\delta_i X_i > 0)\},$$

where $\delta_i \in \{-1, +1\}$, $1 \leqslant i \leqslant n$. Suppose for definiteness that

$$U_0 \subset \{ \underset{1 \leqslant i \leqslant n}{\&} (X_i < 0)\}.$$

We can assume w.l.o.g. that $h(x) \geqslant 0$ for all $x \in \tilde{\mathbb{Q}}^n$, squaring polynomial h if necessary.

Lemma 5 implies that one can find a vector $\gamma = (\gamma_2, \ldots, \gamma_n) \in \Gamma$ such that any solution from the space F_1^n of the system

$$h - \varepsilon_1 = \left(\frac{\partial h}{\partial X_2}\right)^2 - \frac{\gamma_2}{Nn} \sum_{1 \leqslant j \leqslant n} \left(\frac{\partial h}{\partial X_j}\right)^2 = \ldots = \left(\frac{\partial h}{\partial X_n}\right)^2 - \frac{\gamma_n}{Nn} \sum_{1 \leqslant j \leqslant n} \left(\frac{\partial h}{\partial X_j}\right)^2 = 0 \quad (5)$$

is an isolated point in the variety $\bar{W}^{(\varepsilon_1)}$ consisting of all solutions of the system from the space $\bar{F}_1^n$. Define

$$\alpha_j = \sqrt{\gamma_j/Nn} > 0, \; 2 \leqslant j \leqslant n \quad \text{and} \quad \alpha_1 = \sqrt{1 - \sum_{2 \leqslant j \leqslant n} \alpha_j^2} > 0.$$

Introduce the linear function

$$q(X_1, \ldots, X_n) = \alpha_1 X_1 + \ldots + \alpha_n X_n.$$

Let us pick a certain point $x^{(0)} = (x_1^{(0)}, \ldots, x_n^{(0)}) \in U_0$. Consider the closed simplex

$$S = \{ \underset{1 \leqslant i \leqslant n}{\&} (X_i \leqslant 0) \& q(X_1, \ldots, X_n) \geqslant q(x^{(0)})\}.$$

Then $x^{(0)} \in U_0 \cap S$ and the function q has the maximal value $0 > q_0 \in \tilde{\mathbb{Q}}$ on the limited closed set $U_0 \cap S$. Moreover, q_0 is the maximal value of q on the whole component U_0.

A set

$$A = \{q(X_1, \ldots, X_n) = q_0\} \cap U_0 \subset \tilde{\mathbb{Q}}^n$$

lies in the simplex

$$\left\{ \underset{1 \leqslant i \leqslant n}{\&} (X_i \leqslant 0) \& (q(X_1, \ldots, X_n) = q_0) \right\}$$

and therefore $A \subset \mathscr{D}_0(\tau)$ for an appropriate $\tau \in \mathbb{Q}$. Lemma 3 entails the coincidence of the sets

$$st(\{h = \varepsilon_1\} \cap \mathscr{D}_0(\tau + 1)) = (\{h = 0\} \cap \mathscr{D}_0(\tau + 1)) \subset \tilde{\mathbb{Q}}^n.$$

Let $U_0^{(1)} \subset \tilde{\mathbb{Q}}^n$ be a component of connectivity of the semi-algebraic set $\{h = 0\} \cap \mathscr{D}_0(\tau + 1)$ such that $A^{(1)} = A \cap U_0^{(1)} \neq \phi$. Obviously $U_0^{(1)} \subset U_0$. By virtue of lemma 1(b) there exist components of connectivity $W_1, \ldots, W_e$ of the semi-algebraic set

$$(\{h = \varepsilon_1\} \cap \mathscr{D}_0(\tau + 1)) \subset F_1^n$$

such that the equality $st(W_1 \cup \ldots \cup W_e) = U_0^{(1)}$ is true.

The function q has the maximal value $q_1 \in F_1$ on the closed limited semi-algebraic set $(W_1 \cup \ldots \cup W_e) \subset \mathscr{D}_0(\tau + 1)$ in view of the transfer principle (see section 1). Introduce the semi-algebraic set

$$B = \{q(X_1, \ldots, X_n) = q_1\} \cap (W_1 \cup \ldots \cup W_e).$$

Let us show that $st(B) \subset A^{(1)}$. For this purpose it is sufficient to check for any point $y^{(1)} \in B$ that $q(st(y^{(1)})) = q_0$. Indeed, pick a point $x \in A^{(1)} \subset U_0^{(1)}$, then there is a point $y \in W_1 \cup \ldots \cup W_e$ such that $st(y) = x$. Hence,

$$q_0 = q(x) = st(q(y)) \leqslant st(q(y^{(1)})) = st(q_1)$$

taking into account the definition of the set B. On the other hand, $st(y^{(1)}) \in U_0^{(1)} \subset U_0$, therefore $q(st(y^{(1)})) \leqslant q_0$, which was to be shown. One concludes in particular that $B \subset \mathscr{D}_0(\tau + 1/2)$.

The following proposition is well known (see e.g. Thorpe, 1979). Let a polynomial $\rho \in \mathbb{R}[X_1, \ldots, X_n]$ and a point $y^{(2)} \in \{\rho = 0\} \subset \mathbb{R}^n$ be such that the gradient

$$\left(\frac{\partial \rho}{\partial X_1}(y^{(2)}), \ldots, \frac{\partial \rho}{\partial X_n}(y^{(2)}) \right) \neq 0$$

and, apart from that, the linear form q reaches the maximum in the point $y^{(2)}$ on the set $\{\rho = 0\} \cap \mathscr{D}_{y^{(2)}}(\tau^{(2)})$ for a certain $\tau^{(2)} > 0$. Then

$$\left(\frac{\partial \rho}{\partial X_1}(y^{(2)}), \ldots, \frac{\partial \rho}{\partial X_n}(y^{(2)}) \right)$$

is collinear to the vector $(\alpha_1, \ldots, \alpha_n)$. According to the transfer principle this proposition remains also correct if we replace $\mathbb{R}$ by an arbitrary real closed field K.

Apply the proposition to the polynomial $\rho = h - \varepsilon_1$, the field $K = F_1$ and an arbitrary point $y^{(2)} \in B$. Since $y^{(2)} \in B \subset \{h = \varepsilon_1\}$, the gradient

$$\left(\frac{\partial h}{\partial X_1}(y^{(2)}), \ldots, \frac{\partial h}{\partial X_n}(y^{(2)}) \right) \neq 0$$

by virtue of lemma 4(a). As radius $\tau^{(2)}$ one can take $0 < \tau^{(2)} < 1/2$ such that

$$\mathscr{D}_{y^{(2)}}(\tau^{(2)}) \cap (W_1 \cup \ldots \cup W_e) = \mathscr{D}_{y^{(2)}}(\tau^{(2)}) \cap \{h = \varepsilon_1\} \subset \mathscr{D}_0(\tau + 1).$$

The proposition implies that the gradient

$$\left(\frac{\partial h}{\partial X_1}(y^{(2)}), \ldots, \frac{\partial h}{\partial X_n}(y^{(2)}) \right)$$

is collinear to the vector $(\alpha_1, \ldots, \alpha_n)$. Therefore, the point $y^{(2)}$ satisfies the system (5). Besides that, $y^{(2)}$ is an isolated point in the variety $\bar{W}^{(\varepsilon_1)} \subset \bar{F}_1^n$ (see above) and the point $st(y^{(2)}) \in A^{(1)} \subset A \subset U_0$, since $y^{(2)} \in B$.

As in section 2, applying the algorithm from proposition 2, one can produce a null-dimensional component $\bar{W}_1^{(\varepsilon_1)}$ of the variety $\bar{W}^{(\varepsilon_1)}$ irreducible over the field $\mathbb{Q}(\varepsilon_1)$ such that the point $y^{(2)} = (y_1^{(2)}, \ldots, y_n^{(2)}) \in \bar{W}_1^{(\varepsilon_1)}$. The algorithm constructs polynomials $\Phi_1, \ldots, \Phi_n \in \mathbb{Q}[\varepsilon_1][Z]$ for which $\Phi_i(y_i^{(2)}) = 0$, $1 \leqslant i \leqslant n$, furthermore, the bounds $\deg_{\varepsilon_1, Z}(\Phi_i) \leqslant \mathscr{P}(d^n)$ and $l(\Phi_i) \leqslant M\mathscr{P}(d^n)$ are fulfilled. Finally, $(st(\Phi_i))(st(y_i^{(2)})) = 0$ (remark that $st(y_i^{(2)})$ is definable, taking into account that the point $y^{(2)} \in B \subset \mathscr{D}_0(\tau + 1/2)$). Hence, $|st(y_i^{(2)})| \leqslant \exp(M\mathscr{P}(d^n))$ (see e.g. Lang, 1965; also Heindel, 1971), i.e. the point

$$st(y^{(2)}) \in U_0 \cap \mathscr{D}_0(\exp(M\mathscr{P}(d^n))),$$

which completes the proof of the lemma.

Let the polynomials $h_1, \ldots, h_k \in \mathbb{Z}[X_1, \ldots, X_n]$ satisfy the bounds $\deg h_i < d$; $l(h_i) \leqslant M$, $1 \leqslant i \leqslant k$. Consider an arbitrary component of connectivity W of the semi-algebraic set

$$\{(h_1 \geqslant 0) \& \ldots \& (h_k \geqslant 0)\} \subsetneqq \tilde{\mathbb{Q}}^n.$$

The following lemma generalises lemma 9 and estimates real solutions of a system of polynomial inequalities.

LEMMA 10. *For a suitable natural number*

$$R \leqslant \exp((M + \log k)\mathscr{P}(d^n))$$

both $\mathscr{D}_0(R) \cap W \neq \phi$ and $\mathscr{D}_0(R) \setminus W \neq \phi$ are valid.

PROOF. By virtue of lemma 9, for a suitable natural number

$$R \leqslant \exp((M + \log k)\mathscr{P}(d^n)),$$

if a polynomial $h \in \mathbb{Z}[X_1, \ldots, X_{n+1}]$ satisfies the bounds $\deg h \leqslant 2d$ and

$$l(h) \leqslant M + n \log d + \log k,$$

then any component of connectivity of the semi-algebraic set $\{h = 0\} \subset \tilde{\mathbb{Q}}^n$ has non-empty intersection with the ball $\mathscr{D}_0(R)$. We claim that this R is the one desired in the lemma.

At first we show that $W \cap \mathscr{D}_0(R) \neq \phi$. Take a set of indices $I \subset \{1, \ldots, k\}$ maximal with respect to inclusion, such that there exists a point $z \in W$ for which $h_i(z) = 0$ for every $i \in I$. Consider the component of connectivity $U_1 \subset \tilde{\mathbb{Q}}^n$ of the semi-algebraic set $\{\sum_{i \in I} h_i^2 = 0\}$ that contains the point $z \in U_1$ (if $I = \phi$ then $U_1 = \tilde{\mathbb{Q}}^n$). Lemma 9 entails that $U_1 \cap \mathscr{D}_0(R) \neq \phi$ taking into account that

$$l\left(\sum_{i \in I} h_i^2\right) \leqslant M + n \log d + \log k.$$

We assert that $U_1 \subset W$ (in the case when $I = \{1, \ldots, k\}$ the latter inclusion is trivial, so w.l.o.g. we assume further that $I \subsetneqq \{1, \ldots, k\}$). Suppose the contrary. Since U_1 is connected, there exists a point $z_0 \in U_1$ such that for any $\tau > 0$ the intersection $U_1 \cap \mathscr{D}_0(\tau)$

contains points from $U_1 \cap W$ as well as points not belonging to W. (In the case of the field $\mathbb{R}$, the latter statement is obvious, for an arbitrary real closed field one has to make use of the transfer principle.) Then $z_0 \in W$, since the semi-algebraic set

$$\{(h_1 \geqslant 0) \& \ldots \& (h_k \geqslant 0)\} \supset W$$

is determined by only unstrict inequalities and, hence, W is closed in the topology generated by the base consisting of all open balls.

On the other hand, there exists an index $j_0 \notin I$ for which $h_{j_0}(z_0) = 0$. Otherwise, for an appropriate $\tau_0 > 0$ and every point $v \in \mathscr{D}_{z_0}(\tau_0)$, the inequality $h_j(v) > 0$ is true for each $j \notin I$. Consider the component of connectivity U_2 of the intersection $U_1 \cap \mathscr{D}_{z_0}(\tau_0)$ which contains the point $z_0 \in U_2$. Then $U_2 \subset W$ taking into account that W is a component of connectivity of the set $\{(h_1 \geqslant 0) \& \ldots \& (h_k \geqslant 0)\}$. For a suitable τ_1 satisfying the inequalities $\tau_0 > \tau_1 > 0$, any component of connectivity (except U_2) of the intersection $U_1 \cap \mathscr{D}_{z_0}(\tau_0)$ has no common points with the ball $\mathscr{D}_{z_0}(\tau_1)$. This leads to a contradiction to the assumption that the intersection $U_1 \cap \mathscr{D}_{z_0}(\tau_1)$ has at least one point not belonging to W.

Thus, we have proved the existence of $j_0 \notin I$ such that $h_{j_0}(z_0) = 0$. This contradicts to the maximality of the set I of indices, therefore

$$W \cap \mathscr{D}_0(R) \supset U_1 \cap \mathscr{D}_0(R) \neq \phi.$$

From the above one can infer that if U is a component of connectivity of some non-empty semi-algebraic set

$$\{(h_1 > 0) \& \ldots \& (h_m > 0) \& (h_{m+1} \geqslant 0) \& \ldots \& (h_k \geqslant 0)\} \subset \tilde{\mathbb{Q}}^n,$$

then for a suitable natural number

$$R_1 \leqslant \exp\left((M + k)\mathscr{P}((md)^n)\right),$$

the intersection $U \cap \mathscr{D}_0(R_1) \neq \phi$. Indeed, introduce a new variable X_{n+1} and a semi-algebraic set

$$\{(X_{n+1}h_1 \ldots h_m - 1 \geqslant 0) \& (h_1 \geqslant 0) \& \ldots \& (h_m \geqslant 0) \& (h_{m+1} \geqslant 0) \& \ldots \& (h_k \geqslant 0)\} \subset \tilde{\mathbb{Q}}^{n+1}$$

(cf. beginning of the section). There exists a unique component of connectivity $U^{(1)}$ of the latter semi-algebraic set such that $\pi_2(U^{(1)}) = U$, where the linear projection π_2 is defined by the formula

$$\pi_2(X_1, \ldots, X_n, X_{n+1}) = (X_1, \ldots, X_n).$$

According to what we proved above, $U^{(1)} \cap \mathscr{D}_0(R_1) \neq \phi$ is fulfilled taking into account the bound

$$l(X_{n+1}h_1 \ldots h_m - 1) \leqslant M + kn \log d.$$

Hence, $U \cap \mathscr{D}_0(R_1) \neq \phi$.

Assume now that $\mathscr{D}_0(R) \subset W$. Then $h_i(x) \geqslant 0$ is valid for each point $x \in \mathscr{D}_0(R)$ and each $1 \leqslant i \leqslant k$. We can suppose w.l.o.g. that the polynomial h_1 is not non-negative everywhere on $\tilde{\mathbb{Q}}^n$. (Otherwise, if $h_i \geqslant 0$ on $\tilde{\mathbb{Q}}^n$ for each $1 \leqslant i \leqslant k$, then

$$\{(h_1 \geqslant 0) \& \ldots \& (h_k \geqslant 0)\} = \tilde{\mathbb{Q}}^n.)$$

Then the intersection $\{h_1 < 0\} \cap \mathscr{D}_0(R) \neq \phi$ by virtue of what we proved above. This contradicts to the assumption and completes the proof of the lemma.

Introduce now the polynomial

$$g_1 = (f_1 + \varepsilon_1) \ldots (f_{k+1} + \varepsilon_1) - \varepsilon_1^{k+1} \in \mathbb{Z}[\varepsilon_1][X_1, \ldots, X_n].$$

Apply the algorithm from section 2 (see lemma 8) to this polynomial and to the natural number $R+1$, where $R \leqslant \exp(M\mathscr{P}(kd)^n)$ is obtained from lemma 10, in which the polynomials $f_1, \ldots, f_{k+1}$ are taken as $h_1, \ldots, h_k$. As a result, the algorithm produces a representative set $\mathscr{S} \subset F_1^n$ for the semi-algebraic set $\{g_1 = 0\} \cap \mathscr{D}_0(R+1)$. The set $\mathscr{S}$ is a union of classes of points conjugate over the field $\mathbb{Q}(\varepsilon_1)$ (see lemma 8). Any class of conjugate points from the set $\mathscr{S}$ is given by a polynomial $\psi \in \mathbb{Z}[\varepsilon_1][Z]$ irreducible over field $\mathbb{Q}(\varepsilon_1)$ and by expressions for the coordinates

$$\zeta_i(\eta) = \sum_j \rho_i^{(j)} \eta^j \ (1 \leqslant i \leqslant n),$$

where $\rho_i^{(j)} \in \mathbb{Q}(\varepsilon_1)$, $0 \leqslant j < \deg_Z(\psi)$ and $\eta \in F_1$, $\psi(\eta) = 0$. Apart from that, the following bounds are valid:

$$\deg_{\varepsilon_1, Z}(\psi), \deg_{\varepsilon_1}(\zeta_i(\eta)) \leqslant \mathscr{P}((kd)^n) \quad \text{and} \quad l(\psi), l(\zeta_i(\eta)) \leqslant M\mathscr{P}(kd)^n.$$

Apply the algorithm from lemma 7 in the case of the field $H = \mathbb{Q}$ to each class of conjugate points $(\zeta_1(\eta), \ldots, \zeta_n(\eta))$ from the set $\mathscr{S}$. Thus, a finite set $\mathscr{R} \subset \bar{\mathbb{Q}}^n$ is obtained which contains the standard part of every point from the set $\mathscr{S}$, provided that the standard part is definable. Analogously, the set $\mathscr{R}$ is a union of classes of points conjugate over the field $\mathbb{Q}$ (see lemma 7).

Fix a certain class of conjugate points from $\mathscr{R}$ which is given by polynomial $\Phi \in \mathbb{Q}[Z]$ irreducible over the field $\mathbb{Q}$ and by expressions for the coordinates

$$\chi_i(\omega) = \sum_j \beta_i^{(j)} \omega^j,$$

where $\beta_i^{(j)} \in \mathbb{Q}$, $1 \leqslant i \leqslant n$, $0 \leqslant j < \deg_Z(\Phi)$ and $\omega \in \tilde{\mathbb{Q}}$, $\Phi(\omega) = 0$. Besides that, the following bounds are true:

$$\deg_Z(\Phi) \leqslant \mathscr{P}(kd)^n) \quad \text{and} \quad l(\Phi), l(\chi_i(\omega)) \leqslant M\mathscr{P}((kd)^n)$$

(see the proof of lemma 7). Write

$$q_e(\omega) = f_e(\chi_1(\omega), \ldots, \chi_n(\omega)), \quad 1 \leqslant e \leqslant k+1$$

for suitable polynomials $q_e \in \mathbb{Q}[Z]$. If the polynomial Φ divides the polynomial q_e for some $1 \leqslant e \leqslant k+1$, then $q_e(\omega) = 0$ and vice versa. Select all such polynomials q_e. Assume that $q_{e_0}(\omega) \neq 0$ for a certain $1 \leqslant e_0 \leqslant k+1$. Following, e.g. Heindel (1971), one can find a polynomial p_5 such that the inequalities

$$|\omega'' - \omega'|, |\omega' - \omega_1| > 2^{(-Mp_5((kd)^n))} = C$$

are correct for all roots $\omega_1 \in \tilde{\mathbb{Q}}$ of polynomial q_{e_0} and each pair of distinct real roots $\omega' \neq \omega''$ of polynomial the Φ. Here we make use of the inequalities

$$\deg_Z(q_e) \leqslant \mathscr{P}((kd)^n) \quad \text{and} \quad)l(q_e) \leqslant M\mathscr{P}((kd)^n).$$

Following, e.g. Heindel (1971), the algorithm yields for every real root $\omega_0 \in \tilde{\mathbb{Q}}$ of the polynomial Φ its rational approximation $\omega_2 \in \mathbb{Q}$ which satisfies the inequality $|\omega_0 - \omega_2| < C/2$. Hence, an interval $(\omega_2 - C/2, \omega_2 + C/2) \subset \tilde{\mathbb{Q}}$ with the rational endpoints contains only one root ω_0 of the polynomial Φ. After that the algorithm evaluates $q_{e_0}(\omega_2)$ for all indices $1 \leqslant e_0 \leqslant k+1$ such that $q_{e_0}(\omega) \neq 0$ (observe that the latter is valid iff $q_{e_0}(\omega_0) \neq 0$). If $q_{e_0}(\omega_2) > 0$ is fulfilled for an index e_0, then the algorithm includes the point $(\chi_1(\omega_0), \ldots, \chi_n(\omega_0))$ into the set $\mathscr{T}'$, where $\mathscr{T}' \subset \mathscr{R}$ is the required representative set of the semi-algebraic set

$$V = \{(f_1 \geqslant 0) \& \ldots \& (f_{k+1} \geqslant 0)\} \subset \tilde{\mathbb{Q}}^n$$

(see the beginning of the present section). This completes the description of the algorithm.

Now we shall show that $\mathcal{T}'$ is really a representative set for the set V. Furthermore, we determine bounds on the parameters of the points from $\mathcal{T}'$ and on the running time of the algorithm which produces $\mathcal{T}'$. These bounds are similar to the corresponding bounds in the theorem (see the introduction). Let us check at first that $\mathcal{R} \cap V = \mathcal{T}'$. Let a point $(\chi_1(\omega_0), \ldots, \chi_n(\omega_0)) \in \mathcal{T}'$. Assume that $q_{e_0}(\omega_0) \neq 0$ for a certain $1 \leqslant e_0 \leqslant k+1$ (see above). Suppose that $q_{e_0}(\omega_0) < 0$. Then there exists a root $\omega_3 \in \tilde{\mathbb{Q}}$ of the polynomial q_{e_0} which lies between ω_0 and ω_2 since $q_{e_0}(\omega_2) > 0$. Therefore $|\omega_0 - \omega_3| < C/2$, so the supposition leads to a contradiction taking into account that $\Phi(\omega_0) = 0$. Hence,

$$f_{e_0}(\chi_1(\omega_0), \ldots, \chi_n(\omega_0)) = q_{e_0}(\omega_0) > 0$$

for all $1 \leqslant e_0 \leqslant k+1$ such that $q_{e_0}(\omega_0) \neq 0$, i.e. $(\chi_1(\omega_0), \ldots, \chi_n(\omega_0)) \in V$, thus $\mathcal{T}' \subset \mathcal{R} \cap V$.

Conversely, let a point

$$\chi = (\chi_1(\omega_0), \ldots, \chi_n(\omega_0)) \in \mathcal{R} \cap V.$$

Then

$$\omega_0 = \sum_{1 \leqslant j \leqslant n} \lambda_j' \chi_j(\omega_0)$$

for the appropriate natural numbers λ_j' (see lemma 7), therefore $\omega_0 \in \tilde{\mathbb{Q}}$. Assume that the point $\chi \notin \mathcal{T}'$. Then one can find an index $1 \leqslant e_0 \leqslant k+1$ such that $q_{e_0}(\omega_0) \neq 0$ and $q_{e_0}(\omega_2) < 0$ (see the construction of $\mathcal{T}'$ above). On the other hand, $q_{e_0}(\omega_0) > 0$ since $\chi \in V$. Hence, there exists a real root of the polynomial q_{e_0} between the numbers ω_0 and ω_2, which leads to a contradiction similar to the above.

Let us consider now an arbitrary component of connectivity $V^{(1)}$ of the semi-algebraic set V and prove that $V^{(1)} \cap \mathcal{T}' \neq \phi$. Remark that the set $V^{(1)}$ is closed (see the proof of lemma 10 above). Lemma 10 implies that $\mathcal{D}_0(R) \cap V^{(1)} \neq \phi$ and $\mathcal{D}_0(R) \backslash V^{(1)} \neq \phi$. Therefore, one can find a certain point

$$x \in (\partial V^{(1)}) \cap \mathcal{D}_0(R) \subset V^{(1)} \cap \mathcal{D}_0(R).$$

Denote by $V^{(2)}$ the unique component of connectivity of the semi-algebraic set $V \cap \mathcal{D}_0(R+1)$ that contains the point $x \in V^{(2)}$. Observe that $V^{(2)} \subset V^{(1)}$ in view of the connectivity of $V^{(1)}$.

Introduce the semi-algebraic set

$$W = \{(f_1 + \varepsilon_1 > 0) \& \ldots \& (f_{k+1} + \varepsilon_1 > 0) \& (g_1 \geqslant 0)\} \subset F_1^n.$$

Lemma 2(d) entails the existence of a point $z \in W \cap \{g_1 = 0\}$ such that each coordinate of the point $(z - x)$ is infinitesimal relatively to the field $\mathbb{Q}$. In particular, the point $z \in \mathcal{D}_0(R+1)$. According to lemma 2(a),

$$st(W \cap \mathcal{D}_0(R+1)) = (V \cap \mathcal{D}_0(R+1)).$$

Pick all components of connectivity $W^{(1)}, \ldots, W^{(e)}$ of the semi-algebraic set $W \cap \mathcal{D}_0(R+1)$, for which

$$st(W^{(j)}) \subset V^{(2)}, \quad 1 \leqslant j \leqslant e.$$

Then by virtue of lemma 1(b),

$$V^{(2)} = st(W^{(1)} \cup \ldots \cup W^{(e)}).$$

Taking into account that $st(z) = x$, we deduce by lemma 1(b) that the point $z \in W^{(1)} \cup \ldots \cup W^{(e)}$. Let w.l.o.g. $z \in W^{(1)}$.

Denote by U the unique component of connectivity of the semi-algebraic set $\{g_1 = 0\} \cap \mathcal{D}_0(R+1)$ that contains the point $z \in U$. One can infer that $U \subset W \cap \mathcal{D}_0(R+1)$ by lemma 2(c) since $z \in U \cap W^{(1)}$. This implies the inclusion $U \subset W^{(1)}$ in view of the connectivity of U. Hence, $st(U) \subset V^{(2)}$.

Involving lemma 8 from section 2, the algorithm described above produces a representative set $\mathscr{S}$ for the semi-algebraic set $\{g_1 = 0\} \cap \mathscr{D}_0(R+1)$. In particular, there is a certain point $u \in U \cap \mathscr{S}$. Then $st(u) \in V^{(2)} \subset V^{(1)}$ by what we proved above. Besides that, $st(u) \in \mathscr{R}$ according to the construction of $\mathscr{R}$ (cf. lemma 7). Thus, the point $st(u) \in \mathscr{T}' \cap V^{(1)}$. So we have shown that $\mathscr{T}'$ is a representative set for the set V.

To complete the proof of the theorem (see the introduction), it remains to estimate the running time of the algorithm and the parameters of the points (and their number) in the set $\mathscr{T}'$. For producing the representative set $\mathscr{S}$ for the set $\{g_1 = 0\} \cap \mathscr{D}_0(R+1)$ the algorithm works within time $\mathscr{P}(M(kd)^{n^2})$ by lemma 8. The running time of the algorithm from lemma 7, which constructs the set $\mathscr{R}$, can be estimated by the same bound. Thereby the algorithm yields a rational approximation $\omega_2 \in \mathbb{Q}$ to a real root $\omega_0 \in \tilde{\mathbb{Q}}$ of the polynomial Φ such that $|\omega_0 - \omega_2| < C/2$. This requires no more time than $\mathscr{P}(M(kd)^n)$ (see e.g. Heindel, 1971). Also, within the latter time-bound one can evalute $q_{e_0}(\omega_2)$. Thus, the time-bound desired in the theorem is guaranteed.

The number of points in the set $\mathscr{S}$ does not exceed $\mathscr{P}((kd)^{n^2})$, according to lemma 8. Hence, the number of points in the set $\mathscr{T}'$ also is bounded by $\mathscr{P}((kd)^{n^2})$ in view of the construction in the proof of lemma 7. The estimations for $\deg_Z(\Phi)$ and $l(\Phi)$, $l(\chi_i(\omega))$ were obtained earlier. Based on Heindel (1971), we get the inequalities $l(C)$, $l(\omega_2) \leqslant M\mathscr{P}((kd)^n)$. This entails the bounds on $l(b_1)$, $l(b_2)$, where $b_1 = \omega_2 - C/2$, $b_2 = \omega_2 + C/2 \in \mathbb{Q}$ (see above). This completes the proof of the theorem.

In conclusion we make a remark that the theorem allows a direct generalisation to input polynomials $f_1, \ldots, f_k \in H[X_1, \ldots, X_n]$, where the field H is, for example, of the kind $H = \mathbb{Q}(\varepsilon_1, \ldots, \varepsilon_s)$, where ε_{i+1} is infinitesimal relatively to the field $\mathbb{Q}(\varepsilon_1, \ldots, \varepsilon_i)$ for $0 \leqslant i < s$. The method of the proof and all the above constructions are the same, only the form of the necessary bounds has to be changed respectively.

4. General Outline of the Algorithm

Before implementing the algorithm one has to fulfil the following routine work (fortunately, this has only to be done once). Namely, one has to determine the bounds in the proofs of propositions 1, 2, lemmas 9, 10 "hidden" in the denotation $\mathscr{P}$ and, as a result, to obtain a specified polynomial $p \in \mathbb{Z}[Z]$ with non-negative coefficients satisfying the following property (cf. lemma 10).

Let some polynomials $h_1, \ldots, h_k \in \mathbb{Z}[X_1, \ldots, X_n]$ satisfy the bounds $\deg h_i < d$; $l(h_i) \leqslant M$, $1 \leqslant i \leqslant k$. Let us set the integer $R = 3^{((M + [\log k])p(d^n))}$. Then for each component of connectivity W of the semi-algebraic set

$$\{(h_1 \geqslant 0) \& \ldots \& (h_k \geqslant 0)\} \subsetneqq \tilde{\mathbb{Q}}^n$$

both $\mathscr{D}_0(R) \cap W \neq \phi$ and $\mathscr{D}_0(R) \backslash W \neq \phi$ are valid.

Now we proceed to exposing an outline of the algorithm.

Let an input system of inequalities

$$f_1 > 0, \ldots, f_m > 0, f_{m+1} \geqslant 0, \ldots, f_k \geqslant 0$$

be given (see (1)), where the polynomials $f_1, \ldots, f_k \in \mathbb{Z}[X_2, \ldots, X_{n-1}]$ satisfy the bounds

$$\deg_{X_2, \ldots, X_{n-1}}(f_i) < d; \quad l(f_i) \leqslant M, \quad 1 \leqslant i \leqslant k$$

(see (2)). Introduce a new variable X_1 and a polynomial $f_{k+1} = X_1 f_1 \ldots f_m - 1$. Denote by π_1 a linear projection defined by the formula

$$\pi_1(X_1, \ldots, X_{n-1}) = (X_2, \ldots, X_{n-1}).$$

The algorithm then yields a representative set $\mathcal{T}' \subset \tilde{\mathbb{Q}}^{n-1}$ for the system

$$f_1 \geqslant 0, \ldots, f_m \geqslant 0, f_{m+1} \geqslant 0, \ldots, f_k \geqslant 0, f_{k+1} \geqslant 0.$$

Then as the representative set $\mathcal{T}$ required in the theorem, the algorithm takes $\mathcal{T} = \pi_1(\mathcal{T}')$.

As at the beginning of the present section we take R corresponding to the polynomials $f_1, \ldots, f_{k+1}$.

Let us introduce new variables ε_1, ε, X_n and a polynomial

$$g_1 = (f_1 + \varepsilon_1) \ldots (f_{k+1} + \varepsilon_1) - \varepsilon_1^{k+1} \in \mathbb{Z}[\varepsilon_1][X_1, \ldots, X_{n-1}].$$

Consider now the polynomial

$$g = g_1^2 + (X_1^2 + \ldots + X_n^2 - (R+1))^2 \in \mathbb{Z}[\varepsilon_1][X_1, \ldots, X_n].$$

Define $N = (8kd)^n$ and introduce a family $\Gamma \subset \mathbb{Z}^{n-1}$ consisting of N^{n-1} integer vectors $\Gamma = \{\gamma = (\gamma_2, \ldots, \gamma_n)\}$, where each γ_i ranges independently over all values $1, \ldots, N$. The algorithm looks over all elements of Γ. Let us fix a vector $\gamma = (\gamma_2, \ldots, \gamma_n) \in \Gamma$, and consider the following system of equations where

$$\Delta = \sum_{1 \leqslant j \leqslant n} (\partial g / \partial X_j)^2 \text{ (cf. (4n))}:$$

$$g - \varepsilon = \left(\frac{\partial g}{\partial X_2} \right)^2 - \frac{\gamma_2}{Nn} \Delta = \ldots = \left(\frac{\partial g}{\partial X_n} \right)^2 - \frac{\gamma_n}{Nn} \Delta = 0. \tag{6}$$

Denote by $\bar{V}^{(\varepsilon)} \subset \bar{F}^n$ the variety of solutions of the latter system. Now the algorithm from proposition 2 is applied to system (6) and, as a result, components $\bar{V}_j^{(\varepsilon)}$ of the variety

$$\bar{V}^{(\varepsilon)} = \bigcup_j \bar{V}_j^{(\varepsilon)}.$$

irreducible over the field $\mathbb{Q}(\varepsilon_1, \varepsilon)$ are produced. After that the algorithm selects all the null-dimensional components.

Then the algorithm, at first for each null-dimensional component, yields a certain finite family $\mathcal{R}_1 \subset \bar{F}_1^n$ containing the standard part (relatively to ε) of every point (for which the standard part is definable) from this component. The procedure for obtaining such an $\mathcal{R}_1$ is described in details in section 2 (see lemma 7). Then the algorithm selects the points from $\mathcal{R}_1$ lying in the semi-algebraic set $\{g = 0\} \cap F_1^n$ (see section 2). Collecting the points produced for all vectors $\gamma \in \Gamma$ and for all null-dimensional components $\bar{V}_j^{(\varepsilon)}$, the algorithm gets a representative set $\mathcal{S}' \subset F_1^n$ for the equation $g = 0$ (see lemma 8). Denote by π a linear projection defined by a formula

$$\pi(X_1, \ldots, X_n) = (X_1, \ldots, X_{n-1}).$$

The algorithm produces the set $\mathcal{S} = \pi(\mathcal{S}') \subset F_1^{n-1}$.

After that the algorithm applies the analogous procedure to the points from the set $\mathcal{S}$. As a result the algorithm obtains a finite set $\mathcal{R} \subset \tilde{\mathbb{Q}}^{n-1}$ containing the standard part (relative to ε_1) of every point (for which the standard part is definable) from the set $\mathcal{S}$ (see section 3).

Finally, the algorithm constructs the required set $\mathcal{T}' \subset \mathcal{R}$ as a subset of $\mathcal{R}$ consisting of all those points from $\mathcal{R}$ that belong to the space $\tilde{\mathbb{Q}}^{n-1}$ (i.e. have real coordinates) and satisfy inequalities $f_1 \geqslant 0, \ldots, f_k \geqslant 0, f_{k+1} \geqslant 0$ (see section 3). At last, set $\mathcal{T} = \pi_1(\mathcal{T}')$.

References

Chistov, A. L., Grigor'ev, D. Yu. (1982). *Polynomial-time factoring of multivariable polynomials over a global field.* Preprint LOMI E–5–82, Leningrad.

Chistov, A. L., Grigor'ev, D. Yu. (1983*a*). *Subexponential-time solving systems of algebraic equations. I.* Preprint LOMI E-9-83, Leningrad.

Chistov, A. L., Grigor'ev, D. Yu. (1983*b*). *Subexponential-time solving systems of algebraic equations. II.* Preprint LOMI E-10-83, Leningrad.

Chistov, A. L. (1984). Polynomial-time factoring of polynomials and finding compounds of a variety within subexponential time. *Notes of Sci. Seminars of Leningrad Department of Math. Steklov Inst.* **137**, 124–188 (in Russian, English transl. to appear in J. Soviet Math.).

Chistov, A. L., Grigor'ev, D. Yu. (1984). Complexity of quantifier elimination in the theory of algebraically closed fields. *Springer Lec. Notes Comp. Sci.* **176**, 17–31.

Collins, G. E. (1975). Quantifier elimination for real closed fields by cylindrical algebraic decomposition. *Springer Lec. Notes Comp. Sci.* **33**, 134–183.

Grigor'ev, D. Yu. (1984). Factoring multivariable polynomials over a finite field and solving systems of algebraic equations. *Notes of Sci. Seminars of Leningrad Department of Math. Steklov Inst.* **137**, 20–79 (in Russian, English transl. to appear in J. Soviet Math.).

Grigor'ev, D. Yu. (1987). Computational complexity in polynomial algebra. Proc. Intern. Congress of Mathematicians, Berkeley.

Heindel, L. E. (1971). Integer arithmetic algorithms for polynomial real zero determination. *J. Assoc. Comp. Mach.* **18/4**, 533–548.

Heintz, J. (1983). Definability and fast quantifier elimination in algebraically closed field. *Theor. Comp. Sci.* **24**, 239–278.

Khachian, L. G. (1979). A polynomial algorithm in linear programming. *Soviet Math. Doklady* **20/1**, 191–194.

Lang, S. (1965). *Algebra.* New York: Addison-Wesley.

Lenstra, A. R. (1984). Factoring multivariable polynomials over algebraic number fields. *Springer Lec. Notes Comp. Sci.* **176**, 389–396.

Milnor, J. (1964). On the Betti numbers of real varieties. *Proc. Amer. Math. Soc.* **15/2**, 275–280.

Milnor, J. (1965). *Topology from the Differentiable Viewpoint.* University Press of Virginia.

Mignotte, M. (1974). An inequality about factors of polynomials. *Math. Comput.* **28/128**, 1153–1157.

Shafarevich, I. R. (1974). *Basic Algebraic Geometry.* Berlin: Springer-Verlag.

Tarski, A. (1951). *A Decision Method for Elementary Algebra and Geometry.* University of California Press.

Thorpe, J. A. (1979). *Elementary Topics in Differential Geometry.* Berlin: Springer-Verlag.

Vorobjov, N. N., Jr. (1984). Bounds of real roots of a system of algebraic equations. *Notes of Sci. Seminars of Leningrad Department of Math. Steklov Inst.* **137**, 7–19 (in Russian, English transl. to appear in J. Soviet Math.).

Vorobjov, N. N., Jr., Grigor'ev, D. Yu. (1985). Finding real solutions of systems of algebraic inequalities within the subexponential time. *Soviet Math. Dokl.* **32** (1), 316–320.

Wüthrich, H. R. (1976). Ein Entscheidungsverfahren für die Theorie der reell-abgeschlossenen Körper. In: Komplexität von Entschiedungs Problemen, (Specker, E, Strassen, V., Eds). *Springer Lec. Notes Comp. Sci.* **43**, 138–162.

J. Symbolic Computation (1988) **5**, 65–108

Complexity of Deciding Tarski Algebra

D. Yu GRIGOR'EV

*Leningrad Department of Steklov Mathematical Institute
of the Academy of Sciences of the USSR,
Fontanka embankment* 27, *Leningrad* 191011, *USSR*

(*Received* 20 *March* 1985)

Let a formula of Tarski algebra contain k atomic subformulas of the kind $(f_i \geqslant 0)$, $1 \leqslant i \leqslant k$, where the polynomials $f_i \in \mathbb{Z}[X_1, \ldots, X_n]$ have degrees $\deg(f_i) < d$, let 2^M be an upper bound for the absolute value of every coefficient of the polynomials f_i, $1 \leqslant i \leqslant k$, let $a \leqslant n$ be the number of quantifier alternations in the prenex form of the formula. A decision method for Tarski algebra is described with the running time polynomial in $M(kd)^{(O(n))^{4a-2}}$. Previously known decision procedures have a time complexity polynomial in $(Mkd)^{2^{O(n)}}$.

Introduction

The decidability of the first order theory of real closed fields (or, in other words, Tarski algebra) was proved for the first time in Tarski (1951) (see also Seidenberg, 1954; Cohen 1969). Moreover, quantifier elimination methods for Tarski algebra were suggested in these papers. The running time, however, of the method from Tarski (1951) cannot be bounded by any finite tower of exponential functions. Collins (1975) (see also Monk, 1974; Wüthrich, 1976) proposes a quantifier elimination procedure for Tarski algebra with a running time bounded by $\mathscr{L}^{5^{O(n)}}$, where $\mathscr{L}$ denotes the size of an input formula and n is the number of variables occurring in the formula. In the present paper we present a decision algorithm for Tarski algebra that works within time $\mathscr{L}^{(O(n))^{4a-2}}$, where $a \leqslant n$ denotes the number of quantifier alternations in the input formula which can be assumed w.l.o.g. to be in the prenex form.

So the running time of the algorithm described is essentially better than that of the algorithms known earlier in the case of a small number a of quantifier alternations. One can observe that Fischer-Rabin (1974) prove an exponential lower bound on the complexity of deciding Tarski algebra for a sequence of formulas in which the order of growth of the number a of quantifier alternations is the same as the order of growth of the number n of variables. Thus, the number a of quantifier alternations makes the most essential contribution to the complexity of the decision problem. Further, the result of Fischer–Rabin was generalized in Berman (1980) to alternating Turing machines. Besides, EXPSPACE-completeness of the decision problem of Tarski algebra was proved in Mayr & Meyer (1982) and Ben-Or *et al.* (1984). The time bound of the algorithm from the latter paper is the same as in Collins (1975).

The present paper continues (Grigor'ev & Vorobjov, 1987) (see also Vorobjov & Grigor'ev, 1985) and uses its main result (see below, theorem 1 in section 1). One can even consider the present paper as a generalization of Grigor'ev & Vorobjov (1987), where an algorithm is described for finding real solutions of a given system of polynomial inequalities (i.e. $a = 1$ in the notations adopted in the present paper). In section 3 a certain construction from Chistov & Grigor'ev (1984) is involved. Chistov & Grigor'ev (1984) present a quantifier elimination method for the first-order theory of the algebraically closed fields with time-bound $\mathscr{L}^{(O(n))^{2a+1}}$.

Let

$$\exists X_{1,1} \ldots \exists X_{1,s_1} \forall X_{2,1} \ldots \forall X_{2,s_2} \ldots \exists X_{a,1} \ldots \exists X_{a,s_a}(P) \tag{1}$$

be a formula of Tarski algebra, where P is a quantifier-free formula with atomic subformulas $(f_i \geqslant 0)$, $1 \leqslant i \leqslant k$, and $f_i \in \mathbb{Z}[X_{1,1}, \ldots, X_{1,s_1}, \ldots, X_{a,1}, \ldots, X_{a,s_a}]$. We denote by $n = s_1 + \ldots + s_a$ the number of all the variables and by $a \leqslant n$ the number of quantifier alternations (in the presentation of the formula (1) a is odd, but this is not essential). The degrees $\deg_{X_{1,1}, \ldots, X_{a,s_a}}(f_i) < d$ and the absolute value of each integer coefficient of a polynomial f_i is supposed to be less than 2^M, $1 \leqslant i \leqslant k$. For a rational function $g \in \mathbb{Q}(X_1, \ldots, X_n)$ we denote by $l(g)$ the maximum of the bit lengths of the coefficients in relatively prime polynomials $g_1, g_2 \in \mathbb{Z}[X_1, \ldots, X_n]$, where $g = g_1/g_2$. In this notation $l(f_i) \leqslant M$, $1 \leqslant i \leqslant k$.

A decision method for Tarski algebra is an algorithm determining for any formula (1), whether it is valid. By validity of a formula we mean here validity over the field $\mathbb{R}$ of real numbers. On the other hand, one can consider validity over an arbitrary real closed field F (see e.g. Lang, 1965) by virtue of the following statement expressing the fact that all the real closed fields are elementary equivalent (Tarski, 1951) and that any extension of real closed fields is elementary.

Transfer principle. Let a formula (1) be such that the polynomials $f_i \in F[X_{1,1}, \ldots, X_{a,s_a}]$ for some real closed field F, and let $F_1 \supset F$ be any other real closed field containing F. Then the truth values of the formula over the fields F and F_1, respectively, are the same.

The main purpose of the present paper is to prove the following theorem (see also Grigor'ev, 1985; Grigor'ev, 1987).

THEOREM. *One can design a decision algorithm for Tarski algebra which determines the truth value of a formula of the kind* (1) *within a time polynomial in* $M(kd)^{(O(n))^{4a-2}}$.

The latter estimate does not exceed $\mathscr{L}^{(O(n))^{4a-2}}$ where $\mathscr{L}$ denotes the size of the formula (1) (cf. Grigor'ev & Vorobjov, 1987).

We shall utilise the notation $g_1 \leqslant \mathscr{P}(g_2, \ldots, g_m)$ for the functions $g_1 > 0, \ldots, g_m > 0$ if for appropriate natural numbers p, q the inequality $g_1 \leqslant p(g_2 \ldots g_m)^q$ holds.

Let us mention that the running time of the algorithm from Collins (1975) (and also Wüthrich, 1976) can be bounded by $\mathscr{P}((Mkd)^{2^{O(n)}})$.

Further, for the proof of the theorem we need the algorithms from Chistov & Grigor'ev (1982, 1983a, b), Chistov (1984), Grigor'ev (1984), also Chistov & Grigor'ev (1984) on polynomial factoring and on solving systems of algebraic equations. Now we formulate exactly these results. Taking into account that only zero characteristic fields are considered in the present paper and in order to avoid some swelling of formulas due to inseparable fields extensions we restrict ourselves herein to the zero characteristic case.

Thus, consider a ground field $F = \mathbb{Q}(T_1, \ldots, T_e)[\eta]$, where the elements $T_1, \ldots, T_e$ are algebraically independent over $\mathbb{Q}$, the element η is algebraic over the field $\mathbb{Q}(T_1, \ldots, T_e)$, denote by

$$\varphi = \sum_{0 \leqslant i \leqslant \deg_Z(\varphi)} (\varphi_i^{(1)}/\varphi^{(2)})Z^i \in \mathbb{Q}(T_1, \ldots, T_e)[Z]$$

its minimal polynomial over $\mathbb{Q}(T_1, \ldots, T_e)$ with the leading coefficient $lc_Z(\varphi) = 1$, where $\varphi_i^{(1)}$, $\varphi^{(2)} \in \mathbb{Z}[T_1, \ldots, T_e]$ and the degree $\deg(\varphi^{(2)})$ is the least possible. Any polynomial $f \in F[X_1, \ldots, X_n]$ can be uniquely represented in a form

$$f = \sum_{0 \leqslant i < \deg_Z(\varphi); i_1, \ldots, i_n} (a_{i, i_1, \ldots, i_n}/b)\eta^i X_1^{i_1} \ldots X_n^{i_n},$$

where $a_{i, i_1, \ldots, i_n}, b \in \mathbb{Z}[T_1, \ldots, T_e]$ and the degree $\deg(b)$ is the least possible. Let

$$\deg_{T_j}(f) = \max_{i, i_1, \ldots, i_n} \{\deg_{T_j}(a_{i, i_1, \ldots, i_n}), \deg_{T_j}(b)\}.$$

Let $\deg_{X_m}(f) < \tau$, $\deg_{T_j}(f) < \tau_2$, $\deg_{T_j}(\varphi) < \tau_1$, $\deg_Z(\varphi) < \tau_1$, $l(f) \leqslant M_2$, $l(\varphi) \leqslant M_1$ for all $1 \leqslant m \leqslant n$, $1 \leqslant j \leqslant e$. As a size $L_1(f)$ of the polynomial f we consider in proposition 1 a value $\tau^{n+e}\tau_2^e\tau_1 M_2$ and analogously $L_1(\varphi) = \tau_1^{e+1}M_1$.

PROPOSITION 1 (Chistov & Grigor'ev, 1982; Chistov, 1984; Grigor'ev, 1984; Chistov & Grigor'ev, 1984). *One can factor a polynomial f over F within time polynomial in the sizes $L_1(f)$, $L_1(\varphi)$. Furthermore, for any divisor $f_1|f$ where a polynomial $f_1 \in F[X_1, \ldots, X_n]$ has some coefficients equal to 1, the following bounds are true:*

$$\deg_{T_j}(f_1) \leqslant \tau_2 \mathscr{P}(\tau, \tau_1), \qquad l(f_1) \leqslant (M_1 + M_2 + e\tau_2 + n)\mathscr{P}(\tau, \tau_1).$$

For the cases when the field F is finite or F is a finite extension of $\mathbb{Q}$, other polynomial-time algorithms for factoring are described in Lenstra (1984).

Now we proceed to the problem of solving systems of algebraic equations. Let an input system $f_1 = \ldots = f_k = 0$ be given, where the polynomials $f_1, \ldots, f_k \in F[X_1, \ldots, X_n]$. Let $\deg_{X_1, \ldots, X_n}(f_i) < d$, $\deg_{T_1, \ldots, T_e, Z}(\varphi) < d_1$, $\deg_{T_1, \ldots, T_e}(f_i) < d_2$, $l(f_i) \leqslant M_2$ for all $1 \leqslant i \leqslant k$. As size of the system in proposition 2 we consider the value

$$L = kd^n d_1 d_2^e M_2 + d_1^{e+1}M_1.$$

The variety $\mathscr{W} \subset \bar{F}^n$ of all roots (defined over the algebraic closure $\bar{F}$ of the field F) of the system $f_1 = \ldots = f_k = 0$ is decomposable in a union of components

$$\mathscr{W} = \bigcup_\alpha W_\alpha$$

defined and irreducible over the field F. The algorithm from proposition 2 finds the components W_α and outputs every W_α in two ways: by its general point (see below) and, on the other hand, by a certain system of algebraic equations such that W_α coincides with the variety of all roots of this system.

Let $W \subset \bar{F}^n$ be a closed variety of dimension $\dim W = n - m$, defined and irreducible over F. Denote by $t_1, \ldots, t_{n-m}$ some algebraically independent elements over F. A general point of the variety W can be given by the following fields isomorphism:

$$F(t_1, \ldots, t_{n-m})[\theta] \simeq F(X_1, \ldots, X_n) = F(W), \tag{*}$$

where the element θ is algebraic over the field $F(t_1, \ldots, t_{n-m})$, denote by $\Phi(Z) \in F(t_1, \ldots, t_{n-m})[Z]$ its minimal polynomial over $F(t_1, \ldots, t_{n-m})$ with leading coefficient $lc_Z(\Phi) = 1$. The elements $X_1, \ldots, X_n$ are considered herein as the rational (coordinate) functions on the variety W. Under the isomorphism (*) $t_i \to X_{j_i}$ for suitable $1 \leqslant j_1 < \ldots < j_{n-m} \leqslant n$ where $1 \leqslant i \leqslant n - m$. Besides, θ is an image under isomorphism (*) of an appropriate linear function $\sum_{1 \leqslant i \leqslant n} c_i X_i$ where c_i are integers. The algorithm from proposition 2 represents the isomorphism (*) by the integers $c_1, \ldots, c_n$ together with the images of the coordinate functions $X_1, \ldots, X_n$ in the field $F(t_1, \ldots, t_{n-m})[\theta]$. Sometimes, in the formulation of proposition 2, we identify a rational function with its image under the isomorphism.

PROPOSITION 2 (Chistov & Grigor'ev, 1983a, b; Chistov, 1984; Grigor'ev, 1984; Chistov & Grigor'ev, 1984). *An algorithm can be designed which produces a general point of every component W_α and constructs a certain family of polynomials $\psi_\alpha^{(1)}, \ldots, \psi_\alpha^{(N)} \in F[X_1, \ldots, X_n]$ such that W_α coincides with the variety of all roots of the system $\Psi_\alpha^{(1)} = \ldots = \Psi_\alpha^{(N)} = 0$. Denote $n - m = \dim W_\alpha, \theta_\alpha = \theta, \Phi_\alpha = \Phi$ (see (*)). Then $\deg_Z(\Phi_\alpha) \leqslant \deg W_\alpha \leqslant d^m$, for all j, s the degrees*

$$\deg_{T_1, \ldots, T_e, t_1, \ldots, t_{n-m}}(\Phi_\alpha), \deg_{T_1, \ldots, T_e, t_1, \ldots, t_{n-m}}(X_j), \deg_{T_1, \ldots, T_e}(\psi_\alpha^{(s)}) \leqslant d_2 \mathscr{P}(d^m, d_1)$$

and

$$\deg_{X_1, \ldots, X_n}(\psi_\alpha^{(s)}) \leqslant d^{2m}.$$

The number of equations $N \leqslant m^2 d^{4m}$. Furthermore,

$$l(\Phi_\alpha), l(X_j) \leqslant (M_1 + M_2 + (n + e) \log d_2)\mathscr{P}(d^m, d_1)$$

and

$$l(\psi_\alpha^{(s)}) \leqslant (M_1 + M_2 + e \log d_2)\mathscr{P}(d^n, d_1).$$

Finally, the total running time of the algorithm can be bounded by $\mathscr{P}(M_1, M_2, (d^n d_1 d_2)^{n+e}, k)$. Obviously, the latter value does not exceed $\mathscr{P}(L^{\log L})$, in other words, is subexponential in the size.

Let us briefly describe the further contents of the paper. In section 1 we present an algorithm for producing a representative set for the partition of the space $\mathbb{R}^n$ into maximal connected subsets on each of which a given family of polynomials has constant signs (lemma 2).

In Grigor'ev & Vorobjov (1987) (see also Vorobjov & Grigor'ev, 1985) a relevant tool for calculations with the infinitesimals was introduced. It is developed further in section 2. We ascertain some properties of semialgebraic curves over real closed fields; in addition some properties of the decomposition of a semi-algebraic set in its semialgebraic components of connectivity under passage to the limit (or standard part), i.e. when zeros are substituted for infinitesimals (lemmas 3, 4, 5).

In section 3 we describe a construction allowing to reduce a projection of a semialgebraic set along many variables to a projection along two variables (lemma 10). In this connection the question, whether a point belongs to the projection of a semi-algebraic set, is replaced by the question, whether a point infinitely close to the initial one belongs to the projection (lemma 6).

At the beginning of section 4 it is proved that the construction suggested in section 3 has a relevant additional property (lemma 11) which, together with sections 1 and 3, completes the description of the decision algorithm (lemma 13).

An outline and time analysis of the decision algorithm described is carried out in section 5. This completes the proof of the theorem.

As an application of the theorem we give a procedure to compute the dimension of a semialgebraic set within subexponential time in the last section 6.

If we were supplied with either a polynomial-time procedure for eliminating one quantifier in Tarski algebra or with a polynomial-time algorithm for finding the components of connectivity of a given semialgebraic set, then it would be possible to design a quantifier elimination method for Tarski algebra with the same time-bound as in the theorem. This follows from the construction in section 3 (see lemma 10). The author would like to conjecture the existence of a quantifier elimination procedure with the same time-bound as in the theorem.

1. Producing a Representative Set for a Semialgebraic Set

Let polynomials $f_1, \ldots, f_k \in \mathbb{Q}[X_1, \ldots, X_n]$ with rational coefficients be given, in addition $\deg(f_i) = \deg_{X_1, \ldots, X_n}(f_i) < d$ and bit lengths $l(f_i) \leqslant M$, $1 \leqslant i \leqslant k$. Let Π be a certain quantifier-free formula of Tarski algebra with atomic subformulas of the kind $(f_i \geqslant 0)$. Then the set, consisting of all points from $\mathbb{R}^n$ satisfying formula Π, is a semialgebraic set which we denote by $\{\Pi\} \subset \mathbb{R}^n$. The set $V = \{\Pi\} = \bigcup_j V_j$ coincides with the union of its components of connectivity V_j, each of which is also a semialgebraic set (see e.g. Collins, 1975; Wüthrich, 1976; and also Grigor'ev & Vorobjov, 1987). The procedures described in Collins (1975) and Wüthrich (1976) find quantifier-free formulas Π_j of Tarski algebra such that $V_j = \{\Pi_j\}$ within time $(Mkd)^{2^{O(n)}}$.

A finite set $\mathscr{S} \subset V$ is called a representative set for the set V if for every j the intersection $V_j \cap \mathscr{S} \neq \phi$. With the help of procedures from Collins (1975) and Wüthrich (1976) one can produce a representative set for V also within time $(Mkd)^{2^{O(n)}}$. Further, we need the main result from Grigor'ev & Vorobjov (1987) (see also Vorobjov & Grigor'ev, 1985) considerably improving the latter time-bound.

THEOREM 1. *One can design an algorithm which, for an arbitrary semialgebraic set of the sort*

$$\{(f_1 > 0)\& \ldots \&(f_m > 0)\&(f_{m+1} \geqslant 0)\& \ldots \&(f_k \geqslant 0)\} \subset \mathbb{R}^n,$$

produces a representative set $\mathscr{S}$ with a number of points not greater than $\mathscr{P}((kd)^{n^2})$; the running time of the algorithm does not exceed $\mathscr{P}(M, (kd)^{n^2})$. Furthermore, for every point $(\xi_1, \ldots, \xi_n) \in \mathscr{S}$ the algorithm constructs a polynomial $\Phi \in \mathbb{Q}[Z]$ irreducible over the field $\mathbb{Q}$, and expressions

$$\xi_i = \xi_i(\theta) = \sum_{0 \leqslant j < \deg(\Phi)} \alpha_j^{(i)} \theta^j,$$

where $\alpha_j^{(i)} \in \mathbb{Q}$, $1 \leqslant i \leqslant n$, $0 \leqslant j < \deg(\Phi)$ and $\theta \in \mathbb{R}$, $\Phi(\theta) = 0$. Moreover, the algorithm yields a pair of rational numbers $b_1, b_2 \in \mathbb{Q}$ such that in the interval $(b_1, b_2) \subset \mathbb{R}$ θ is the unique real root of the polynomial Φ. Moreover,

$$\theta = \sum_{1 \leqslant i \leqslant n} \lambda_i \xi_i(\theta)$$

for suitable natural numbers $1 \leqslant \lambda_i \leqslant \deg(\Phi)$. Finally, the following bounds are valid:

$$\deg(\Phi) \leqslant \mathscr{P}((kd)^n); \qquad l(\Phi), l(\xi_i(\theta)), l(b_1), l(b_2) \leqslant M\mathscr{P}((kd)^n).$$

In particular, the coordinates of the points from $\mathscr{S}$ are real algebraic numbers.

REMARK. Based on theorem 1 and on a result from Heindel (1971), one can find δ-approximations to the points of the representative set $\mathscr{S}$ within time $\mathscr{P}(\log(1/\delta), M, (kd)^{n^2})$.

Following Heintz (1983) we call $(\{f_i\}_{1 \leqslant i \leqslant k})$-cell any nonempty semialgebraic set of the form

$$\left\{ \underset{i \in I}{\&} (f_i = 0)\& \underset{i_1 \in I_1}{\&} (f_{i_1} > 0)\& \underset{i_2 \in I_2}{\&} (f_{i_2} < 0) \right\},$$

where $I \cup I_1 \cup I_2 = \{1, \ldots, k\}$. The next lemma allows us to find all $(\{f_i\}_{1 \leqslant i \leqslant k})$-cells.

LEMMA 1. *One can enumerate all partitions of the set of indices $I \cup I_1 \cup I_2 = \{1, \ldots, k\}$ such that the corresponding semialgebraic set*

$$\{ \underset{i \in I}{\&} (f_i = 0) \& \underset{i_1 \in I_1}{\&} (f_{i_1} > 0) \& \underset{i_2 \in I_2}{\&} (f_{i_2} < 0) \}$$

is nonempty, i.e. is a $(\{f_i\}_{1 \leq i \leq k})$-cell, within time $\mathscr{P}(M, (kd)^{n^2})$. Furthermore, the number of $(\{f_i\}_{1 \leq i \leq k})$-cells does not exceed $\mathscr{P}((kd)^n)$.

PROOF. We shall conduct the enumeration of all $(\{f_i\}_{1 \leq i \leq k})$-cells by recursion on k. Assume that all $(\{f_i\}_{1 \leq i \leq j})$-cells are already selected for a certain $0 \leq j < k$. For each $(\{f_i\}_{1 \leq i \leq j})$-cell $K \subset \mathbb{R}^n$, the algorithm detects with the aid of theorem 1, what sets among the following three:

$$K \cap \{f_{j+1} = 0\}, \quad K \cap \{f_{j+1} > 0\}, \quad K \cap \{f_{j+1} < 0\}$$

are nonempty, i.e. are $(\{f_i\}_{1 \leq i \leq j+1})$-cells. Thus, all $(\{f_i\}_{1 \leq i \leq j+1})$-cells will be enumerated.

Let us now estimate the number of all $(\{f_i\}_{1 \leq i \leq k})$-cells. Any nonempty subset of $\mathbb{C}^n$ of the kind

$$\{ \underset{i \in I}{\&} (f_i = 0) \& \underset{i \notin I}{\&} (f_i \neq 0) \},$$

where $I \subset \{1, \ldots, k\}$ is called a complex $(\{f_i\}_{1 \leq i \leq k})$-cell (see Heintz, 1983).

According to Heintz (1983) the number of complex $(\{f_i\}_{1 \leq i \leq k})$-cells is less or equal to $(1 + \sum_{1 \leq i \leq k} \deg(f_i))^n$. For an arbitrary fixed subset of indices $I \subset \{1, \ldots, k\}$, the number of $(\{f_i\}_{1 \leq i \leq k})$-cells of the sort

$$\{ \underset{i \in I}{\&} (f_i = 0) \& \underset{i_1 \in I_1}{\&} (f_{i_1} > 0) \& \underset{i_2 \in I_2}{\&} (f_{i_2} < 0) \} \subset \mathbb{R}^n$$

for all possible I_1, I_2 does not exceed the number of components of connectivity (in $\mathbb{R}^n$) of the semialgebraic set

$$\{ \underset{i \in I}{\&} (f_i = 0) \& \underset{i \notin I}{\&} (f_i \neq 0) \} \subset \mathbb{R}^n,$$

because on every such component of connectivity all the signs $\mathrm{sgn}(f_i)$, $1 \leq i \leq k$ are constant. The set

$$\{ \underset{i \in I}{\&} (f_i = 0) \& \underset{i \notin I}{\&} (f_i \neq 0) \}$$

coincides with the projection of the semialgebraic set

$$V^{(1)} = \{ \underset{i \in I}{\&} (f_i = 0) \& (Z \prod_{i \notin I} f_i = 1) \} \subset \mathbb{R}^{n+1}$$

along the variable Z. By virtue of Milnor (1964) the number of components of connectivity of the set $V^{(1)}$ is less or equal to $(2kd)^{n+1}$. Therefore, the whole number of $(\{f_i\}_{1 \leq i \leq k})$-cells does not exceed

$$(1 + \sum_{1 \leq i \leq k} \deg(f_i))^n (2kd)^{n+1} \leq \mathscr{P}(kd)^n).$$

This entails also the required in the lemma time-bound in view of theorem 1 and completes the proof of the lemma.

Let the polynomials $g_1, \ldots, g_m \in \mathbb{R}[X_1, \ldots, X_n]$. Denote by $\mathscr{U}(\{g_i\}_{1 \leq i \leq m})$ the partition of the space $\mathbb{R}^n$ into maximal connected subsets on each of which the signs $\mathrm{sgn}(g_i)$ are constant for every $1 \leq i \leq m$ (cf. Wüthrich, 1976). Observe that the family of elements of the partition $\mathscr{U}(\{g_i\}_{1\leqslant i\leqslant m})$ coincides with the family of components of connectivity of all possible $(\{g_i\}_{1\leqslant i\leqslant m})$-cells. A finite set $\mathscr{S}\subset\mathbb{R}^n$ is called a *representative set* for the partition $\mathscr{U}(\{g_i\}_{1\leqslant i\leqslant m})$ if each element of the partition contains at least one point from the set $\mathscr{S}$.

Consider now polynomials

$$g_1,\ldots,g_m\in\mathbb{Q}[Y_1,\ldots,Y_q,X_1,\ldots,X_n].$$

Let $y=(y_1,\ldots,y_q)\in\mathbb{R}^q$ be a real algebraic point given in the following form (cf. theorem 1): $\Phi_1(Z)\in\mathbb{Q}[Z]$ is a polynomial irreducible over $\mathbb{Q}$;

$$y_p=y_p(\theta_1)=\sum_{0\leqslant j<\deg(\Phi_1)}\rho_j^{(p)}\theta_1^j$$

are expressions where $\rho_j^{(p)}\in\mathbb{Q}$, $1\leqslant p\leqslant q$, $0\leqslant j<\deg(\Phi_1)$ and $\theta_1\in\mathbb{R}$, $\Phi_1(\theta_1)=0$; furthermore, a pair of rational numbers $c_1,c_2\in\mathbb{Q}$ is given such that in the interval $(c_1,c_2)\subset\mathbb{R}$, θ is the unique real root of the polynomial Φ_1. Moreover,

$$\theta_1=\sum_{1\leqslant p\leqslant q}\lambda_p^{(1)}y_p(\theta_1)$$

for some natural numbers $\lambda_p^{(1)}$, $1\leqslant p\leqslant q$. In addition, the following bounds are fulfilled: $\deg(g_j)<d$; $l(g_j)\leqslant M$ for every $1\leqslant j\leqslant m$ and $\deg(\Phi_1)<d_1$;

$$l(\Phi_1),\ l(y_p(\theta_1)),\ l(c_1),\ l(c_2)\leqslant M_1$$

for all $1\leqslant p\leqslant q$. Introduce polynomials

$$\hat{g}_j(X_1,\ldots,X_n)=g_j(y_1,\ldots,y_q,X_1,\ldots,X_n)\in\mathbb{R}[X_1,\ldots,X_n],\quad 1\leqslant j\leqslant m.$$

The following lemma allows us to produce a representative set for the partition $\mathscr{U}(\{\hat{g}_j\}_{1\leqslant j\leqslant m})$.

LEMMA 2. *One can design an algorithm which produces a representative set $\mathscr{S}\subset\mathbb{R}^n$ for the partition $\mathscr{U}(\{\hat{g}_j\}_{1\leqslant j\leqslant m})$ within time $\mathscr{P}(M,M_1,q,(mdd_1)^{n^2})$, where $\mathscr{S}$ contains not more than $\mathscr{P}((mdd_1)^{n^2})$ points. Moreover, for every point $(\xi_1,\ldots,\xi_n)\in\mathscr{S}$ the algorithm constructs polynomial $\Phi\in\mathbb{Q}[Z]$ irreducible over $\mathbb{Q}$, and expressions*

$$\xi_i=\xi_i(\theta)=\sum_{0\leqslant j<\deg(\Phi)}\alpha_j^{(i)}\theta^j,$$

where $\alpha_j^{(i)}\in\mathbb{Q}$, $1\leqslant i\leqslant n$, $0\leqslant j<\deg(\Phi)$ and $\theta\in\mathbb{R}$, $\Phi(\theta)=0$; apart from that, the algorithm yields a pair of rational numbers $b_1,b_2\in\mathbb{Q}$ such that in the interval $(b_1,b_2)\subset\mathbb{R}$ θ is the only real root of the polynomial Φ. Furthermore, the expressions

$$y_p=y_p(\theta)=\sum_{0\leqslant j<\deg(\Phi)}\mu_j^{(p)}\theta^j$$

are constructed by the algorithm, where $\mu_j^{(p)}\in\mathbb{Q}$, $1\leqslant p\leqslant q$, $0\leqslant j<\deg(\Phi)$. In addition

$$\theta=\lambda_0\theta_0+\sum_{1\leqslant i\leqslant n}\lambda_i\xi_i$$

for certain natural numbers $1\leqslant\lambda_i\leqslant\deg(\Phi)$, $0\leqslant i\leqslant n$. At last, we have the following bounds:

$$\deg(\Phi)=\mathscr{P}((mdd_1)^n)$$

and

$$l(\Phi),\ l(\xi_i(\theta)),\ l(y_p(\theta)),\ l(b_1),\ l(b_2)\leqslant(M+M_1+q)\mathscr{P}((mdd_1)^n).$$

PROOF. Consider the following family consisting of $(m+2)$ polynomials in $(n+1)$ variables $T, X_1, \ldots, X_n$ with rational coefficients:

$$\tilde{g}_j(T, X_1, \ldots, X_n) = g_j(y_1(T), \ldots, y_q(T), X_1, \ldots, X_n), \quad 1 \leqslant j \leqslant m;$$

$$\tilde{g}_{m+1}(T, X_1, \ldots, X_n) = \Phi_1(T); \quad \tilde{g}_{m+2}(T, X_1, \ldots, X_n) = (T-c_1)(c_2-T).$$

To begin with, let us enumerate all $(\{\tilde{g}_j\}_{1 \leqslant j \leqslant m+2})$-cells based on lemma 1. Next for every $(\{\tilde{g}_j\}_{1 \leqslant j \leqslant m+2})$-cell

$$\left\{ \underset{j \in J}{\&} \, (\tilde{g}_j = 0) \, \& \, \underset{j_1 \in J_1}{\&} \, (\tilde{g}_{j_1} > 0) \, \& \, \underset{j_2 \in J_2}{\&} \, (\tilde{g}_{j_2} < 0) \right\}$$

such that the requirements $\tilde{g}_{m+1} = 0$, $\tilde{g}_{m+2} > 0$ are satisfied, i.e. $(m+1) \in J$, $(m+2) \in J_1$, the algorithm produces a representative set $\mathscr{S}_{J, J_1, J_2} \subset \mathbb{R}^{n+1}$ using theorem 1. Denote by $\pi: \mathbb{R}^{n+1} \to \mathbb{R}^n$ the linear projection defined by the formula $\pi(T, X_1, \ldots, X_n) = (X_1, \ldots, X_n)$ and put

$$\mathscr{S} = \bigcup_{J, J_1, J_2} \pi(\mathscr{S}_{J, J_1, J_2}) \subset \mathbb{R}^n,$$

where the union is taken over all $(\{\tilde{g}_j\}_{1 \leqslant j \leqslant m})$-cells satisfying the requirements stated above. Then $\mathscr{S}$ is a representative set for the partition $\mathscr{U}(\{\hat{g}_j\}_{1 \leqslant j \leqslant m})$, since the partition $\mathscr{U}(\{\hat{g}_j\}_{1 \leqslant j \leqslant m})$ is isomorphic by means of the projection π to the partition of the space $\mathbb{R}^{n+1} \cap \{T = \theta_1\} \simeq \mathbb{R}^n$ formed by the components of connectivity of all $(\{\tilde{g}_j\}_{1 \leqslant j \leqslant m+2})$-cells satisfying the requirements $(m+1) \in J$, $(m+2) \in J_1$, taking into account that the requirements $\tilde{g}_{m+1}(T) = 0$, $\tilde{g}_{m+2}(T) > 0$ imply the equality $T = \theta_1$.

Let a point $(\theta_1, \xi_1, \ldots, \xi_n) \in \mathscr{S}_{J, J_1, J_2}$. According to theorem 1, a polynomial $\Phi \in \mathbb{Q}[Z]$ irreducible over the field $\mathbb{Q}$ corresponds to the point, as well as expressions

$$\xi_i = \xi_i(\theta) = \sum_{0 \leqslant j < \deg(\Phi)} \alpha_j^{(i)} \theta^j, \quad 1 \leqslant i \leqslant n$$

and

$$\theta_1 = \theta_1(\theta) = \sum_{0 \leqslant j < \deg(\Phi)} \alpha_j^{(0)} \theta^j;$$

moreover, the algorithm from theorem 1 yields an interval $(b_1, b_2) \subset \mathbb{R}$ (with rational endpoints $b_1, b_2 \in \mathbb{Q}$) which contains a unique real root $\theta \in (b_1, b_2)$ of the polynomial Φ. Furthermore,

$$\theta = \lambda_0 \theta_1 + \sum_{1 \leqslant i \leqslant n} \lambda_i \xi_i$$

for appropriate natural numbers $1 \leqslant \lambda_i \leqslant \deg(\Phi)$, $0 \leqslant i \leqslant n$. Substituting the constructed expression $\theta_1(\theta)$ in the expressions $y_p(\theta_1)$, $1 \leqslant p \leqslant q$ we obtain the expressions $y_p(\theta)$, $1 \leqslant p \leqslant q$. Observe that $\deg_{T, X_1, \ldots, X_n}(\tilde{g}_j) < dd_1$, $1 \leqslant j \leqslant m+2$. Hence, the number of points in the set $\mathscr{S}_{J, J_1, J_2}$ is not greater than $\mathscr{P}((mdd_1)^{n^2})$ by virtue of theorem 1; on the other hand the number of $(\{\tilde{g}_j\}_{1 \leqslant j \leqslant m+2})$-cells does not exceed $\mathscr{P}((mdd_1)^n)$ in view of lemma 1. Therefore, the set $\mathscr{S}$ contains not more than $\mathscr{P}((mdd_1)^{n^2})$ points.

Theorem 1 entails the bound $\deg(\Phi) \leqslant \mathscr{P}((mdd_1)^n)$. The bit lengths of coefficients are bounded by

$$l(\tilde{g}_j) \leqslant M + M_1 d + d \log(d_1) + q \log(d);$$

from this we obtain the bounds:

$$l(\Phi), l(\xi_i(\theta)), l(\theta_1(\theta)), l(b_1), l(b_2) \leqslant (M + M_1 + q)\mathscr{P}((mdd_1)^n), \quad 1 \leqslant i \leqslant n;$$

thus,

$$l(y_p(\theta)) \leqslant (M + M_1 + q)\mathscr{P}((mdd_1)^n), \quad 1 \leqslant p \leqslant q.$$

Finally, the running time of the algorithm can be estimated by $\mathscr{P}(M, M_1, q, (mdd_1)^{n^2})$ by lemma 1 and theorem 1, this completes the proof of the lemma.

2. Semialgebraic Curves Over Real Closed Fields
(With Infinitesimals)

First of all we shall recall some facts about real closed fields.

Throughout the present section, F denotes a real closed field. Consider the ordered field $F(\varepsilon)$, in which the order is defined by the condition $0 < \varepsilon < \beta$ for all $0 < \beta \in F$, in other words, the element ε is infinitesimal relatively to the elements of the field F. For any ordered field F_1 denote by $\tilde{F}_1 \supset F_1$ (defined uniquely up to an isomorphism of the ordered fields) the real closure of the field F_1 (see e.g. Lang, 1965). For instance, $\tilde{\mathbb{Q}} \subset \mathbb{R}$ is the field of real algebraic numbers.

A Puiseux series (or in other words, power-fractional series) is a series of the kind

$$b = \sum_{i \geq 0} \beta_i \varepsilon^{v_i/\mu},$$

where $0 \neq \beta_i \in F$ for each i, the integers $v_0 < v_1 < \ldots$ increase, the natural number $\mu \geq 1$ may depend on the series. The totality of all Puiseux series (with added zero) forms the field $F((\varepsilon^{1/\infty})) \supset \widetilde{F(\varepsilon)} \supset F(\varepsilon)$ containing the real closure $\widetilde{F(\varepsilon)}$ of the field $F(\varepsilon)$. The order in the field $F((\varepsilon^{1/\infty}))$ (that induces the order in the field $\widetilde{F(\varepsilon)}$) is lexicographical. Furthermore, the field

$$\bar{F}((\varepsilon^{1/\infty})) \supset \overline{F(\varepsilon)} = F(\varepsilon)[\sqrt{-1}]$$

of Puiseux series over the algebraic closure $\bar{F} = F[\sqrt{-1}]$ of the field F contains the algebraic closure $\overline{F(\varepsilon)}$ of the field $F(\varepsilon)$.

If $v_0 < 0$, then the element b is called *infinitely large* (relatively to the field F); if $v_0 > 0$, then b is *infinitesimal*. A vector $(b_1, \ldots, b_n) \in (F((\varepsilon^{1/\infty})))^n$ is called F-finite if every coordinate b_i is not infinitely large. For any F-finite element $b \in F((\varepsilon^{1/\infty}))$, its *standard part* $st(b) \in F$ is defined and is equal to β_0 in cases when $v_0 = 0$, or $st(b) = 0$ when $v_0 > 0$. For an F-finite vector $(b_1, \ldots, b_n) \in (F((\varepsilon^{1/\infty})))^n$, its standard part is

$$st(b_1, \ldots, b_n) = (st(b_1), \ldots, st(b_n)).$$

If each point from a set $V \subset (F((\varepsilon^{1/\infty})))^n$ is F-finite, then its standard part $st(V) \subset F^n$ is defined as the set of standard parts of all points from the set V. In the sequel, we deal with the subfield $\widetilde{F(\varepsilon)}$ of $F((\varepsilon^{1/\infty}))$ and apply to it the notions introduced.

Now we shall demonstrate how the transfer principle can work and show (a known fact) that any semialgebraic set over a real closed field F can be represented uniquely as a union of its components of connectivity, each in its turn being a semialgebraic set. Consider a semialgebraic set $W = \{\Pi\} \subset F^n$, determined by a quantifier-free formula Π of Tarski algebra with the atomic subformulas of the kind $(f \geq 0)$, where the polynomials $f \in F[X_1, \ldots, X_n]$. By a format of the formula Π we shall mean the sum of the number of its variables, the number of atomic subformulas and the degrees of the polynomials f.

In the case of the field $F = \mathbb{R}$, the set W is uniquely representable in a union of its components of connectivity $W = \bigcup_i W_i$, where every W_i is in its turn a semialgebraic set (and connected in the euclidean topology). From e.g. the papers by Collins (1975), and Wüthrich (1976), one can deduce the existence of a function $\mathfrak{N}$ such that if a format of a formula Π is less than $\mathscr{N}$, then the whole number of the components W_i is less than $\mathfrak{N}(\mathscr{N})$

and, moreover, one can find quantifier-free formulas Π_i of Tarski algebra each of the format less than $\mathfrak{N}(\mathcal{N})$ that $W_i = \{\Pi_i\}$. Indeed, the algorithms from Collins (1975) and Wüthrich (1976) allow to produce a cylindrical algebraic decomposition of a semialgebraic set and as a corollary to produce the decomposition on the components of connectivity. For a given format $\mathcal{N}$ of an initial formula (with symbolic coefficients) each of the two algorithms can be represented as a rooted tree (directed outward the root) having vertices either with out-degree one or out-degree three. To the root corresponds the initial formula, to any vertex of the tree with out-degree one corresponds an arithmetic operation; to any vertex with out-degree three corresponds a polynomial. The computation for arbitrary initial formula, with specified coefficients substituted instead of symbolic ones, proceeds along a suitable path of the tree starting from the root, performing the corresponding arithmetic operation in a vertex with out-degree one, and branching in a vertex with out-degree three according to the sign of the corresponding polynomial. This representation as a tree provides the desired function $\mathfrak{N}$.

Thus, for a given $\mathcal{N}$, one can yield a formula $\Omega_{\mathcal{N}}$ of Tarski algebra (for the case of the field $F = \mathbb{R}$), expressing the existence of decomposition of any semialgebraic set $W = \{\Pi\}$, with the format of Π less than $\mathcal{N}$ into less than $\mathfrak{N}(\mathcal{N})$ its components of connectivity $W = \bigcup_i \{\Pi_i\}$, such that the format of every Π_i is less than $\mathfrak{N}(\mathcal{N})$. Moreover, the formula $\Omega_{\mathcal{N}}$ states that for each pair of indices $i \neq j$ the components $\{\Pi_i\}$ and $\{\Pi_j\}$ are "separated", i.e. the following formula of Tarski algebra is valid:

$$\forall ((a_1, \ldots, a_n) \in \{\Pi_i\}) \, \exists z > 0 (\forall (b_1, \ldots, b_n) \in \{\Pi_j\}) \Big(\sum_{1 \leqslant l \leqslant n} (a_l - b_l)^2 \geqslant z \Big).$$

Besides, the formula $\Omega_{\mathcal{N}}$ claims the "connectedness" of every $\{\Pi_i\}$, this means that there do not exist two "separated" semialgebraic subsets of $\{\Pi_i\}$, each determined by a quantifier-free formula of Tarski algebra with the format less than $\mathfrak{N}(\mathfrak{N}(\mathcal{N}))$.

Apart from that, for given $\mathcal{N}, \mathcal{M}$ one can prove (for the case of the field $F = \mathbb{R}$) a formula $\Omega_{\mathcal{N},\mathcal{M}}$ of Tarski algebra expressing the following. If $\{\Pi\}$ (where the format of Π is less than $\mathcal{N}$) can be represented as a union of more than one and less than $\mathcal{M}$ pairwise "separated" semialgebraic sets, each being determined by a quantifier-free formula of Tarski algebra of the format less than $\mathcal{M}$, then $\{\Pi\}$ can be represented as a union of more than one and less than $\mathfrak{N}(\mathcal{N})$ pairwise "separated" semialgebraic "connected" sets, each being determined by a quantifier-free formula of Tarski algebra of the format less than $\mathfrak{N}(\mathcal{N})$.

Applying the transfer principle to all the formulas $\Omega_{\mathcal{N}}, \Omega_{\mathcal{N},\mathcal{M}}$, one concludes that any semialgebraic set (over a real closed field F) can be uniquely represented as a union of its pairwise "separated" "components of connectivity", moreover, each component is semialgebraic and is "connected", i.e. cannot be represented as a union of a finite number pairwise "separated" semialgebraic sets. Below, we utilise the terms connected semialgebraic set and components of connectivity of a semialgebraic set without quotation marks, since the notion of connectedness in any topology will not be considered.

As usual, one can define a *semialgebraic curve* $C \subset F^n$ as a semialgebraic set for which there exists a linear projection on the line (i.e. on F), such that the inverse image of every point under the projection consists of a finite number of points and in addition the latter number is less than a certain number depending only on the curve C. A mapping $\varphi : V_1 \to V_2$ where $V_1 \subset F^n$, $V_2 \subset F^m$ are semialgebraic sets, is called semialgebraic if its graph is a semialgebraic subset in the space F^{m+n}. We shall utilise the terms continuous mapping, open and closed set in the sense of the topology with the base of all open balls. We denote by $\mathscr{D}_x(R)$ the closed ball of radius R with the centre in the point x.

We shall call a set $V \subset F^n$ *monotone* iff there exists a vector $(\delta_1, \ldots, \delta_n) \in \{-1, +1\}^n$ satisfying the following property: for any pair of points

$$v^{(1)} = (v_1^{(1)}, \ldots, v_n^{(1)}), \quad v^{(2)} = (v_1^{(2)}, \ldots, v_n^{(2)}) \in V$$

either

$$\delta_1 v_1^{(1)} \geqslant \delta_1 v_1^{(2)}, \ldots, \delta_n v_n^{(1)} \geqslant \delta_n v_n^{(2)} \quad \text{or} \quad \delta_1 v_1^{(1)} \leqslant \delta_1 v_1^{(2)}, \ldots, \delta_n v_n^{(1)} \leqslant \delta_n v_n^{(2)}$$

are fulfilled.

One can prove the next lemma first for the case of the field $F = \mathbb{R}$ bounding the formats of the constructed semialgebraic sets (and their number) via the formats of the given semialgebraic sets, and after that make use of the transfer principle. The proof for the case $F = \mathbb{R}$ is quite cumbersome and routine, on the other hand, the reader can reproduce it for himself without great difficulties, therefore we omit the proof.

LEMMA 3.
  (a) *The image of any connected semialgebraic curve under the action of a continuous curve on this semialgebraic mapping is also a connected semialgebraic curve.*
  (b) *Any pair of points of a connected semialgebraic set can be joined by a closed connected semialgebraic curve entirely situated in this set and in a certain ball.*
  (c) *One can represent any closed semialgebraic curve as a union of a finite number of monotone closed semialgebraic curves.*

Let the elements $\varepsilon_1 > \varepsilon_2 > \ldots > \varepsilon_m > 0$ be such that the element ε_{i+1} is infinitesimal relatively to the field $F(\varepsilon_1, \ldots, \varepsilon_i)$ for each $0 \leqslant i < m$. For every element $\alpha \in \widetilde{F(\varepsilon_1, \ldots, \varepsilon_m)}$ one can uniquely define its standard part $st(\alpha) \in F$ (provided that it exists) by recursion on m. We denote $F_m = \widetilde{F(\varepsilon_1, \ldots, \varepsilon_m)}$.

LEMMA 4.
  (a) *Given a closed connected semialgebraic curve $C \subset \mathscr{D}_0(R) \subset F^n$, where $R \in F$, one can find a closed connected semialgebraic curve $C^{(\varepsilon)} \subset \mathscr{D}_0(R) \subset F_m^n$ such that $C^{(\varepsilon)} \supset st(C^{(\varepsilon)}) = C$ and, furthermore, for a certain quantifier-free formula Π of Tarski algebra both $C = \{\Pi\} \subset F^n$ and $C^{(\varepsilon)} = \{\Pi\} \subset F_m^n$ are true;*
  (b) *Let $W \subset \mathscr{D}_0(R) \subset F_m^n$ be a connected semialgebraic set where $R \in F$. Assume that the set $st(W) \subset V \subset F^n$ for a certain semialgebraic set V. Then $st(W) \subset V_1$ for a suitable component of connectivity V_1 of the set V.*

REMARK. Under the conditions of item (b) it is apparently possible to prove that $st(W) \subset F^n$ is a semialgebraic set, but since we shall not need this further, we do not dwell on its proof.

PROOF. (a) By virtue of lemma (3c), one can represent C as a union of monotone closed curves and after that decompose each monotone curve on the components of connectivity. Thus, we obtain a representation $C = \bigcup_i C_i$ of C as a union of monotone closed connected curves C_i. Let $C_i = \{\Pi_i\}$ for an appropriate quantifier-free formula Π_i of Tarski algebra. We set

$$C_i^{(\varepsilon)} = \{\Pi_i\} \subset F_m^n, \quad \Pi = \bigvee_i \Pi_i \quad \text{and} \quad C^{(\varepsilon)} = \{\Pi\} = \bigcup_i C_i^{(\varepsilon)}.$$

Evidently $C_i^{(\varepsilon)} \supset C_i$ and $st(C_i^{(\varepsilon)}) \supset C_i$. The transfer principle implies the inclusion $C_i^{(\varepsilon)} \subset \mathcal{D}_0(R)$, since it is equivalent to the formula $\forall\, x(\Pi_i(x) \Rightarrow \|x\| \leqslant R)$ of Tarski algebra.

We fix a curve C_i and we claim that $C_i^{(\varepsilon)} \subset F_m^n$ is a monotone closed connected semialgebraic curve and besides that $C_i = st(C_i^{(\varepsilon)})$. From this the statement of item (a) will be concluded easily.

The closedness of semialgebraic set $C_i^{(\varepsilon)}$ can be inferred from the transfer principle. Now we are going to prove that $C_i^{(\varepsilon)}$ is a monotone curve. Consider a linear projection π of the n-dimensional space on the line which is given by the formula

$$\pi(x_1, \ldots, x_n) = \sum_{1 \leqslant j \leqslant n} \delta_j x_j,$$

where $\delta_j = \pm 1$ are taken from the definition of monotonicity of curve C_i. In view of lemma (3a), the image $\pi(C_i) \subset F$ is a connected semialgebraic set, therefore, $\pi(C_i)$ is an interval. Moreover, in the case of the field $F = \mathbb{R}$, the image $\pi(C_i)$ of the compact set C_i is also a compact, in particular $\pi(C_i)$ is a closed interval. The transfer principle entails that $\pi(C_i) = [\gamma_1, \gamma_2]$ is a closed interval in the case of an arbitrary real closed field F. Moreover, the mapping $\pi: C_i \to [\gamma_1, \gamma_2]$ is bijective, since C_i is monotone. Henceforth, $\pi(C_i^{(\varepsilon)}) = [\gamma_1, \gamma_2] \subset F_m$ according to the transfer principle, furthermore, the mapping $\pi: C_i^{(\varepsilon)} \to [\gamma_1, \gamma_2]$ is bijective. In particular, we deduce that $C_i^{(\varepsilon)}$ is a semialgebraic curve and, in addition, it is monotone again by the transfer principle.

Now we shall check that $st(C_i^{(\varepsilon)}) \subset C_i$. Indeed, let a point $x = (x_1, \ldots, x_n) \in C_i^{(\varepsilon)}$; then there exists a point

$$y = (y_1, \ldots, y_n) \in C_i \subset C_i^{(\varepsilon)}$$

such that

$$\pi(y) = st(\pi(x)) \in [\gamma_1, \gamma_2] \subset F$$

(see above). The elements $\delta_j(x_j - y_j) \in F_m$ are either non-negative for all $1 \leqslant j \leqslant n$ or non-positive for all $1 \leqslant j \leqslant n$, since $C_i^{(\varepsilon)}$ is monotone. On the other hand,

$$st\Big(\sum_{1 \leqslant j \leqslant n} \delta_j(x_j - y_j)\Big) = st(\pi(x)) - \pi(y) = 0,$$

therefore,

$$st(x_j - y_j) = \delta_j\, st(\delta_j(x_j - y_j)) = 0$$

for all $1 \leqslant j \leqslant n$, i.e. $st(x) = y$, that proves the inclusion $st(C_i^{(\varepsilon)}) \subset C_i$.

At last, consider a semialgebraic mapping $\pi^{-1}: [\gamma_1, \gamma_2] \to C_i^{(\varepsilon)}$. The mapping is continuous by virtue of monotonicity of $C_i^{(\varepsilon)}$, henceforth, the curve $C_i^{(\varepsilon)}$ is connected in view of lemma (3a). Now we shall show that $C^{(\varepsilon)}$ is connected. Suppose the contrary. Then for some family I of indices the intersection

$$\Big(\bigcup_{i \in I} C_i^{(\varepsilon)}\Big) \cap \Big(\bigcup_{i \notin I} C_i^{(\varepsilon)}\Big) = \phi.$$

Taking into account that the curve C is connected and that curves C_i are closed, we deduce the existence of a point

$$x \in \Big(\bigcup_{i \in I} C_i\Big) \cap \Big(\bigcup_{i \notin I} C_i\Big) \subset \Big(\bigcup_{i \in I} C_i^{(\varepsilon)}\Big) \cap \Big(\bigcup_{i \notin I} C_i^{(\varepsilon)}\Big).$$

The obtained contradiction proves the connectivity of $C^{(\varepsilon)}$.

(b) For every component of connectivity V_i of the set V, consider the following semialgebraic set $U_i \subset F^n$ (cf. lemma 1 in Grigor'ev & Vorobjov, 1987). A point $x \in U_i$ iff there exist such $\tau_i^{(x)} \geqslant 0$, $\tau_{1i}^{(x)} > 0$ that the intersection $\mathcal{D}_x(\tau_i^{(x)}) \cap V_i \neq \phi$, and for each $j \neq i$ the intersection $\mathscr{D}_x(\tau_i^{(x)} + \tau_{1i}^{(x)}) \cap V_j = \phi$. Now we shall establish various properties of the set U_i, which is semialgebraic since it is represented by some formula of Tarski algebra.

(α) The inclusion $V_i \subset U_i$ is correct. Indeed, for any point $y \in V_i$, one can set $\tau_i^{(y)} = 0$ and can take $\tau_{1i}^{(y)}$ from the definition of components of connectivity (see above).

(β) The set U_i is open. Let a point $x \in U_i$, we shall check that the ball $\mathscr{D}_x(\tau_{1i}^{(x)}/3) \subset U_i$. Actually, we assert that for any point $z \in \mathscr{D}_x(\tau_{1i}^{(x)}/3)$, one can take $\tau_i^{(z)} = \tau_i^{(x)} + \tau_{1i}^{(x)}/3$ and $\tau_{1i}^{(z)} = \tau_{1i}^{(x)}/3$. There is such a point $y \in V_i$ that the distance $\|y - x\| \leqslant \tau_i^{(x)}$, henceforth,

$$\|y - z\| \leqslant \|y - x\| + \|x - z\| \leqslant \tau_i^{(z)}.$$

Besides that, for any point $z_1 \in V_j$, where $j \neq i$ the inequalities

$$\|z_1 - z\| \geqslant \|z_1 - x\| - \|z - x\| > \tau_i^{(z)} + \tau_{1i}^{(z)}$$

are true.

(γ) For $i \neq j$ the intersection $U_i \cap U_j = \phi$. Suppose, on the contrary, that a certain point $x \in U_i \cap U_j$. There exist points $y_i \in V_i$, $y_j \in V_j$, for which $\|x - y_i\| \leqslant \tau_i^{(x)}$, $\|x - y_j\| \leqslant \tau_j^{(x)}$. On the other hand, $\|x - y_j\| > \tau_i^{(x)} + \tau_{1i}^{(x)}$ and $\|x - y_i\| > \tau_j^{(x)} + \tau_{1j}^{(x)}$, this leads to the contradiction.

Let $U_i = \{\Pi_i\}$ for each i for suitable quantifier-free formulas Π_i of Tarski algebra. Consider semialgebraic sets $U_i^{(\varepsilon)} = \{\Pi_i\} \subset F_m^n$, for which properties ($\alpha$), ($\beta$), ($\gamma$) are fulfilled according to the transfer principle.

Let an arbitrary point $w \in W$; then the point $st(w) \in V_{i_0}$ (which is defined since $\|w\| \leqslant R$) for an appropriate i_0. By virtue of (α), (β) the ball $\mathscr{D}_{st(w)}(\tau) \subset U_{i_0}$ for a suitable $0 < \tau \in F$. In view of the transfer principle, the inclusion $\mathscr{D}_{st(w)}(\tau) \subset U_{i_0}^{(\varepsilon)}$ is valid in the space F_m^n. Therefore, $w \in \mathscr{D}_{st(w)}(\tau) \subset U_{i_0}^{(\varepsilon)}$. Thus,

$$W \subset \bigcup_i U_i^{(\varepsilon)}.$$

Since set W is connected, one can infer that $W \subset U_{i_1}^{(\varepsilon)}$ for a certain i_1, taking into account properties (β), (γ) of sets $U_i^{(\varepsilon)}$. This implies according to the proved above that $st(W) \subset V_{i_1}$. The lemma is proved.

Consider polynomials

$$p_i \in F[\varepsilon_1, \ldots, \varepsilon_m][X_1, \ldots, X_n], \quad 1 \leqslant i \leqslant k.$$

Write p_i in the form

$$p_i = \sum_{j_1, \ldots, j_m} p_i^{(j_1, \ldots, j_m)} \varepsilon^{j_1} \ldots \varepsilon_m^{j_m},$$

where $p_i^{(j_1, \ldots, j_m)}$ are polynomials in $F[X_1, \ldots, X_n]$.

LEMMA 5.
 (a) *Let $W \subset F^n$ be an element of the partition $\mathscr{U}(\{p_i^{(j_1, \ldots, j_m)}\}_{1 \leqslant i \leqslant k; j_1, \ldots, j_m})$ (see section 1), and let $C \subset W \cap \mathscr{D}_0(R)$ be a closed connected semialgebraic curve, where $R \in F$. Furthermore, let $C^{(\varepsilon)} \subset \mathscr{D}_0(R) \subset F_m^n$ be a closed connected semialgebraic curve such that $C^{(\varepsilon)} \supset st(C^{(\varepsilon)}) = C$ and such that for a suitable quantifier-free formula Π of Tarski algebra equalities $C = \{\Pi\} \subset F^n$, $C^{(\varepsilon)} = \{\Pi\} \subset F_m^n$ are fulfilled (cf. lemma (4a)). Then $C^{(\varepsilon)} \subset W^{(\varepsilon)}$ for some unique element $W^{(\varepsilon)} \subset F_m^n$ of the partition $\mathscr{U}(\{p_i\}_{1 \leqslant i \leqslant k})$.*
 (b) *For any element W of the partition $\mathscr{U}(\{p_i^{(j_1, \ldots, j_m)}\}_{1 \leqslant i \leqslant k; j_1, \ldots, j_m})$ there is a unique element $W^{(\varepsilon)}$ of the partition $\mathscr{U}(\{p_i\}_{1 \leqslant i \leqslant k})$ such that $W \subset W^{(\varepsilon)} \cap F^n$.*
 (c) *Let $g_1, \ldots, g_k$ be polynomials in $F[X_1, \ldots, X_n]$, and let $W_1 \subset F^n$ be a certain component of connectivity of the semialgebraic set $\{(g_1 \geqslant 0) \& \ldots \& (g_k \geqslant 0)\} \subset F^n$.*

Then there exists a unique element $W_1^{(\varepsilon)} \subset F_3^n$ of the partition

$$\mathcal{U}(\{g_1+\varepsilon_1, \ldots, g_k+\varepsilon_1, (g_1+\varepsilon_1)\ldots(g_k+\varepsilon_1)-\varepsilon_3\})$$

which contains $W_1 \subset W_1^{(\varepsilon)}$.

PROOF. (a) Taking into account that the curve $C^{(\varepsilon)}$ is connected, it is sufficient to show that $C^{(\varepsilon)}$ is situated entirely in some $(\{p_i\}_{1 \leqslant i \leqslant k})$-cell (see section 1).

Assume that the polynomial $p_i^{(j'_1, \ldots, j'_m)}$ vanishes on the set W for some $j'_1, \ldots, j'_m$. Since the curve $C = \{\Pi\} \subset W$, the following statement is correct: if a point $x \in F^n$ satisfies the formula Π, then $p_i^{(j'_1, \ldots, j'_m)}(x) = 0$. The transfer principle entails that the polynomial $p_i^{(j'_1, \ldots, j'_m)}$ vanishes also on the curve $C^{(\varepsilon)} = \{\Pi\} \subset F_m^n$.

We introduce a lexicographical order on multi-indices setting $(j'_1, \ldots, j'_m) \prec (j_1, \ldots, j_m)$ if $j'_m = j_m, \ldots, j'_l = j_l$ and $j'_{l-1} < j_{l-1}$ for a certain $1 < l \leqslant m+1$. Let us denote $\mathrm{least}(p_i) = p_i^{(j_1, \ldots, j_m)}$ iff the sign $\mathrm{sgn}(p_i^{(j_1, \ldots, j_m)})$ is different from zero on the set W (otherwise, if there is no such multi-index $(j_1, \ldots, j_m)$, then we set $\mathrm{least}(p_i) = 0$) and in addition the sign $(p_i^{(j'_1, \ldots, j'_m)})$ is equal to zero for each multi-index $(j'_1, \ldots, j'_m) \prec (j_1, \ldots, j_m)$. We recall that sign $\mathrm{sgn}(p_i^{(j''_1, \ldots, j''_m)})$ is constant on W for every multi-index $(j''_1, \ldots, j''_m)$. Note that polynomial $p_i^{(j'_1, \ldots, j'_m)}$ vanishes on the curve $C^{(\varepsilon)}$, provided that $(j'_1, \ldots, j'_m) \prec (j_1, \ldots, j_m)$ by virtue of the facts proved above.

For any point $c \in C^{(\varepsilon)}$ we claim the equality

$$\mathrm{sgn}(p_i(c)) = \mathrm{sgn}((\mathrm{least}(p_i))(st(c))), \quad 1 \leqslant i \leqslant k.$$

In the case when $\mathrm{least}(p_i) = 0$, both sides of the claimed equality are zeros according to the ascertained above. If $\mathrm{least}(p_i) = p_i^{(j_1, \ldots, j_m)} \neq 0$, then $p_i^{(j_1, \ldots, j_m)}(st(c)) \neq 0$ since $st(c) \in C \subset W$. Therefore, taking into account the equalities

$$p_i(c) = p_i^{(j_1, \ldots, j_m)}(c)\varepsilon_1^{j_1} \ldots \varepsilon_m^{j_m} + w_1 \varepsilon_1^{j_1} \ldots \varepsilon_m^{j_m}$$

$$= p_i^{(j_1, \ldots, j_m)}(st(c))\varepsilon_1^{j_1} \ldots \varepsilon_m^{j_m} + w_2 \varepsilon_1^{j_1} \ldots \varepsilon_m^{j_m}$$

for suitable infinitesimals $w_1, w_2 \in F_m$ (relatively to the field F), one can deduce the desired equality

$$\mathrm{sgn}(p_i(c)) = \mathrm{sgn}(p_i^{(j_1, \ldots, j_m)}(st(c))) \neq 0.$$

Finally, for an arbitrary pair of points $c_1, c_2 \in C^{(\varepsilon)}$, points $st(c_1), st(c_2) \in C \subset W$. So the definition of the partition $\mathcal{U}(\{p_i^{(j_1, \ldots, j_m)}\}_{1 \leqslant j \leqslant k; j_1, \ldots, j_m})$ implies

$$\mathrm{sgn}((\mathrm{least}(p_i))(st(c_1))) = \mathrm{sgn}((\mathrm{least}(p_i))(st(c_2)))$$

for all $1 \leqslant i \leqslant k$, therefore, $\mathrm{sgn}(p_i(c_1)) = \mathrm{sgn}(p_i(c_2))$; $1 \leqslant i \leqslant k$ in view of the fact claimed above. Thus, arbitrary points $c_1, c_2 \in C^{(\varepsilon)}$ belong to the same $(\{p_i\}_{1 \leqslant i \leqslant k})$-cell, which was to be shown.

(b) Suppose that there are points $x_1 \in W_1^{(\varepsilon)} \cap W$, $x_2 \in W_2^{(\varepsilon)} \cap W$ for two distinct elements $W_1^{(\varepsilon)} \neq W_2^{(\varepsilon)}$ of the partition $\mathcal{U}(\{p_i\}_{1 \leqslant i \leqslant k})$. According to lemma 3(b), there exists a closed connected semialgebraic curve $C \subset W \cap \mathcal{D}_0(R)$ for some $R \in F$ which contains points $x_1, x_2 \in C$. By virtue of lemma (4a) one can find a closed connected semialgebraic curve $C^{(\varepsilon)} \subset \mathcal{D}_0(R) \subset F_m^n$ such that $C^{(\varepsilon)} \supset st(C^{(\varepsilon)}) = C$ and, furthermore, $C = \{\Pi\} \subset F^n$, $C^{(\varepsilon)} = \{\Pi\} \subset F_m^n$ for a suitable quantifier-free formula Π of Tarski algebra. Then $C^{(\varepsilon)} \subset W^{(\varepsilon)}$ for some unique element $W^{(\varepsilon)}$ of the partition $\mathcal{U}(\{p_i\}_{1 \leqslant i \leqslant k})$ in view of part (a) of the present lemma, in particular $x_1, x_2 \in W^{(\varepsilon)}$, which contradicts the hypothesis.

(c) One can uniquely decompose $W_1 = \bigcup_j W_{1j}$, where W_{1j} is an element of the partition $\mathscr{U}(\{g_i\}_{1 \leqslant i \leqslant k})$ for every j. Taking into account the equality of the partitions

$$\mathscr{U}(\{g_i\}_{1 \leqslant i \leqslant k}) = \mathscr{U}(\{g_i\}_{1 \leqslant i \leqslant k} \cup \{\prod_{i \in I} g_i\}_{I \subset \{1,\ldots,k\}}),$$

one deduces from part (b) of the present lemma that for each W_{1j} there is a unique element $W_{1j}^{(\varepsilon)} \subset F_3^n$ of the partition

$$\mathscr{U}(g_1 + \varepsilon_1, \ldots, g_k + \varepsilon_1, (g_1 + \varepsilon_1) \ldots (g_k + \varepsilon_1) - \varepsilon_3),$$

which contains $W_{1j} \subset W_{1j}^{(\varepsilon)}$.

Suppose that for some indices $j_1 \neq j_2$, elements $W_{1j_1}^{(\varepsilon)} \neq W_{1j_2}^{(\varepsilon)}$ are distinct. Let us pick out points $x_1 \in W_{1j_1}, x_2 \in W_{1j_2}$. According to lemma (3b), one can find a closed connected semialgebraic curve $C \subset W_1 \cap \mathscr{D}_0(R)$ for a suitable $R \in F$ such that $x_1, x_2 \in C$. By virtue of lemma (4a) there exists a quantifier-free formula Π of Tarski algebra which satisfies the following conditions: $C = \{\Pi\} \subset F^n$, the closed connected semialgebraic curve $C^{(\varepsilon)} = \{\Pi\} \subset \mathscr{D}_0(R) \subset F_3^n$ and $C^{(\varepsilon)} \supset st(C^{(\varepsilon)}) = C$. For any point $c \in C \subset W_1$, the inequalities $g_i(c) \geqslant 0$, $1 \leqslant i \leqslant k$ are valid. So for every point $c^{(\varepsilon)} \in C^{(\varepsilon)}$, the inequalities $g_i(c^{(\varepsilon)}) \geqslant 0$, $1 \leqslant i \leqslant k$ are also correct in view of the transfer principle, therefore $g_i(c^{(\varepsilon)}) + \varepsilon_1 > 0$, $1 \leqslant i \leqslant k$ and

$$(g_1(c^{(\varepsilon)}) + \varepsilon_1) \ldots (g_k(c^{(\varepsilon)}) + \varepsilon_1) - \varepsilon_3 \geqslant \varepsilon_1^k - \varepsilon_3 > 0.$$

Since the curve $C^{(\varepsilon)}$ is connected, the inclusion $C^{(\varepsilon)} \subset W_1^{(\varepsilon)}$ is fulfilled for an appropriate element $W_1^{(\varepsilon)} \subset F_3^n$ of the partition $\mathscr{U}(g_1 + \varepsilon_1, \ldots, g_k + \varepsilon_1, (g_1 + \varepsilon_1) \ldots (g_k + \varepsilon_1) - \varepsilon_3)$, in particular, $x_1, x_2 \in W_1^{(\varepsilon)}$. We have obtained a contradiction which completes the proof of the lemma.

3. Projections of a Semialgebraic Set

Let a formula of Tarski algebra be given by

$$\exists X_1 \ldots \exists X_{s-1}((f_1 > 0) \& \ldots \& (f_m > 0) \& (f_{m+1} \geqslant 0) \& \ldots \& (f_{k-1} \geqslant 0)), \qquad (2)$$

where $f_1, \ldots, f_{k-1} \in \mathbb{Q}[Z_1, \ldots, Z_n, X_1, \ldots, X_{s-1}]$ are polynomials with $\deg(f_i) < d$, $l(f_i) \leqslant M$, $1 \leqslant i \leqslant k-1$. We add a new variable X_s, let $f_k = X_s f_1 \ldots f_m - 1$, and consider the formula

$$\exists X_1 \ldots \exists X_{s-1} \exists X_s((f_1 \geqslant 0) \& \ldots \& (f_{k-1} \geqslant 0) \& (f_k \geqslant 0)) \qquad (3)$$

equivalent to formula (2). We introduce one more variable Z_0, denote

$$f_0 = Z_0 - X_1^2 - \ldots - X_s^2$$

and consider the formula

$$\exists Z_0 \exists X_1 \ldots \exists X_{s-1} \exists X_s((f_0 \geqslant 0) \& (f_1 \geqslant 0) \& \ldots \& (f_{k-1} \geqslant 0) \& (f_k \geqslant 0)), \qquad (4)$$

which is equivalent to formulas (2) and (3).

In the present section we describe an algorithm which constructs (see lemma 10 below) a formula of the first-order theory of a certain real closed field F_3. The formula is of the form $\exists T(P_1)$, where P_1 is a quantifier-free formula with the coefficients in the field F_3, and is such that $\exists T(P_1)$ is equivalent (over the field $\tilde{\mathbb{Q}}$) to the formula

$$\exists X_1 \ldots \exists X_{s-1} \exists X_s((f_0 \geqslant 0) \& (f_1 \geqslant 0) \& \ldots \& (f_{k-1} \geqslant 0) \& (f_k \geqslant 0)), \qquad (5)$$

i.e. both formulas determine the same (semialgebraic) set in the space $\tilde{\mathbb{Q}}^{n+1}$; in other words a point $(z_0, z_1, \ldots, z_n) \in \tilde{\mathbb{Q}}^{n+1}$ satisfies (5) iff it satisfies the formula $\exists T(P_1)$. So, the algorithm reduces a projection along many variables (see formula (2)) to a projection along two variables Z_0, T, taking into account that formula (2) (and (4)) is equivalent to the formula $\exists\, Z_0 \in \tilde{\mathbb{Q}}\;\exists\, T \in F_3(P_1)$ (being not a formula of the first-order theory!)

We begin with the construction of the formula $\exists\, T(P_1)$. Let the element $\varepsilon_1 > 0$ be infinitesimal relatively to the field $\mathbb{Q}$, let the element $\varepsilon_2 > 0$ be infinitesimal relatively to $\mathbb{Q}(\varepsilon_1)$ and let the element $\varepsilon_3 > 0$ be infinitesimal relatively to $\mathbb{Q}(\varepsilon_1, \varepsilon_2)$. We denote $F_1 = \widetilde{\mathbb{Q}(\varepsilon_1)}$, $F_2 = \widetilde{\mathbb{Q}(\varepsilon_1, \varepsilon_2)}$, $F_3 = \widetilde{\mathbb{Q}(\varepsilon_1, \varepsilon_2, \varepsilon_3)}$ (see section 2). Consider the polynomial

$$g = (f_0 + \varepsilon_1)(f_1 + \varepsilon_1) \ldots (f_k + \varepsilon_1) - \varepsilon_3 \in \mathbb{Q}[\varepsilon_1, \varepsilon_3][Z_0, Z_1, \ldots, Z_n, X_1, \ldots, X_s]$$

and the formula

$$\exists\, X_1 \ldots \exists\, X_s((g = 0)\,\&\,(f_0 + \varepsilon_1 > 0)\,\&\,\ldots\,\&\,(f_k + \varepsilon_1 > 0)). \tag{6}$$

Consider points $z = (z_0, z_1, \ldots, z_n) \in \tilde{\mathbb{Q}}^{n+1}$ and $0 \neq v \in \tilde{\mathbb{Q}}^{n+1}$, and let $\tilde{z} = z + \varepsilon_2 v \in F_2^{n+1}$. We denote the semialgebraic sets

$$V = \{(f_0 \geqslant 0)\,\&\,\ldots\,\&\,(f_k \geqslant 0)\} \subset \tilde{\mathbb{Q}}^{s+n+1},$$

$$V^{(\varepsilon)} = \{(f_0 + \varepsilon_1 > 0)\,\&\,\ldots\,\&\,(f_k + \varepsilon_1 > 0)\} \subset F_3^{s+n+1}.$$

LEMMA 6.

(a) *For an arbitrary $0 < R \in \tilde{\mathbb{Q}}$, the following three sets coincide:*

$$V \cap \mathscr{D}_0(R) = st(V^{(\varepsilon)} \cap \mathscr{D}_0(R))$$
$$= st(V^{(\varepsilon)} \cap \{g \geqslant 0\} \cap \mathscr{D}_0(R)) \subset (V^{(\varepsilon)} \cap \{g \geqslant 0\} \cap \mathscr{D}_0(R)) \subset F_3^{s+n+1};$$

(b) *The sign $\mathrm{sgn}\,(f_i + \varepsilon_1)$ is constant on any component of connectivity of the semialgebraic set $\{g \geqslant 0\} \subset F_3^{s+n+1}$ for each $0 \leqslant i \leqslant k$;*

(c) *Formula (5) is true at point z iff formula (6) is valid at point $\tilde{z}$. Moreover, if formula (5) is true at point z and some point $(z, x) \in \tilde{\mathbb{Q}}^{s+n+1}$ belongs to a certain component of connectivity V_1 of the set V, and by the same token (cf. part (a) of the present lemma) the point (z, x) belongs to a corresponding component of connectivity $V_1^{(\varepsilon)}$ of the set $\{g \geqslant 0\}$, then there exists a point $(\tilde{z}, \tilde{x}) \in V_1^{(\varepsilon)}$ such that $g(\tilde{z}, \tilde{x}) = 0$ and such that point $(z, st(\tilde{x})) = st(\tilde{z}, \tilde{x}) \in \tilde{\mathbb{Q}}^{s+n+1}$ is defined.*

PROOF. (a) (see also lemma 2 from Grigor'ev & Vorobjov, 1987). Let a point $u \in V^{(\varepsilon)} \cap \mathscr{D}_0(R)$, then $\|st(u)\| \leqslant R$ and $f_i(st(u)) \geqslant 0$ for every $0 \leqslant i \leqslant k$; this entails the inclusion $st(V^{(\varepsilon)} \cap \mathscr{D}_0(R)) \subset V \cap \mathscr{D}_0(R)$. Now consider a point $w \in V \cap \mathscr{D}_0(R)$, then $f_i(w) + \varepsilon_1 \geqslant \varepsilon_1 > 0$, $0 \leqslant i \leqslant k$ and $g(w) \geqslant \varepsilon_1^{k+1} - \varepsilon_3 > 0$, henceforth,

$$(V \cap \mathscr{D}_0(R)) \subset st(V^{(\varepsilon)} \cap \{g \geqslant 0\} \cap \mathscr{D}_0(R)) \cap (V^{(\varepsilon)} \cap \{g \geqslant 0\} \cap \mathscr{D}_0(R)).$$

(b) (see also lemma 2 from Grigor'ev & Vorobjov, 1987). Let $W_1 \subset \{g \geqslant 0\}$ be a component of connectivity of the semialgebraic set $\{g \geqslant 0\}$. If the sign $\mathrm{sgn}\,(f_i + \varepsilon_1)$ is not constant on W_1, then there exists a point $u \in W_1$ such that $f_i(u) + \varepsilon_1 = 0$. The inequality $g(u) = -\varepsilon_3 < 0$ leads to contradiction.

(c) Assume that formula (6) is valid at point $\tilde{z}$, then there exists a point $(\tilde{z}, \tilde{x}) \in V^{(\varepsilon)} \cap \{g = 0\}$. Since $f_0(\tilde{z}, \tilde{x}) + \varepsilon_1 > 0$, i.e. $\|\tilde{x}\|^2 < z_0 + 2\varepsilon_1$, one concludes that the point $st(\tilde{x})$ is defined and $(z, st(\tilde{x})) \in V$ is fulfilled according to part (a) of this lemma, i.e. formula (5) is true at point z.

Conversely, suppose that formula (5) is valid at point z, then some point $(z, x) \in V$ belongs to an appropriate component of connectivity V_1 of the set V, whence $(z, x) \in V_1^{(\varepsilon)}$

for a suitable component of connectivity $V_1^{(\varepsilon)}$ of the set $V^{(\varepsilon)} \cap \{g \geqslant 0\}$ (see item (a) of this lemma). Note that $V_1 \subset V_1^{(\varepsilon)}$ by virtue of lemma 5(c). Since $f_i(z, x) + \varepsilon_1 \geqslant \varepsilon_1,\ 0 \leqslant i \leqslant k$ and $g(z, x) \geqslant \varepsilon_1^{k+1} - \varepsilon_3$, there exists such a natural number q that for any point $(z', x') \in \mathcal{D}_{(z, x)}(\varepsilon_1^q)$ the inequalities $f_i(z', x') + \varepsilon_1 \geqslant \varepsilon_1/2,\ 0 \leqslant i \leqslant k$ are correct; so $g(z', x') \geqslant (\varepsilon_1/2)^{k+1} - \varepsilon_3 > 0$, i.e. $\mathcal{D}_{(z, x)}(\varepsilon_1^q) \subset V_1^{(\varepsilon)}$. Obviously, $(\tilde{z}, x) \in \mathcal{D}_{(z, x)}(\varepsilon_1^q)$.

On the s-dimensional plane $\mathcal{L} = \{(\tilde{z}, x') : x' \in F_3^s\} \subset F_3^{s+n+1}$ we consider an arbitrary ray $\gamma \subset \mathcal{L}$ with endpoint $(\tilde{z}, x)$ and with a rational directing vector. The intersection $K_1 = \gamma \cap \{g < 0\}$ is not empty. Indeed, otherwise $\gamma \subset V_1^{(\varepsilon)}$ in view of part (b) of the present lemma; so for any point $(\tilde{z}, x') \in \gamma$ the inequality $\|x'\|^2 < z_0 + 2\varepsilon_1$ is fulfilled, which leads to a contradiction. The set K_1 is semialgebraic and, henceforth, K_1 is the union of a finite number of intervals. We denote by $(\tilde{z}, \tilde{x}) \in \gamma$ one of the endpoints of these intervals which is the nearest to the point $(\tilde{z}, x)$. Then $g(\tilde{z}, \tilde{x}) = 0$ since in an arbitrary neighbourhood (on the ray γ) of the point $(\tilde{z}, \tilde{x})$ there is a point in which the polynomial g has a negative value as well as a point in which g has a positive value.

The closed interval $J \subset \gamma$ with endpoints $(\tilde{z}, x)$ and $(\tilde{z}, \tilde{x})$ is contained in $V_1^{(\varepsilon)}$, according to part (b) of the present lemma, in particular $(\tilde{z}, \tilde{x}) \in V_1^{(\varepsilon)}$, this entails $\|\tilde{x}\|^2 < z_0 + 2\varepsilon_1$ and that the formula (6) is true at point $\tilde{z}$. It remains to show that $st(\tilde{z}, \tilde{x}) \in V_1$. The set $st(J) \subset \tilde{\mathbb{Q}}^{s+n+1}$ is well defined and coincides with the closed interval with endpoints (z, x) and $(z, st(\tilde{x}))$. Indeed, let $(u_1, \ldots, u_s) \in \mathbb{Q}^s$ be the directing vector of the ray γ. Then $\tilde{x} = x + \alpha(u_1, \ldots, u_s)$ for a certain $0 < \alpha \in F_3$ being $\mathbb{Q}$-finite by the facts proved above. For any $0 \leqslant \beta \leqslant \alpha,\ \beta \in F_3$, the equality

$$st(x + \beta(u_1, \ldots, u_s)) = x + st(\beta)(u_1, \ldots, u_s)$$

is correct. This implies that $st(J)$ is a closed interval. Finally, $st(J) \subset V$ by virtue of part (a) of the present lemma, henceforth, $st(J) \subset V_1$, which completes the proof of the lemma.

Now we return to the description of the algorithm. It involves the following construction from Grigor'ev & Vorobjov (1987) (see also Vorobjov & Grigor'ev, 1985). We denote by $\Gamma \subset \mathbb{Z}^{s-1}$ the family consisting of vectors of the kind $\gamma = (\gamma_2, \ldots, \gamma_s) \in \Gamma$, where $\gamma_i,\ 2 \leqslant i \leqslant s$ run independently over all integers from 1 up to $N_2 = (2(k+1)d)^s$. For a point $z^{(2)} \in F_2^{n+1}$ we denote by $g(z^{(2)})$ the polynomial

$$g(z^{(2)}, X_1, \ldots, X_s) \in F_2[\varepsilon_3][X_1, \ldots, X_s].$$

The following lemma can be inferred from lemmas 4, 5 in Grigor'ev & Vorobjov (1987) (see there the corollary after lemma 5; cf. also Vorobjov & Grigor'ev, 1985).

LEMMA 7. *Let a point $z^{(2)} \in F_2^{n+1}$.*

(a) *For each component of connectivity V_2 of the variety $\{g(z^{(2)}) = 0\} \subset F_3^s$, provided that V_2 is situated in a certain ball, and for every vector $\gamma = (\gamma_2, \ldots, \gamma_s) \in \Gamma$ the system of equations*

$$g(z^{(2)}) = \left(\frac{\partial g(z^{(2)})}{\partial X_2}\right)^2 - \frac{\gamma_2}{N_2 s} \sum_{1 \leqslant i \leqslant s} \left(\frac{\partial g(z^{(2)})}{\partial X_i}\right)^2 = \cdots$$

$$= \left(\frac{\partial g(z^{(2)})}{\partial X_s}\right)^2 - \frac{\gamma_s}{N_2 s} \sum_{1 \leqslant i \leqslant s} \left(\frac{\partial g(z^{(2)})}{\partial X_i}\right)^2 = 0 \qquad (7)$$

has a root in V_2.

(b) *There exists a vector $\gamma = (\gamma_2, \ldots, \gamma_s) \in \Gamma$ such that any solution of system (7), which belongs to the space F_3^s, is an isolated point of the algebraic variety consisting of all solutions of system (7) in the space $\bar{F}_3^s$.*

We recall that $\bar{F}_3 = F_3[\sqrt{-1}]$ is the algebraic closure of F_3. Lemma 7(b) entails, in particular, that for a relevant $\gamma \in \Gamma$, system (7) has only finite number of solutions in the space F_3^s.

In order to verify formula (6) in a certain point $z^{(2)}$, it is sufficient to test, whether there exists a component of connectivity V_3 of the variety $\{g(z^{(2)}) = 0\} \subset F_3^s$ and a point $x \in V_3$ from some representative set for the variety $\{g(z^{(2)}) = 0\}$ (considered below) such that the inequalities $f_i(z^{(2)})(x) + \varepsilon_1 > 0$, $0 \leqslant i \leqslant k$ hold, taking into account that signs $\operatorname{sgn}(f_i(z^{(2)}) + \varepsilon_1)$, $0 \leqslant i \leqslant k$ are constant on V_3 according to lemma 6(b). As a representative set of the points x, the algorithm will take solutions of system (7) in the space F_3^s, this family of points x suffices in view of lemma 7(a) and the observation that if $\operatorname{sgn}(f_0(z^{(2)})(x^{(1)}) + \varepsilon_1) > 0$ for points $x^{(1)} \in V_3$ belonging to some component of connectivity V_3 of the variety $\{g(z^{(2)}) = 0\} \subset F_3^s$, then $V_3 \subset \mathcal{D}_0((z_0 + 2\varepsilon_1)^{1/2})$.

Let us fix for the time being a vector $\gamma \in \Gamma$ and denote $h_1 = g$,

$$h_j = \left(\frac{\partial g}{\partial X_j}\right)^2 - \frac{\gamma_j}{N_2 s} \sum_{1 \leqslant i \leqslant s} \left(\frac{\partial g}{\partial X_i}\right)^2 \in \mathbb{Q}[\varepsilon_1, \varepsilon_3][Z_0, \ldots, Z_n, X_1, \ldots, X_s]; \quad 2 \leqslant j \leqslant s.$$

We introduce a new variable X_0 and polynomials

$$\bar{h}_j = X_0^{\deg x_1, \ldots, x_s(h_j)} h_j(Z_0, \ldots, Z_n, X_1/X_0, \ldots, X_s/X_0), \quad 1 \leqslant j \leqslant s$$

homogeneous with respect to variables $X_0, X_1, \ldots, X_s$. Consider a certain point $z^{(3)} \in \bar{F}_3^{n+1}$, one more new variable Y and the following system of equations homogeneous with respect to variables $X_0, \ldots, X_s$ (cf. section 5 in Chistov & Grigor'ev, 1983b; Chistov & Grigor'ev, 1984):

$$\bar{h}_j(z^{(3)}) - Y X_j^{\deg X_0, \ldots, X_s(\bar{h}_j)} = 0; \quad 1 \leqslant j \leqslant s \tag{8}$$

over a field $\bar{F}_3(Y)$.

Besides, consider a system of equations

$$\bar{h}_j - Y X_j^{\deg x_0, \ldots, x_s(\bar{h}_j)} = 0; \quad 1 \leqslant j \leqslant s \tag{8'}$$

in the variables $X_0, \ldots, X_s$ over a field $\mathbb{Q}(\varepsilon_1, \varepsilon_3)(Y, Z_0, \ldots, Z_n)$. Further, we need some similar statements about both systems (8) and (8'). Thus, we consider an arbitrary field F (of zero characteristic), a vector $\mathscr{Z} \in F^{n+1}$ and a system

$$\bar{h}_j(\mathscr{Z}) - Y X_j^{\deg x_0, \ldots, x_s(\bar{h}_j)} = 0; \quad 1 \leqslant j \leqslant s. \tag{8''}$$

Define the field $H_1 = F(Y)$. Later we consider two cases: $F = \bar{F}_3$ and $\mathscr{Z} = z^{(3)}$ or $F = \mathbb{Q}(\varepsilon_1, \varepsilon_3)(Z_0, \ldots, Z_n)$ and $\mathscr{Z} = (Z_0, \ldots, Z_n)$.

PROPOSITION 3. *System (8') has a finite nonzero number of solutions in the projective space $\mathbb{P}^s(\bar{H}_1)$ (see section 5 in Chistov & Grigor'ev, 1983b and also Chistov & Grigor'ev, 1984).*

PROOF. We define $T_1 = 1/Y$, then $H_1 = F(Y_1)$ and system (8'') is equivalent to a system

$$Y_1 \bar{h}_j(\mathscr{Z}) - X_j^{\deg x_0, \ldots, x_s(\bar{h}_j)} = 0; \quad 1 \leqslant j \leqslant s$$

(i.e. the varieties of solutions of the latter system and system (8'') in the space $\mathbb{P}^s(\bar{H}_1)$ coincide). The latter system has a finite nonzero number of solutions iff its u-resultant $\mathscr{R}(Y_1) \in F[u_0, \ldots, u_s][Y_1]$ does not vanish identically. On the other hand, $\mathscr{R}(0) \neq 0$, taking into account that $\mathscr{R}(0)$ equals to the u-resultant of the system $X_j^{\deg x_0, \ldots, x_s(\bar{h}_j)} = 0$; $1 \leqslant j \leqslant s$ having a finite number of solutions in $\mathbb{P}^s(\bar{F})$. This completes the proof of the proposition.

Consider now system (8″) in the variables $X_0, \ldots, X_s, Y$ over the field F and the variety of its solutions $U_F \subset \bar{F}^{s+2}$.

PROPOSITION 4. *The irreducible (over F) components $\mathscr{V}_t \subset \bar{F}^{s+2}$ of the variety U_F that are not situated in any union of a finite number of hyperplanes of the kind $\{Y = \beta\}$, where $\beta \in \bar{F}$, correspond bijectively to the classes of solutions conjugate over the field H_1 in the space $\mathbb{P}^s(\bar{H}_1)$ of the system (8″). Under this correspondence, if $J \subset F[Y, X_0, \ldots, X_s]$ is an ideal of a component $\mathscr{V}_t$, then $H_1 \otimes_F J \subset H_1[X_0, \ldots, X_s]$ is an ideal of the corresponding conjugate class. Furthermore, $\dim \mathscr{V}_t = 2$ for each t (see section 5 in Chistov & Grigor'ev, 1983b and also Chistov & Grigor'ev, 1984).*

PROOF. Indeed, denote by J_{H_1} (resp. J_F) the ideal, generated by the polynomials $\bar{h}_j(\mathscr{L}) - Y X_j^{\deg_{X_0, \ldots, X_s}(\bar{h}_j)}$ for $1 \leqslant j \leqslant s$ in the H_1-algebra $\Lambda_{H_1} = H_1[X_0, \ldots, X_s]$ (resp. in F-algebra $\Lambda_F = F[Y, X_0, \ldots, X_s]$). Then

$$\Lambda_{H_1} = H_1 \otimes_F \Lambda_F = (F[Y] \backslash \{0\})^{-1} \Lambda_F.$$

Therefore, there is a bijective correspondence preserving the inclusion relation between prime ideals $\mathscr{I}_{H_1}^{(i)} \subset \Lambda_{H_1}$, and on the other hand, prime ideals $\mathscr{I}_F^{(i)} \subset \Lambda_F$ such that $\mathscr{I}_F^{(i)} \cap F[Y] = \{0\}$; besides, under this correspondence $\mathscr{I}_{H_1}^{(i)} = H_1 \otimes_F \mathscr{I}_F^{(i)}$ (see Lang, 1965). Apart from that, to every prime ideal $\mathscr{I}_F^{(i)}$ such that $\mathscr{I}_F^{(i)} \cap F[Y] = \{0\}$ corresponds a variety $\mathscr{V}_{\mathscr{I}_F^{(i)}} \subset \bar{F}^{s+2}$ irreducible over F, which is not situated in any union of a finite number of hyperplanes of the kind $\{Y = \beta\}$ for $\beta \in F$ (and conversely). Denote by $\mathscr{V}_{\mathscr{I}_{H_1}^{(i)}} \subset \mathbb{P}^s(\bar{H}_1)$ a variety irreducible over H_1 corresponding to a homogeneous prime ideal $\mathscr{I}_{H_1}^{(i)}$, provided that $\mathscr{I}_{H_1}^{(i)} \neq (X_0, \ldots, X_s)$. Let $\mathscr{I}_{H_1}^{(i)} \supset J_{H_1}$ (resp. $\mathscr{I}_F^{(i)} \supset J_F$) be a certain prime ideal homogeneous (resp. homogeneous relatively to $X_0, \ldots, X_s$) and minimal among prime ideals containing J_{H_1} (resp. J_F). In other words, the factor ideal $\mathscr{I}_{H_1}^{(i)}/J_{H_1}$ (resp. $\mathscr{I}_F^{(i)}/J_F$) is a minimal prime ideal in the factorring Λ_{H_1}/J_{H_1} (resp. Λ_F/J_F). Hence, $\mathscr{V}_{\mathscr{I}_{H_1}^{(i)}}$ (resp. $\mathscr{V}_{\mathscr{I}_F^{(i)}}$) is an irreducible (over H_1, resp. F) component of the variety of roots of system (8″) in the space $\mathbb{P}^s(\bar{H}_1)$ (resp. $\bar{F}^{s+2}$), taking into account that $\mathscr{I}_{H_1}^{(i)} \neq (X_0, \ldots, X_s)$ since system (8″) has at least one root in the space $\mathbb{P}^s(\bar{H}_1)$ by proposition 3. This entails the correspondence claimed in proposition 4 between components $\mathscr{V}_t$ and, on the other hand, classes of solutions of system (8″) conjugate over H_1. Finally, for each $\mathscr{V}_t$ there is a suitable prime ideal $\mathscr{I}_F^{(i)}$ such that $\mathscr{V}_t = \mathscr{V}_{\mathscr{I}_F^{(i)}}$, therefore

$$\dim \mathscr{V}_t = \deg tr_F(\Lambda_F/\mathscr{I}_F^{(i)}) = \deg tr_F((H_1 \otimes_F \Lambda_F)/(H_1 \otimes_F \mathscr{I}_F^{(i)}))$$

$$= \deg tr_F(\Lambda_{H_1}/\mathscr{I}_{H_1}^{(i)}) = \deg tr_{H_1}(\Lambda_{H_1}/\mathscr{I}_{H_1}^{(i)}) + 1 = \dim \mathscr{V}_{\mathscr{I}_{H_1}^{(i)}} + 2 = 2$$

(the latter equality follows from proposition 3). The proposition is proved.

PROPOSITION 5. *The variety*

$$\left(\bigcup_t \mathscr{V}_t \right) \cap \{Y = 0\} \subset \bar{F}^{s+1}$$

considered as a subvariety of the space $\bar{F}^{s+1}$ with coordinates $X_0, \ldots, X_s$ is a union of a finite number of lines passing through the origin of coordinates. Besides, for every isolated solution $(x_1, \ldots, x_s) \in \bar{F}^s$ of system $h_1(\mathscr{L}) = \ldots = h_s(\mathscr{L}) = 0$ (obtained from system (7) by replacing the point $z^{(2)}$ by the point $\mathscr{L}$) its cone (which is a line) $\{(\lambda, \lambda x_1, \ldots, \lambda x_s)_{\lambda \in F}\} \subset \bar{F}^{s+1}$ is a component of the variety

$$\left(\bigcup_t \mathscr{V}_t \right) \cap \{Y = 0\}$$

irreducible over the field $\bar{F}$.

PROOF. For each t, by virtue of proposition 4, $\dim \mathscr{V}_t = 2$. Apart from that, $\mathscr{V}_t \cap \{Y = 0\} \subsetneq \mathscr{V}_t$. Therefore, $\dim(\mathscr{V}_t \cap \{Y = 0\}) = 1$ according to the theorem on the dimension of intersection (see Shafarevich, 1974). Furthermore, the variety $\mathscr{V}_t \cap \{Y = 0\}$ is homogeneous since $\mathscr{V}_t$ is a component of the variety, consisting of solutions of system (8″), which is homogeneous relatively to the coordinates $X_0, \ldots, X_s$. Thus, $\mathscr{V}_t \cap \{Y = 0\}$ is a union of a finite number of lines passing through the origin of coordinates.

Taking into account that $\mathrm{con} = \{(\lambda, \lambda x_1, \ldots, \lambda x_s)_{\lambda \in \bar{F}}\} \subset U_F \cap \{Y = 0\}$, one can infer that there exists a certain irreducible component $\mathscr{V} \subset U_F$ of the variety U_F which contains $\mathrm{con} \subset \mathscr{V}$. If $\mathscr{V} = \mathscr{V}_t$ for some t, then con is a line being a component of $\mathscr{V}_t \cap \{Y = 0\}$ and the required statement of the proposition is valid. Otherwise, $\mathscr{V}$ is situated in a union of a finite number of hyperplanes of the kind $\{Y = \beta\}$, therefore, $\mathscr{V} \subset \{Y = 0\}$ because of the irreducibility of $\mathscr{V}$. Denote by $\mathscr{W} \subset \mathbb{P}^s(\bar{F})$ a projective irreducible variety such that the cone $\mathrm{con}(\mathscr{W}) = \mathscr{V}$. According to theorem on the dimension of intersections (see Shafarevich, 1974), $\dim \mathscr{V} \geqslant 2$, hence $\dim \mathscr{W} \geqslant 1$. On the other hand, the point $(1 : x_1 : \ldots : x_s) \in \mathscr{W}$ is an isolated solution of a system $\bar{h}_1(\mathscr{L}) = \ldots = \bar{h}_s(\mathscr{L}) = 0$. This leads to a contradiction with the fact that $\mathscr{W}$ also satisfies the latter system. The proposition is proved.

In the sequel we shall make use of the following construction from Lazard (1981). Let $g_0, \ldots, g_{k-1} \in F[X_0, \ldots, X_s]$ be homogeneous polynomials of degrees $\delta_0 \geqslant \delta_1 \geqslant \ldots \geqslant \delta_{k-1}$, respectively. Introduce new variables $u_0, \ldots, u_s$ algebraically independent over a field $F(X_0, \ldots, X_s)$. Consider a polynomial $g_k = X_0 u_0 + \ldots + X_s u_s$ and set

$$D = \sum_{0 \leqslant i \leqslant s} \delta_i - s,$$

where $\delta_k = \ldots = \delta_s = 1$ if $k \leqslant s$. Denote by $\mathscr{B}_i$ (resp. $\mathscr{B}$) a space of homogeneous polynomials in the variables $X_0, \ldots, X_s$ of degree $D - \delta_i$ (resp. D) over the field $F(u_0, \ldots, u_s)$. Consider a linear mapping $\mathscr{A} : \mathscr{B}_0 \oplus \ldots \oplus \mathscr{B}_k \to \mathscr{B}$ over the field $F(u_0, \ldots, u_s)$ given by the formula

$$\mathscr{A}(b_0, \ldots, b_k) = \sum_{0 \leqslant i \leqslant k} g_i b_i.$$

Denote by

$$\rho_i = \binom{n + D - \delta_i}{n}, \qquad \tau = \binom{n + D}{n}$$

binomial coefficients. One can write an arbitrary element $b = (b_0, \ldots, b_k) \in \mathscr{B}_0 \oplus \ldots \oplus \mathscr{B}_k$ in the form

$$b = (b_{0,1}, \ldots, b_{0,\rho_0}, b_{1,1}, \ldots, b_{1,\rho_1}, \ldots, b_{k,1}, \ldots, b_{k,\rho_k}),$$

where $b_{i,1}, \ldots, b_{i,\rho_i}$ are the coefficients of the polynomial b_i, provided that some numeration of monomials of degree $D - \delta_i$ is fixed. Similarly, one can write elements from the space $\mathscr{B}$. In a chosen coordinate system the mapping $\mathscr{A}$ has a matrix A of size $\tau \times (\sum_{0 \leqslant i \leqslant k} \rho_i)$. The matrix A can be uniquely represented in a form $A = (A^{(n)}, A^{(f)})$, where $A^{(n)}$ (we call it the number part of A) contains $\sum_{0 \leqslant i \leqslant k-1} \rho_i$ columns and $A^{(f)}$ (we call it the formal part of A) contains ρ_k columns; furthermore, the entries of $A^{(n)}$ belong to the field F, the entries of $A^{(f)}$ are linear forms in variables $u_0, \ldots, u_s$ over F.

The next proposition is a certain effective version of Hilbert's Nullstellensatz.

PROPOSITION 6. (Lazard, 1981). *A system $g_0 = \ldots = g_{k-1} = 0$ has no roots in $\mathbb{P}^s(\bar{F})$ iff the ideal $(g_0, \ldots, g_{k-1}) \supset (X_0, \ldots, X_s)^D$.*

The proof of the next proposition is based on the latter one.

PROPOSITION 7 (Lazard, 1981).

(a) *A system $g_0 = \ldots = g_{k-1} = 0$ has a finite number of roots in $\mathbb{P}^s(\bar{F})$ iff rank rank $(A) = \tau$; in parts (b), (c), (d) we suppose that rank rank $(A) = \tau$.*

(b) *All $\tau \times \tau$ minors (by a minor we mean the determinant of a submatrix) of matrix A generate a principal ideal whose generator $R \in F[u_0, \ldots, u_s]$ is their greatest common divisor.*

(c) *a form R homogeneous relatively to the variables $u_0, \ldots, u_s$ is a product*

$$R = \prod_{1 \leqslant i \leqslant D_2} L_i^{e_i}$$

of linear forms

$$L_i = \sum_{0 \leqslant j \leqslant s} \xi_j^{(i)} u_j, \quad 1 \leqslant i \leqslant D_2$$

with coefficients from $\bar{F}$, furthermore, $(\xi_0^{(i)} : \ldots : \xi_s^{(i)}) \in \mathbb{P}^s(\bar{F})$ is a root of the system $g_0 = \ldots = g_{k-1} = 0$ with the multiplicity e_i.

(d) *Let Δ be a nonsingular $\tau \times \tau$ submatrix of A containing rank $(A^{(n)})$ columns in the number part $A^{(n)}$ (obviously, such a submatrix exists). Then the determinant $\det \Delta$ equals to R up to a factor from F^*, besides,*

$$\deg R = D_1 = \sum_{1 \leqslant i \leqslant D_2} e_i = \tau - \text{rank}(A^{(n)}).$$

We shall apply later this proposition in the case when $k = n$. In this case, R coincides with the classical u-resultant of the system of the polynomials $g_0, \ldots, g_{n-1}$. We shall make use of the suggested explicit form of R.

The algorithm constructs a matrix A with entries in the ring

$$\mathbb{Q}[\varepsilon_1, \varepsilon_3][Y, Z_0, \ldots, Z_n, u_0, \ldots, u_s]$$

corresponding to system (8′) considered in the variables $X_0, \ldots, X_s$. Denote by $A_{(z^{(3)})}$ (resp. $A_{\mathscr{Z}}$) a matrix corresponding to system (8) (resp. (8″)) and obtained from matrix A by substituting the coordinates of the vector $z^{(3)}$ (resp. $\mathscr{Z}$) instead of the variables $Z_0, \ldots, Z_n$. Denote by

$$R \in \mathbb{Q}[\varepsilon_1, \varepsilon_3, Y, Z_0, \ldots, Z_n, u_0, \ldots, u_s] \text{ (resp. } R_{(z^{(3)})} \in \bar{F}_3[Y, u_0, \ldots, u_s],$$

resp. $R_{\mathscr{Z}} \in F[Y, u_0, \ldots, u_s])$ the u-resultant of system (8′) (resp. (8), resp. (8″)) (see proposition 7(b)). u-Resultants do not vanish because of proposition 3. One can assume w.l.o.g. that $Y \nmid R$, $Y \nmid R_{(z^{(3)})}$, $Y \nmid R_{\mathscr{Z}}$, otherwise one divides the respective polynomial by the highest possible power of the variable Y.

Consider arbitrary polynomials $\Xi_0, \ldots, \Xi_\mu \in F[Y, X_0, \ldots, X_s]$ homogeneous in the variables $X_0, \ldots, X_s$ such that the variety $\bigcup_t \mathscr{V}_t$ (see proposition 4) coincides with the variety of all roots in the space $\bar{F}^{s+2}$ of the system $\Xi_0 = \ldots = \Xi_\mu = 0$. By virtue of proposition 4, the variety of all roots of the latter system over the field $\bar{H}_1$ coincides with the variety $U_{H_1} \subset \mathbb{P}^s(\bar{H}_1)$ of all roots of system (8″). Owing to proposition 3, U_{H_1} has a finite number of points; denote by $\hat{A}_{\mathscr{Z}}$ a matrix with τ_1 rows corresponding to the system $\Xi_0 = \ldots = \Xi_\mu = 0$ and by $0 \neq \hat{R}_{\mathscr{Z}} \in F[Y, u_0, \ldots, u_s]$ its u-resultant. Again, dividing $\hat{R}_{\mathscr{Z}}$ by the highest possible power of Y, we can sssume w.l.o.g. that $Y \nmid \hat{R}_{\mathscr{Z}}$.

Proposition 7(c) entails that

$$\hat{R}_{\mathscr{L}} = \prod_j (L_j^{(1)})^{\hat{\gamma}_j}, \qquad R_{\mathscr{L}} = \prod_j (L_j^{(1)})^{\gamma_j},$$

where the linear forms $L_j^{(1)} = \xi_0^{(j)} u_0 + \ldots + \xi_s^{(j)} u_s$ correspond bijectively to the points $(\xi_0^{(j)} : \ldots : \xi_s^{(j)}) \in U_{H_1}$ of the variety U_{H_1}. Furthermore, the integers $\hat{\gamma}_j, \gamma_j$ are positive. Hence, the relations $R_{\mathscr{L}} | (\hat{R}_{\mathscr{L}})^{\hat{\gamma}}$ and $\hat{R}_{\mathscr{L}} | (R_{\mathscr{L}})^{\gamma}$ are valid in the ring $H_1[u_0, \ldots, u_s]$ for relevant integers $\hat{\gamma}, \gamma$. Therefore,

$$R_{\mathscr{L}}(0, u_0, \ldots, u_s) | (\hat{R}_{\mathscr{L}}(0, u_0, \ldots, u_s))^{\hat{\gamma}} \quad \text{and} \quad \hat{R}_{\mathscr{L}}(0, u_0, \ldots, u_s) | (R_{\mathscr{L}}(0, u_0, \ldots, u_s))^{\gamma}$$

are true in the ring $F[u_0, \ldots, u_s]$.

Consider a system

$$\Xi_0(0, X_0, \ldots, X_s) = \ldots = \Xi_\mu(0, X_0, \ldots, X_s) = 0.$$

It has a finite number of roots $\mathscr{W} \subset \mathbb{P}^s(\bar{F})$ and herewith the cone

$$\operatorname{con}(\mathscr{W}) = \left(\bigcup_t \mathscr{V}_t \right) \cap \{ Y = 0 \} \subset \bar{F}^{s+1}$$

(see proposition 5). The matrix $\hat{A}_{\mathscr{L}}(0)$, corresponding to the latter system is obtained from the matrix $\hat{A}_{\mathscr{L}}$ by substituting 0 instead of Y. Let $\Delta(0)$ be a certain nonsingular $\tau_1 \times \tau_1$ submatrix of the matrix $\hat{A}_{\mathscr{L}}(0)$, containing the maximal possible number of columns in the number part of $\hat{A}_{\mathscr{L}}(0)$, then

$$\det(\Delta(0)) = \prod_i L_i^{\hat{c}_i}$$

is the u-resultant of the system

$$\Xi_0(0, X_0, \ldots, X_s) = \ldots = \Xi_\mu(0, X_0, \ldots, X_s) = 0,$$

where the linear forms L_i correspond bijectively to the points of the set $\mathscr{W}$ according to proposition 7(b). Denote by Δ the $\tau_1 \times \tau_1$ submatrix of the matrix $\hat{A}_{\mathscr{L}}$ formed by the same columns as the matrix $\Delta(0)$. Then $\det(\Delta) \neq 0$ and, by proposition 7(b), $\hat{R}_{\mathscr{L}} | \det(\Delta)$ in the ring $H_1[u_0, \ldots, u_s]$. Hence, $\hat{R}_{\mathscr{L}}(0, u_0, \ldots, u_s) | \det(\Delta(0))$. Thus

$$R_{\mathscr{L}}(0, u_0, \ldots, u_s) = \prod_i L_i^{c_i}$$

is fulfilled for suitable integers c_i. Our next purpose is to show that each $c_i > 0$.

Denote by $\mathscr{J}_F^{(t)} \subset F[Y, X_0, \ldots, X_s]$ the prime ideal defining the component $\mathscr{V}_t$, and by $\mathscr{J}_{H_1}^{(t)} \subset H_1[X_0, \ldots, X_s]$ the prime ideal defining the class of points of the variety U_{H_1} conjugate over the field H_1, that correspond to each other by proposition 4. Introduce a polynomial

$$E(Y, X_0, \ldots, X_s) = R_{\mathscr{L}}\left(Y, -\sum_{1 \leqslant i \leqslant s} u_i X_i, u_1 X_0, \ldots, u_s X_0\right) = \sum_I E_I u^I,$$

where the polynomials $E_I \in F[Y, X_0, \ldots, X_s]$ and $u^I = u_1^{I_1} \ldots u_s^{I_s}$ is a monomial respective to a multi-index $I = (I_1, \ldots, I_s)$. Let a point $(\xi_0 : \ldots : \xi_s) \in U_{H_1}$ and

$$L_{j_0}^{(1)}(u_0, \ldots, u_s) = \xi_0 u_0 + \ldots + \xi_s u_s$$

be the corresponding linear form. Consider a polynomial

$$E_{j_0}^{(1)}(X_0, \ldots, X_s) = L_{j_0}^{(1)}\left(-\sum_{1 \leqslant i \leqslant s} u_i X_i, u_1 X_0, \ldots, u_s X_0\right)$$

$$= -\xi_0 \sum_{1 \leqslant i \leqslant s} u_i X_i + \xi_1 u_1 X_0 + \ldots + \xi_s u_s X_0 \in \bar{H}_1[X_0, \ldots, X_s, u_1, \ldots, u_s].$$

Obviously,

$$E(Y, X_0, \ldots, X_s) = \prod_j (E_j^{(1)}(X_0, \ldots, X_s))^{\gamma_j}.$$

Then

$$0 = E_{j_0}^{(1)}(\xi_0, \ldots, \xi_s) \in \bar{H}_1[u_1, \ldots, u_s],$$

and, conversely, if $E_{j_0}^{(1)}(\zeta_0, \ldots, \zeta_s) = 0$ and some of the $\zeta_0, \ldots, \zeta_s$ does not vanish, then the points

$$(\xi_0 : \ldots : \xi_s) = (\zeta_0 : \ldots : \zeta_s) \in \mathbb{P}^s(\bar{H}_1)$$

coincide. Therefore

$$0 = \prod_j (E_j^{(1)}(\hat{\xi}_0, \ldots, \hat{\xi}_s))^{\gamma_j} = E(Y, \hat{\xi}_0, \ldots, \hat{\xi}_s)$$

for every point $(\hat{\xi}_0 : \ldots : \hat{\xi}_s) \in U_{H_1}$. This implies that a polynomial $E_I \in \mathscr{J}_{H_1}^{(t)}$ for each I, t. Hence, by virtue of proposition 4, $E_I \in \mathscr{J}_F^{(t)}$. This means that the polynomial $E(Y, X_0, \ldots, X_s)$ vanishes on the variety $(\bigcup_t \mathscr{V}_t)$. Therefore, the polynomial $E(0, X_0, \ldots, X_s)$ vanishes on the variety

$$\left(\bigcup_t \mathscr{V}_t\right) \cap \{Y = 0\}.$$

By the above arguments, this entails that, for any point of the variety

$$\left(\bigcup_t \mathscr{V}_t\right) \cap \{Y = 0\}$$

and its corresponding linear form L_{i_0},

$$L_{i_0}^t | R_{\mathscr{Z}}(0, u_0, \ldots, u_s) = \prod_i L_i^{c_i}.$$

This completes the proof of the following

PROPOSITION 8. *Let $R_{\mathscr{Z}}$ be the u-resultant of system (8″). Then*

$$R_{\mathscr{Z}}(0, u_0, \ldots, u_s) = \prod_i L_i^{c_i}$$

for appropriate positive integers c_i, where the linear forms $L_i = \zeta_0 u_0 + \ldots + \zeta_s u_s$ correspond bijectively to the points $(\zeta_0 : \ldots : \zeta_s) \in \mathscr{W}$ of the set $\mathscr{W} \subset \mathbb{P}^s(\bar{F})$ such that its cone

$$\mathrm{con}(\mathscr{W}) = \left(\bigcup_t \mathscr{V}_t\right) \cap \{Y = 0\}.$$

Applying lemma 7(a), (b), proposition 5 and proposition 8 to a point $\mathscr{Z} = z^{(2)} \in F_2^{n+1}$ one can infer the following

COROLLARY. *A point $z^{(2)} \in F_2^{n+1}$ satisfies formula (6) iff there exist a vector*

$$\gamma = (\gamma_2, \ldots, \gamma_s) \in \Gamma$$

and a linear form $\hat{L}_i = \xi_0^{(i)} u_0 + \ldots + \xi_s^{(i)} u_s$ such that $\hat{L}_i | R_{z^{(2)}}(0, u_0, \ldots, u_s)$ and $\xi_0^{(i)} \neq 0$. Furthermore, the point $(\xi_1^{(i)}/\xi_0^{(i)}, \ldots, \xi_s^{(i)}/\xi_0^{(i)}) \in F_3^s$ belongs to the space F_3^s. Finally, the inequalities

$$f_i(z^{(2)}, \xi_1^{(i)}/\xi_0^{(i)}, \ldots, \xi_s^{(i)}/\xi_0^{(i)}) + \varepsilon_1 > 0; \quad 0 \leqslant i \leqslant k$$

are fulfilled.

Our next aim is to contruct a polynomial

$$\psi_0(Z_0, \ldots, Z_n) \in \mathbb{Q}[\varepsilon_1, \varepsilon_3][Z_0, \ldots, Z_n, u_0, \ldots, u_s]$$

and a polynomial

$$P(Z_0, \ldots, Z_n) \in \mathbb{Q}[\varepsilon_1, \varepsilon_3][Z_0, \ldots, Z_n]$$

such that

$$\psi_0(Z_0, \ldots, Z_n) = \lambda \prod_{i \in I} L_i^{c_i} | R(0, Z_0, \ldots, Z_n, u_0, \ldots, u_s)$$

is a product (up to a factor $\lambda \in \overline{F_3(Z_0, \ldots, Z_n)}{}^*$) of all linear forms $L_i = \zeta_0^{(i)} u_0 + \ldots + \zeta_s^{(i)} u_s$, $i \in I$ which are factors of the polynomial $R(0, Z_0, \ldots, Z_n, u_0, \ldots, u_s)$ such that $\overline{\zeta_j^{(i)} \in \mathbb{Q}(\varepsilon_1, \varepsilon_3)(Z_0, \ldots, Z_n)}$, $\zeta_0^{(i)} \neq 0$ and, besides,

$$\psi_0(z^{(3)}) = \hat{\lambda} \prod_{i \in J} \hat{L}_i^{c_i} | R_{z^{(3)}}(0, u_0, \ldots, u_s)$$

is a product (up to a factor $\hat{\lambda} \in \bar{F}_3^*$) of all linear forms

$$\hat{L}_i = (\xi_0^{(i)} u_0 + \ldots + \xi_s^{(i)} u_s) | R_{z^{(3)}}(0, u_0, \ldots, u_s), \quad i \in J$$

such that $\xi_j^{(i)} \in \bar{F}_3$, $\xi_0^{(i)} \neq 0$, provided that $P(z^{(3)}) \neq 0$, where a point $z^{(3)} \in \bar{F}_3^{n+1}$ (cf. the construction from Chistov & Grigor'ev, 1984).

Now the algorithm applies the Gaussian algorithm to the matrix A. Let ρ steps of Gaussian algorithm be already carried out and let $(i_0, j_0), \ldots, (i_{\rho-1}, j_{\rho-1})$ be a sequence of leading entries, herewith $j_0 < \ldots < j_{\rho-1}$, and $i_\alpha \neq i_\beta$ when $\alpha \neq \beta$. By the current step the matrix A is reduced by elementary transformations to a matrix $A^{(\rho)} = (a_{ij}^{(\rho)})$ (at the beginning $A^{(0)} = A$) with entries from the field $\mathbb{Q}(\varepsilon_1, \varepsilon_3)(Z_0, \ldots, Z_n, u_0, \ldots, u_s)$.

Moreover, $a_{ij}^{(\rho)} = 0$ if either $j_{\beta-1} < j < j_\beta$ for a certain $0 \leqslant \beta \leqslant \rho - 1$, or $j = j_\beta$ for some $0 \leqslant \beta \leqslant \rho - 1$ and $i \neq i_\alpha$ for all $0 \leqslant \alpha \leqslant \rho - 1$, or $j = j_\beta$ for some $0 \leqslant \beta \leqslant \rho - 1$ and $i = i_\alpha$ for some $0 \leqslant \beta < \alpha \leqslant \rho - 1$. The next leading entry (i_ρ, j_ρ) is picked out so that j_ρ is the least possible index such that $j_{\rho-1} < j_\rho$ and $a_{i_\rho, j_\rho}^{(\rho)} \neq 0$. For each i different from all $i_0, \ldots, i_\rho$ set the entry $a_{ij}^{(\rho+1)} = a_{ij}^{(\rho)} - a_{i_\rho, j}^{(\rho)} a_{i, j_\rho}^{(\rho)} / a_{i_\rho, j_\rho}^{(\rho)}$ (an elementary transformation over rows with the leading entry (i_ρ, j_ρ)). This completes the description of applying Gaussian algorithm to the matrix A and producing matrices $A = A^{(0)}, \ldots, A^{(\tau-1)}$, where τ is the number of rows in the matrix A, taking into account that $\operatorname{rank}(A) = \tau$ by virtue of proposition 3 and proposition 7(a).

Let $i \neq i_\alpha$ for all $0 \leqslant \alpha \leqslant \rho - 1$ and $j \neq j_\beta$ for all $0 \leqslant \beta \leqslant \rho - 1$. Denote by $\Delta_{ij}^{(\rho)}$ a submatrix of the matrix A, formed by the rows $i_0, \ldots, i_{\rho-1}, i$ and by the columns $j_0, \ldots, j_{\rho-1}, j_{(\rho)}$. It is well known (see e.g. Heintz, 1983) that $a_{ij} = \det(\Delta_{ij}^{(\rho)}) / \det(\Delta_{i_{\rho-1}, j_{\rho-1}}^{(\rho-1)})$. This statement, as usual, guarantees that the Gaussian algorithm can be realised within the available time (see section 5 below).

After carrying out the Gaussian algorithm, the algorithm under description calculates polynomials

$$\psi_1 = a_{i_0, j_0}^{(0)} \ldots a_{i_{\tau-1}, j_{\tau-1}}^{(\tau-1)} = \det(\Delta_{i_{\tau-1}, j_{\tau-1}}^{(\tau-1)})$$

and

$$P_1 = \prod_{0 \leqslant \rho \leqslant \tau-1} (a_{i_\rho, j_\rho}^{(\rho)})^{\tau-\rho} = \prod_{0 \leqslant \rho \leqslant \tau-1} \det(\Delta_{i_\rho, j_\rho}^{(\rho)})$$

from the ring $\mathbb{Q}[\varepsilon_1, \varepsilon_3][Y, Z_0, \ldots, Z_n, u_0, \ldots, u_s]$. Denote by ρ_0 the unique number such that the entry $a_{i_{\rho 0-1}, j_{\rho 0-1}}^{(0)}$ belongs to the number part of the matrix A and the entry $a_{i_{\rho 0}, j_{\rho 0}}^{(0)}$ belongs to the formal part of A. Because of the choice of entry $a_{i_{\rho 0}, j_{\rho 0}}^{(\rho 0)}$ with the least possible $j_{\rho 0}$ one deduces that the rank of the number part of A equals ρ_0. Therefore, proposition 7(d) implies the coincidence of the polynomial ψ_1 with the u-resultant R of system (8') up to a factor from $(\mathbb{Q}(\varepsilon_1, \varepsilon_3)(Y, Z_0, \ldots, Z_n))^*$. Observe that if

$$0 \neq P_1(z^{(3)}) \in \bar{F}_3[Y, u_0, \ldots, u_s],$$

then the polynomial $\psi_1(z^{(3)})$ coincides with the u-resultant $R_{z^{(3)}}$ of system (8) up to a factor from $(\bar{F}_3(Y))^*$ again according to proposition 7(d).

Let us write
$$\psi_1 = \sum_{jo \leqslant j} \psi_1^{(j)} Y^j,$$

where the polynomials
$$\psi_1^{(j)}(Z_0, \ldots, Z_n) \in \mathbb{Q}[\varepsilon_1, \varepsilon_3][Z_0, \ldots, Z_n, u_0, \ldots, u_s],$$

furthermore, $\psi_1^{(jo)} \neq 0$. Then the polynomial
$$R(0, Z_0, \ldots, Z_n, u_0, \ldots, u_s) \in \mathbb{Q}[\varepsilon_1, \varepsilon_3][Z_0, \ldots, Z_n, u_0, \ldots, u_s]$$

coincides with the polynomial $\psi_1^{(jo)}$ up to a factor from $(\mathbb{Q}\varepsilon_1, \varepsilon_3)(Z_0, \ldots, Z_n))^*$. If
$$0 \neq \psi_1^{(jo)}(z^{(3)}) \in \bar{F}_3[u_0, \ldots, u_s]$$

and $P_1(z^{(3)}) \neq 0$, then the polynomial
$$R_{z^{(3)}}(0, u_0, \ldots, u_s) \in \bar{F}_3[u_0, \ldots, u_s]$$

coincides with the polynomial $\psi_1^{(jo)}(z^{(3)})$ up to a factor from $\bar{F}_3^*$.

We write
$$\psi_1^{(jo)} = \sum_{m \leqslant mo} \psi_1^{(jo, m)} u_0^m,$$

where the polynomials
$$\psi_1^{(jo, m)}(Z_0, \ldots, Z_n) \in \mathbb{Q}[\varepsilon_1, \varepsilon_3][Z_0, \ldots, Z_n, u_1, \ldots, u_s],$$

furthermore, $\psi_1^{(jo, mo)} \neq 0$. Then a polynomial $\psi_1^{(jo, mo)}$ coincides up to a factor from $\overline{F_3(Z_0, \ldots, Z_n)}^*$ with the product
$$\prod_{i \notin I} L_i^{c_i} | R(0, Z_0, \ldots, Z_n, \ldots, Z_n, u_0, \ldots, u_s)$$

of all linear forms
$$L_i = \zeta_0^{(i)} u_0 + \ldots + \zeta_s^{(i)} u_s, \quad i \notin I$$

being factors of $R(0, Z_0, \ldots, Z_n, u_0, \ldots, u_s)$ such that $\zeta_0^{(i)} = 0$; $\zeta_j^{(i)} \in \overline{\mathbb{Q}(\varepsilon_1, \varepsilon_3)(Z_0, \ldots, Z_n)}$ (see proposition 8). If
$$0 \neq \psi_1^{(jo, mo)}(z^{(3)}) \in \bar{F}_3[u_1, \ldots, u_s]$$

and $P_1(z^{(3)}) \neq 0$, then a polynomial $\psi_1^{(jo, mo)}(z^{(3)})$ coincides up to a factor from $\bar{F}_3^*$ with the product $\prod_{i \notin J} \hat{L}_i^{c_i} | R_{z^{(3)}}(0, u_0, \ldots, u_s)$ of all linear forms
$$\hat{L}_i = \xi_0^{(i)} u_0 + \ldots + \xi_s^{(i)} u_s$$

being factors of $R_{z^{(3)}}(0, u_0, \ldots, u_s)$ such that $\xi_0^{(i)} = 0$; $\xi_j^{(i)} \in \bar{F}_3$ (see again proposition 8). Hence, the desired product
$$\prod_{i \in I} L_i^{c_i} | R(0, Z_0, \ldots, Z_n, u_0, \ldots, u_s)$$

of all linear forms
$$L_i = \zeta_0^{(i)} u_0 + \ldots + \zeta_s^{(i)} u_s, \quad i \in I$$

such that $\zeta_0^{(i)} \neq 0$, coincides up to a factor from $\overline{F_3(Z_0, \ldots, Z_n)}^*$ with a quotient
$$\psi_1^{(jo)} / \psi_1^{(jo, mo)} \in \mathbb{Q}(\varepsilon_1, \varepsilon_3)(Z_0, \ldots, Z_n)[u_0, u_1, \ldots, u_s],$$

hence
$$\psi_1^{(jo, m)} / \psi_1^{(jo, mo)} \in \mathbb{Q}(\varepsilon_1, \varepsilon_3)(Z_0, \ldots, Z_n)[u_1, \ldots, u_s]$$

for each $m \leqslant m_0$. Similarly, the product

$$\prod_{i \in J} \hat{L}_i^{\hat{c}_i} \mid R_{z^{(3)}}(0, u_0, \ldots, u_s)$$

of all linear forms

$$\hat{L}_i = \xi_0^{(i)} u_0 + \ldots + \xi_s^{(i)} u_s, \quad i \in J$$

such that $\xi_0^{(i)} \neq 0$; $\xi_j^{(i)} \in \bar{F}_3$ coincides up to a factor from $\bar{F}_3^*$ with a quotient

$$(\psi_1^{(jo)}/\psi_1^{(jo, mo)})(z^{(3)}) \in \bar{F}_3[u_0, u_1, \ldots, u_s].$$

Hence,

$$(\psi_1^{(jo, m)}/\psi_1^{(jo, mo)}) \in \bar{F}_3[u_1, \ldots, u_s]$$

for each $m \leqslant m_0$, provided that $\psi_1^{(jo, mo)}(z^{(3)}) \neq 0$ and $P_1(z^{(3)}) \neq 0$.

After that the algorithm calculates the quotient $\psi_1^{(jo)}/\psi_1^{(jo, mo)}$ factoring both polynomials $\psi_1^{(jo)}, \psi_1^{(jo, mo)}$ over the field $\mathbb{Q}(\varepsilon_1, \varepsilon_3)(Z_0, \ldots, Z_n)$ using proposition 1. The quotient $(\psi_1^{(jo)}/\psi_1^{(jo, mo)})(z^{(3)})$ is obtained by substituting in the quotient $(\psi_1^{(jo)}/\psi_1^{(jo, mo)})$ the coordinates of the point $z^{(3)}$ instead of the variables $Z_0, \ldots, Z_n$, provided that $\psi_1^{(jo, mo)}(z^{(3)}) \neq 0$, $P_1(z^{(3)}) \neq 0$. We represent $\psi_1^{(jo)}/\psi_1^{(jo, mo)} = \psi_0/P_2$ for a certain polynomial $P_2 \in \mathbb{Q}[\varepsilon_1, \varepsilon_3][Z_0, \ldots, Z_n]$ of the least possible degree and $\psi_0 \in \mathbb{Q}[\varepsilon_1, \varepsilon_3][Z_0, \ldots, Z_n, u_0, \ldots, u_s]$. If $P_2(z^{(3)}) \neq 0$, $\psi_1^{(jo, mo)}(z^{(3)}) \neq 0$, $P_1(z^{(3)}) \neq 0$, then $\psi_0(z^{(3)})$ coincides with $(\psi_1^{(jo)}/\psi_1^{(jo, mo)})(z^{(3)})$ up to a factor from $\bar{F}_3^*$ and thus, with the product

$$\prod_{i \in J} \hat{L}_i^{\hat{c}_i} \mid R_{z^{(3)}}(0, u_0, \ldots, u_s)$$

of all linear forms

$$\hat{L}_i = \xi_0^{(i)} u_0 + \ldots + \xi_s^{(i)} u_s, \quad i \in J$$

such that $\xi_0^{(i)} = 0$, $\xi_j^{(i)} \in \bar{F}_3$ up to a factor from $\bar{F}_3^*$. So, the polynomial ψ_0 is constructed.

In order to produce a polynomial P, represent

$$P_1 = \sum_{K^{(1)}} \delta_{K^{(1)}}^{(1)}(Z_0, \ldots, Z_n) u_0^{K_0^{(1)}} \ldots u_s^{K_s^{(1)}} Y^{K_{s+1}^{(1)}},$$

$$\psi_1^{(jo, mo)} = \sum_{K^{(2)}} \delta_{K^{(2)}}^{(2)}(Z_0, \ldots, Z_n) u_1^{K_1^{(2)}} \ldots u_s^{K_s^{(2)}},$$

where

$$K^{(1)} = (K_0^{(1)}, \ldots, K_{s+1}^{(1)}), \quad K^{(2)} = (K_1^{(2)}, \ldots, K_s^{(2)})$$

are multi-indices and the polynomials

$$\delta_{K^{(1)}}^{(1)}, \delta_{K^{(2)}}^{(2)} \in \mathbb{Q}[\varepsilon_1, \varepsilon_3][Z_0, \ldots, Z_n].$$

Pick out some multi-indices $K^{(1)}, K^{(2)}$ for which $\delta_{K^{(1)}}^{(1)} \neq 0$ and $\delta_{K^{(2)}}^{(2)} \neq 0$. Finally, we put the polynomial $P = \delta_{K^{(1)}}^{(1)} \delta_{K^{(2)}}^{(2)} P_2$. If $P(z^{(3)}) \neq 0$, then $P_2(z^{(3)}) \neq 0$, $\psi_1^{(jo, mo)}(z^{(3)}) \neq 0$, $P_1(z^{(3)}) \neq 0$. This completes the construction of polynomials ψ_0, P.

Thus, based on the above corollary, we have proved the following

LEMMA 8. *A point $z^{(2)} \in F_2^{n+1}$ such that $P(z^{(2)}) \neq 0$ satisfies formula (6) iff there exist a vector $\gamma = (\gamma_2, \ldots, \gamma_s) \in \Gamma$ and a linear form $\hat{L}_i = \xi_0^{(i)} u_0 + \ldots + \xi_s^{(i)} u_s$ such that $\hat{L}_i \mid \psi_0(z^{(2)})$ (therefore $\xi_0^{(i)} \neq 0$). Furthermore, the point $(\xi_1^{(i)}/\xi_0^{(i)}, \ldots, \xi_s^{(i)}/\xi_0^{(i)}) \in F_3^s$ belongs to the space F_3^s, and, finally, the inequalities*

$$f_i(z^{(2)}, \xi_1^{(i)}/\xi_0^{(i)}, \ldots, \xi_s^{(i)}/\xi_0^{(i)}) + \varepsilon_1 > 0, \quad 0 \leqslant i \leqslant k$$

are fulfilled.

With the aid of proposition 1 (see introduction) the algorithm factorises the polynomial ψ_0 over the field $\mathbb{Q}$. Since ψ_0 is a product of linear forms, its irreducible factors $\Omega \in \mathbb{Q}[\varepsilon_1, \varepsilon_3, Z_0, \ldots, Z_n, u_0, \ldots, u_s]$ correspond bijectively to classes of linear forms L_v, which are conjugate over the field $H_2 = \mathbb{Q}(\varepsilon_1, \varepsilon_3, Z_0, \ldots, Z_n)$, moreover, the product of all linear forms L_v from the considered class corresponding to Ω, equals to Ω up to a factor from H_2.

Now fix Ω for the time being and define $D_1 = \deg_{u_0,\ldots,u_s}(\Omega)$. Let a linear form

$$L_v = (\zeta_0^{(v)} u_0 + \ldots + \zeta_s^{(v)} u_s) | \Omega$$

be a factor of Ω, consider a field

$$H_3^{(v)} = H_2(\zeta_1^{(v)}/\zeta_0^{(v)}, \ldots, \zeta_s^{(v)}/\zeta_0^{(v)})$$

being a finite extension of H_2. Let $\delta : H_3^{(v)} \to \bar{H}_2$ be any field embedding over H_2. Then

$$\delta(L_v/\zeta_0^{(v)}) = L_\mu/\zeta_0^{(\mu)} = u_0 + (\zeta_1^{(\mu)}/\zeta_0^{(\mu)})u_1 + \ldots + (\zeta_s^{(\mu)}/\zeta_0^{(\mu)})u_s$$

for a suitable unique index μ since $\delta(L_v/\zeta_0^{(v)})|\Omega$. Therefore, there exist not more than D_1 embeddings, hence the field degree $[H_3^{(v)} : H_2] \leqslant D_1$ (see Lang, 1965). In fact, $[H_3^{(v)} : H_2] = D_1$, taking into account that a polynomial

$$\prod_\delta \delta(L_v/\zeta_0^{(v)}) \in H_2[u_0, \ldots, u_s]$$

is a factor of the polynomial Ω irreducible over H_2, where the product is taken over all embeddings $\delta : H_3^{(v)} \to \bar{H}_2$ over H_2. This entails the existence of integers $1 \leqslant \lambda_i \leqslant D_1$, $1 \leqslant i \leqslant s$ such that the element

$$\theta^{(v)} = \sum_{1 \leqslant i \leqslant s} \lambda_i(\zeta_i^{(v)}/\zeta_0^{(v)})$$

is primitive in the field $H_2[\theta^{(v)}] = H_3^{(v)}$ by virtue of the theorem on primitive elements (see Lang, 1965).

The algorithm considers all s-tuples $(\lambda_1^{(0)}, \ldots, \lambda_s^{(0)})$, where $1 \leqslant \lambda_i^{(0)} \leqslant D_1$, $1 \leqslant i \leqslant s$. For each s-tuple $(\lambda_1^{(0)}, \ldots, \lambda_s^{(0)})$ it checks, whether the element

$$\theta_0^{(v)} = \sum_{1 \leqslant i \leqslant s} \lambda_1^{(0)}(\zeta_i^{(v)}/\zeta_0^{(v)})$$

is primitive over the field H_2, in the following manner. Substitute in the polynomial Ω the vector $(-\lambda_1^{(0)}, \ldots, -\lambda_s^{(0)})$ instead of $(u_1, \ldots, u_s)$. Then a polynomial

$$\Omega_0(u_0) = \Omega(u_0, -\lambda_1^{(0)}, \ldots, -\lambda_s^{(0)})$$
$$= (\prod_v \zeta_0^{(v)})(\prod_v (u_0 - \theta_0^{(v)})) \in \mathbb{Q}[\varepsilon_1, \varepsilon_3][Z_0, \ldots, Z_n, u_0].$$

Since $\Omega_0(\theta_0^{(v)}) = 0$ and $\deg_{u_0}(\Omega_0) = D_1$, the element $\theta_0^{(v)}$ is primitive in the field $H_3^{(v)}$ iff the polynomial Ω_0 is irreducible over the field H_2. The algorithm tests its irreducibility with the help of proposition 1 and thus, finds a primitive element

$$\theta^{(v)} = \sum_{1 \leqslant i \leqslant s} \lambda_i(\zeta_i^{(v)}/\zeta_0^{(v)})$$

and its minimal polynomial

$$\Phi = \Omega_0/(\prod_v \zeta_0^{(v)}).$$

Then the algorithm produces for each element

$$\zeta_i^{(v)}/\zeta_0^{(v)} = (\zeta_i^{(v)}/\zeta_0^{(v)})(\theta^{(v)}) = \sum_{0 \leqslant j \leqslant D_1} (a_j^{(i)}/b)(\theta^{(v)})^j \in H_2[\theta^{(v)}] = H_3^{(v)}$$

its expression via $\theta^{(v)}$, where the polynomials

$$a_j^{(i)}, b \in \mathbb{Q}[\varepsilon_1, \varepsilon_3][Z_0, \ldots, Z_n].$$

For this goal consider a polynomial

$$\Omega_i = \Omega(-\lambda_1, \ldots, -\lambda_{i-1}, -\lambda_i + u_i, -\lambda_{i+1}, \ldots, -\lambda_s)$$

$$= \left(\prod_v \zeta_0^{(v)}\right)\left(\prod_v (u_0 - \theta_0^{(v)} + (\zeta_i^{(v)}/\zeta_0^{(v)})u_i)\right)$$

and factorise it over the field $H_3^{(v)}$ based on proposition 1. Then we obtain a linear factor $u_0 - \theta^{(v)} + (\zeta_i^{(v)}/\zeta_0^{(v)})(\theta^{(v)})u_i$ of Ω_i and as a result the expression $(\zeta_i^{(v)}/\zeta_0^{(v)})(\theta^{(v)})$.

Thus, the algorithm has produced a polynomial

$$\Phi = \sum_{0 \leqslant j \leqslant D_1} (\alpha_j/\beta)T^j \in H_2[T]$$

irreducible over H_2, where $\alpha_j, \beta \in \mathbb{Q}[\varepsilon_1, \varepsilon_3][Z_0, \ldots, Z_n]$ with the leading coefficient $lc_T(\Phi) = 1$. The expressions

$$(\zeta_i^{(v)}/\zeta_0^{(v)})(\theta^{(v)}) = \sum_{0 \leqslant j < D_1} (\alpha_j^{(i)}/b)(\theta^{(v)})^j \in H_2[\theta^{(v)}]$$

and the integers $1 \leqslant \lambda_i \leqslant D_1$, $1 \leqslant i \leqslant s$ satisfy the following properties. For each root $\theta_0 \in \bar{H}_2$ of the polynomial Φ the equality

$$\theta_0 = \sum_{1 \leqslant i \leqslant s} \lambda_i\left(\sum_{0 \leqslant j < D_1} (a_j^{(i)}/b)\theta_0^j\right)$$

holds, furthermore, a linear from

$$L^{(\theta_0)} = \left(u_0 + \sum_{1 \leqslant i \leqslant s} \left(\sum_{0 \leqslant j < D_1} (a_j^{(i)}/b)\theta_0^j\right)u_i\right)\big| \Omega$$

divides the polynomial Ω in the ring $H_2[\theta_0][u_0, \ldots, u_s]$ and $L^{(\theta_0)} = L_v/\zeta_0^{(v)}$ for an appropriate index v; conversely, every linear form $L_v|\Omega$ dividing Ω equals $L_v = \zeta_0^{(v)}L^{(\theta_0)}$ for a suitable root θ_0 of the polynomial Ω. The fields

$$H_2[\theta_0] \simeq H_2[T]/(\Phi) \simeq H_2[\theta^{(v)}] = H_2(\zeta_1^{(v)}/\zeta_0^{(v)}, \ldots, \zeta_s^{(v)}/\zeta_0^{(v)})$$

are isomorphic. Further, we assume that any element $\eta \in H_2[\theta_0]$ is represented in the form

$$\eta = \sum_{0 \leqslant j < D_0} \eta_j \theta_0^j.$$

Then $\Omega = L^{(\theta_0)}\chi^{(\theta_0)}$ for a certain polynomial $\chi^{(\theta_0)} \in H_2[\theta_0][u_0, \ldots, u_s]$ homogeneous in variables $u_0, \ldots, u_s$ of degree $D_1 - 1$. The algorithm finds the coefficients of the polynomial $\chi^{(\theta_0)}$ factoring the polynomial Ω over the field $H_2[\theta_0]$ with the aid of proposition 1. Let us denote by $L^{(T)}, \chi^{(T)} \in H_2[T][u_0, \ldots, u_s]$ the polynomials obtained by replacing θ_0 by the variable T in the polynomials $L^{(\theta_0)}, \chi^{(\theta_0)}$, respectively. One can show that the polynomial

$$(\Omega - L^{(T)}\chi^{(T)}) = \Phi\tau_1 \in (\Phi) \subset H_2[T][u_0, \ldots, u_s]$$

belongs to the principal ideal (Φ). Analogously, taking into account the equality

$$\theta_0 = \sum_{1 \leqslant i \leqslant s} \lambda_i \zeta_i^{(v)}/\zeta_0^{(v)}$$

one deduces

$$T - \sum_{1 \leqslant i \leqslant s} \lambda_i \sum_j (a_j^{(i)}/b)T^j = \phi\tau_2 \in (\Phi).$$

Let us write

$$\chi^{(\theta_0)} = \sum_j (a_{j1}/b_1)\theta_0^j, \qquad \tau_1 = \sum_j (\alpha_{j1}/\beta_1)T^j, \qquad \tau_2 = \sum_j \frac{\alpha_{j2}\,T^j}{\beta_2},$$

where the polynomials $b_1, \beta_1, \beta_2 \in \mathbb{Q}[\varepsilon_1, \varepsilon_3, Z_0, \ldots, Z_n]$ and the polynomials

$$a_{j1}, \alpha_{j1}, \alpha_{j2} \in \mathbb{Q}[\varepsilon_1, \varepsilon_3, Z_0, \ldots, Z_n][u_0, \ldots, u_s].$$

Assume that a certain point $z^{(3)} \in \bar{F}_3^{n+1}$ satisfies the following conditions:

$$((bb_1\beta\beta_1\beta_2)(z^{(3)}) \neq 0) \ \& \ \Big(\bigvee_q (\varphi_q(z^{(3)}) \neq 0)\Big),$$

and let $\theta^{(0)} \in \bar{F}_3$ be one of the roots of the polynomial $\Phi(z^{(3)})(T) \in \bar{F}_3[T]$. Then the linear form $L^{(\theta^{(0)})}(z^{(3)})|\Omega(z^{(3)})$ divides the polynomial $\Omega(z^{(3)})$ in the ring $\bar{F}_3[u_0, \ldots, u_s]$, since $(\Phi\tau_1)(z^{(3)})(\theta^{(0)}) = 0$. So the linear form $L^{(\theta^{(0)})}(z^{(3)})$ is collinear to one of the linear forms $\hat{L}_{\rho_1}$ in the factorisation

$$\Omega(z^{(3)}) = \prod_{\rho_1} \hat{L}_{\rho_1}^{\mu_{\rho_1}}$$

(cf. above the factorisation of $\psi_0(z^{(3)})$).

Let us denote by

$$D = \mathrm{Res}_T(\Phi, \Phi'_T) = (a_2/b_2) \in H_2$$

the discriminant of polynomial Φ, where a_2, b_2 are polynomials in $\mathbb{Q}[\varepsilon_1, \varepsilon_3, Z_0, \ldots, Z_n]$.

LEMMA 9. *Let a point $z^{(3)} \in \bar{F}_3^{n+1}$ satisfy the conditions*

$$((bb_1 b_2 \beta\beta_1\beta_2)(z^{(3)}) \neq 0) \ \& \ (\varphi_1(z^{(3)}) \neq 0) \ \& \ (D(z^{(3)}) \neq 0). \tag{9}$$

Then

$$\Omega(z^{(3)}) = \delta \prod_\varkappa L^{(\theta_\varkappa^{(0)})}(z^{(3)})$$

for an appropriate $0 \neq \delta \in \bar{F}_3$ where the product is taken over all roots $\theta_\varkappa^{(0)} \in \bar{F}_3$ of the polynomial $\phi(z^{(3)})(T)$. Moreover,

$$\theta_\varkappa^{(0)} = \sum_{1 \leq i \leq s} \lambda_i \Big(\sum_j \frac{a_j^{(i)}(z^{(3)})}{b(z^{(3)})}(\theta_\varkappa^{(0)})^j\Big)$$

where $\Phi(z^{(3)})(\theta_\varkappa^{(0)}) = 0$.

PROOF. The linear form $L^{(\theta_\varkappa^{(0)})}(z^{(3)})|\Omega(z^{(3)})$ divides the polynomial $\Omega(z^{(3)})$ for all $\varkappa$ according to the facts proved above, furthermore $\theta_{\varkappa_1}^{(0)} \neq \theta_{\varkappa_2}^{(0)}$ when $\varkappa_1 \neq \varkappa_2$ since $D(z^{(3)}) \neq 0$.
Finally,

$$\theta_\varkappa^{(0)} = \sum_{1 \leq i \leq s} \lambda_i \left(\sum_j \frac{a_j^{(i)}(z^{(3)})}{b(z^{(3)})}(\theta_\varkappa^{(0)})^j\right),$$

in view of the equality $(\Phi\tau_2)(z^{(3)})(\theta_\varkappa^{(0)}) = 0$. Therefore all linear forms $L^{(\theta_\varkappa^{(0)})}(z^{(3)})$ are pairwise distinct for diverse $\varkappa$, and so $\prod_\varkappa L^{(\theta_\varkappa^{(0)})}(z^{(3)})|\Omega(z^{(3)})$, where the degrees of both these polynomials are equal to $\deg_T(\Phi) = \deg_{u_0, \ldots, u_s}(\Omega)$. This completes the proof of the lemma.

The conditions (9) concern the case of a fixed vector $\gamma \in \Gamma$ and an irreducible factor $\Omega|\Psi_0$. We introduce a hypersurface $\mathscr{L}$ consisting of all points $z^{(3)} \in \bar{F}_3^{n+1}$ which do not

94 D. Yu Grigor'ev

satisfy the conditions similar to (9) for at least one vector $\gamma^{(1)} \in \Gamma$ and one irreducible factor $\Omega^{(1)} | \Psi_0$. Then the degree

$$\deg(\mathscr{L}) \leqslant N_1 = (\operatorname{card} \Gamma)((\deg \psi_0)(\deg(bb_1 b_2 \beta\beta_1 \beta_2 D)) \deg \varphi_1)$$

(an estimate on N_1 will be found below in section 5). Let us denote $N = N_1 n + 1$. It is easy to construct (see e.g. section 2 in Chistov & Grigor'ev, 1983a) linear forms $Y_1, \ldots, Y_N$ over integers in the variables $Z_0, \ldots, Z_n$ such that any $(n+1)$ among them are linearly independent (one can set, for example

$$Y_i = \sum_{0 \leqslant j \leqslant n} i^j Z_j).$$

Let us fix a certain point $z^{(3)} \in \bar{F}_3^{n+1}$. We assert the existence of indices $1 \leqslant i_1 < \ldots < i_n \leqslant N$ such that the line

$$(z^{(3)} + \lambda\{Y_{i_1} = \ldots = Y_{i_n} = 0\})_{\lambda \in \bar{F}_3} \subset \bar{F}_3^{n+1}$$

is not situated in the hypersurface $\mathscr{L}$ (cf. section 2 in Chistov & Grigor'ev, 1983a). Arguing by induction on $0 \leqslant j \leqslant n$, we assume that $j < n$ and indices $1 \leqslant i_1 \leqslant \ldots < i_j \leqslant N$ are already found, for which

$$\dim(\mathscr{L} \cap (z^{(3)} + \lambda\{Y_{i_1} = \ldots = Y_{i_j} = 0\})_{\lambda \in \bar{F}_3}) = n - j.$$

There is an index $1 \leqslant i \leqslant N$ such that the linear function $Y_i - Y_i(z^{(3)})$ does not vanish identically on any irreducible component of variety

$$\mathscr{L} \cap (z^{(3)} + \lambda\{Y_{i_1} = \ldots = Y_{i_j} = 0\})_{\lambda \in \bar{F}_3}.$$

Otherwise, on some irreducible component at least $(n+1)$ functions among $Y_1 - Y_1(z^{(3)}), \ldots, Y_N - Y_N(z^{(3)})$ vanish, taking into account that the number of irreducible components is less or equal to

$$\deg(\mathscr{L} \cap (z^{(3)} + \lambda\{Y_{i_1} = \ldots = Y_{i_j} = 0\})_{\lambda \in \bar{F}_3}) \leqslant \deg(\mathscr{L}) \leqslant N_1$$

by virtue of Bezout's inequality (see Shafarevich, 1974; Heintz, 1983). This leads to a contradiction with the property of linear forms $Y_1, \ldots, Y_N$ and proves the existence of the desired index $1 \leqslant i \leqslant N$. Let us add index i to indices $i_1, \ldots, i_j$, reorder them in increasing order and get the indices $1 \leqslant i_1 < \ldots < i_{j+1} \leqslant N$. Therefore, if a point $z^{(3)}$ is defined over the field $\overline{\mathbb{Q}(\varepsilon_1, \varepsilon_3)}$, then the intersection

$$\mathscr{L} \cap (z^{(3)} + \lambda\{Y_{i_1} = \ldots = Y_{i_n} = 0\})_{\lambda \in \bar{F}_3}$$

consists of a finite number of points which are defined over the field $\overline{\mathbb{Q}(\varepsilon_1, \varepsilon_3)}$.

For any sequence of indices $1 \leqslant i_1 < \ldots < i_n \leqslant N$, let us pick out an arbitrary vector $0 \neq v_{i_1, \ldots, i_n} \in \mathbb{Q}^{n+1}$ lying on the line $\{Y_{i_1} = \ldots = Y_{i_n} = 0\} \subset \bar{F}_3^{n+1}$. Furthermore, we require that one of the coefficients of vector $v_{i_1, \ldots, i_n}$ equals 1. Then for any point $z \in \mathbb{Q}^{n+1}$ there exists a sequence of indices $1 \leqslant i_1 < \ldots < i_n \leqslant N$ such that point $z^{(2)} = z + \varepsilon_2 v_{i_1, \ldots, i_n} \in F_2^{n+1}$ does not belong to the hypersurface $\mathscr{L}$, taking into account that the point $z^{(2)}$ is not defined over the field $\overline{\mathbb{Q}(\varepsilon_1, \varepsilon_3)}$. According to lemmas 8, 9 point $z^{(2)} \in F_2^{n+1} \backslash \mathscr{L}$ satisfies formula (6) iff there is a vector $\gamma \in \Gamma$, an irreducible factor $\Omega | \psi_0$ and a root $\theta^{(0)} \in F_3$ of the polynomial $\Phi(z^{(2)})(T) \in \tilde{\mathbb{Q}}(\varepsilon_1, \varepsilon_2, \varepsilon_3)[T]$ for which the inequalities

$$f_i\left(z^{(2)}, \sum_j \frac{a_j^{(1)}(z^{(2)})}{b(z^{(2)})}(\theta^{(0)})^j, \ldots, \sum_j \frac{a_j^{(s)}(z^{(2)})}{b(z^{(2)})}(\theta^{(0)})^j\right) + \varepsilon_1 > 0, \quad 0 \leqslant i \leqslant k$$

are valid. Here, we use the observation that the point

$$\left(\sum_j \frac{a_j^{(1)}(z^{(2)})}{b(z^{(2)})} (\theta^{(0)})^j, \ldots, \sum_j \frac{a_j^{(s)}(z^{(2)})}{b(z^{(2)})} (\theta^{(0)})^j \right) \in \bar{F}_3^s,$$

where $\Phi(z^{(2)})(\theta^{(0)}) = 0$, belongs to the space F_3^s iff $\theta^{(0)} \in F_3$ since

$$\theta^{(0)} = \sum_{1 \leqslant i \leqslant s} \lambda_i \left(\sum_j \frac{a_j^{(i)}(z^{(2)})}{b(z^{(2)})} (\theta^{(0)})^j \right)$$

by virtue of lemma 9. We recall also that lemma 9 implies that the linear form

$$L^{(\theta^{(0)})}(z^{(2)}) = \left(u_0 + \sum_{1 \leqslant i \leqslant s} \left(\sum_j \frac{a_j^{(i)}(z^{(2)})}{b(z^{(2)})} (\theta^{(0)})^j \right) u_i \right) \in F_3[u_0, \ldots, u_s]$$

divides the polynomial $\Omega(z^{(2)})$. Thus, in view of lemma 6(c) we have proved the following lemma, in which we use introduced above notations.

LEMMA 10. *A point $z \in \tilde{\mathbb{Q}}^{n+1}$ satisfies formula (5) iff there exist a sequence of indices $1 \leqslant i_1 < \ldots < i_n \leqslant N$ such that the point $z^{(2)} = z + \varepsilon_2 v_{i_1, \ldots, i_n} \notin \mathscr{L}$. Furthermore, a vector $\gamma \in \Gamma$, a factor Ω of Ψ_0 irreducible over $\mathbb{Q}$ and, finally, a root $T = \theta^{(0)} \in F_3$ of the polynomial $\phi(z^{(2)})(T)$, where the polynomial Φ is associated with the factor Ω, such that the inequalities*

$$f_i \left(z^{(2)}, \sum_j \frac{a_j^{(1)}(z^{(2)})}{b(z^{(2)})} (\theta^{(0)})^j, \ldots, \sum_j \frac{a_j^{(s)}(z^{(2)})}{b(z^{(2)})} (\theta^{(0)})^j \right) + \varepsilon_1 > 0, \quad 0 \leqslant i \leqslant k$$

are fulfilled. Here, the linear form

$$L^{(\theta)} = u_0 + \sum_{1 \leqslant i \leqslant s} \left(\sum_j \frac{a_j^{(i)}}{b} \theta^j \right) u_i$$

being a factor of the polynomial Ω was constructed by the algorithm described above. Furthermore, the point

$$\left(\sum_j \frac{a_j^{(1)}(z^{(2)})}{b(x^{(2)})} (\theta^{(1)})^j, \ldots, \sum_j \frac{a_j^{(s)}(z^{(2)})}{b(z^{(2)})} (\theta^{(1)})^j \right) \in \bar{F}_3^s$$

is a solution of system (7) for every root $\theta^{(1)} \in \bar{F}_3$ of the polynomial $\Phi(z^{(2)})(T)$.

4. Verifying Formulas of Tarski Algebra

In lemma 10 a formula of the kind $\exists\, T(P_1)$ is produced which is equivalent to formula (5) (over $\tilde{\mathbb{Q}}$, i.e. both formulas determine the same set in $\tilde{\mathbb{Q}}^{n+1}$), where P_1 is a quantifier-free formula of Tarski algebra with atomic subformulas of the sort $(g_j \geqslant 0)$, with polynomials $g_j \in \mathbb{Q}[\varepsilon_1, \varepsilon_2, \varepsilon_3][T, Z_0, \ldots, Z_n]$.

Applying the construction from section 2 in Wüthrich (1976) (see also Collins, 1975) to the family of polynomials $\{g_j\}_j$ (which are considered over the real closed field F_3) and to the variable T, one gets a family of polynomials $\{p_e\}_e$, where $p_e \in \mathbb{Q}[\varepsilon_1, \varepsilon_2, \varepsilon_3][Z_0, \ldots, Z_n]$ which satisfies the following property (see theorem 1 in Wüthrich, 1976 and theorem 5 in Collins, 1975). For every element $W^{(\varepsilon)} \subset F_3^{n+1}$ of the partition $\mathscr{U}(\{p_e\}_e)$ (see section 1) there exists a sequence of semialgebraic functions $q_1, \ldots, q_t : W^{(\varepsilon)} \to F_3$ continuous on $W^{(\varepsilon)}$ (with respect to the topology with the base of all open balls), such that $q_1 < \ldots < q_t$ and the partition, formed by the components of connectivity of the intersection of all elements of the partition $\mathscr{U}(\{g_j\}_j)$ with the cylinder $F_3 \times W^{(\varepsilon)} \subset F_3^{n+2}$, coincides with the following partition of the cylinder:

$$F_3 \times W^{(\varepsilon)} = \{T < q_1(Z_0, \ldots, Z_n)\} \cup \{T > q_t(Z_0, \ldots, Z_n)\} \cup$$

$$\bigcup_{1 \leqslant i \leqslant t-1} \{q_i(Z_0, \ldots, Z_n) < T < q_{i+1}(Z_0, \ldots, Z_n)\} \cup$$

$$\bigcup_{1 \leqslant i \leqslant t} \{T = q_i(Z_0, \ldots, Z_n)\}.$$

As above (see section 2), one bounds the formats of the semialgebraic functions $q_1, \ldots, q_t$ via the formats of $\{p_e\}_e$ in the case of the field $\mathbb{R}$ and then spread these bounds and the properties of $q_1, \ldots, q_t$ to the field F_3 according to the transfer principle.

Let us denote

$$p_e = \sum_{j_1, j_2, j_3} p_e^{(j_1, j_2, j_3)} \varepsilon_1^{j_1} \varepsilon_2^{j_2} \varepsilon_3^{j_3},$$

where the polynomials $p_e^{(j_1, j_2, j_3)} \in \mathbb{Q}[Z_0, \ldots, Z_n]$ (cf. section 2). We denote by $\pi_1 : \tilde{\mathbb{Q}}^{s+n+1} \to \tilde{\mathbb{Q}}^{n+1}$ the linear projection defined by the formula

$$\pi_1(Z_0, \ldots, Z_n, X_1, \ldots, X_s) = (Z_0, \ldots, Z_n).$$

In the following lemma we use the notations from the previous section.

LEMMA 11. *Let a point* $(z^{(0)}, x^{(0)}) = (z_0^{(0)}, \ldots, z_n^{(0)}, x_1^{(0)}, \ldots, x_s^{(0)})$ *belong to some component of connectivity* V_1 *of the semialgebraic set*

$$V = \{(f_0 \geqslant 0) \& \ldots \& (f_k \geqslant 0)\} \subset \tilde{\mathbb{Q}}^{s+n+1}$$

and the point $z^{(0)} = \pi_1(z^{(0)}, x^{(0)}) \in W$, *where* $W \subset \tilde{\mathbb{Q}}^{n+1}$ *is a certain element of the partition* $\mathscr{U}(\{p_e^{(j_1, j_2, j_3)}\}_{e, j_1, j_2, j_3})$. *Then the projection* $\pi_1(V_1) \supset W$.

PROOF. In view of the properties of the vectors $\{v_{i_1, \ldots, i_n}\}$ constructed in the previous section, there exist indices $1 \leqslant i_1 < \ldots < i_n \leqslant N$ such that $z^{(2)} = z^{(0)} + \varepsilon_2 v_{i_1, \ldots, i_n} \notin \mathscr{L}$ (cf. lemma 10). According to lemma 5(c) there is a unique element $V_2^{(\varepsilon)} \subset F_3^{s+n+1}$ of the partition $\mathscr{U}(g, f_0 + \varepsilon_1, \ldots, f_k + \varepsilon_1)$ which contains the component $V_1 \subset V_2^{(\varepsilon)}$. Lemma 6(a) entails the inclusion $V \subset \{g \geqslant 0\} \cap V^{(\varepsilon)}$, where

$$V^{(\varepsilon)} = \{(f_0 + \varepsilon_1 > 0) \& \ldots \& (f_k + \varepsilon_1 > 0)\} \subset F_3^{s+n+1},$$

therefore $V_2^{(\varepsilon)} \subset \{g \geqslant 0\} \cap V^{(\varepsilon)}$. Let us denote by $V_1^{(\varepsilon)}$ the unique component of connectivity of the semialgebraic set $\{g \geqslant 0\} \cap V^{(\varepsilon)}$ which contains the set $V_2^{(\varepsilon)} \subset V_1^{(\varepsilon)}$. Thus $V_1^{(\varepsilon)} \supset V_1$.

By virtue of lemma 6(c) one can find a point $(z^{(2)}, \tilde{x}) \in V_1^{(\varepsilon)}$ such that $g(z^{(2)}, \tilde{x}) = 0$ and $(z^{(0)}, st(\tilde{x})) \in V_1$. Let $V_3^{(\varepsilon)}$ be the unique component of connectivity of the variety $\{g = 0\} \subset F_3^{s+n+1}$, which contains the point $(z^{(2)}, \tilde{x}) \in V_3^{(\varepsilon)}$. Then $V_3^{(\varepsilon)} \subset V_1^{(\varepsilon)}$ in view of lemma 6(b). Consider the unique component of connectivity $V_4^{(\varepsilon)}$ of the semialgebraic set

$$\{(z^{(2)}, y) : y \in F_3^s, g(z^{(2)}, y) = 0\} \subset F_3^{s+n+1},$$

which contains point $(z^{(2)}, \tilde{x}) \in V_4^{(\varepsilon)}$. Obviously, $V_4^{(\varepsilon)} \subset V_3^{(\varepsilon)}$. Note that

$$V_4^{(\varepsilon)} \subset \mathscr{D}_{(z^{(2)}, 0)}((z_0^{(0)} + 2\varepsilon_1)^{1/2}),$$

since $f_0(u) + \varepsilon_1 > 0$ for any point $u \in V_4^{(\varepsilon)} \subset V_1^{(\varepsilon)}$.

Lemma 7(a), (b) implies the existence of a vector $\gamma \in \Gamma$ such that system (7) has a solution $x = (x_1, \ldots, x_s) \in F_3^s$ where point $(z^{(2)}, x) \in V_4^{(\varepsilon)}$. Moreover, point x is isolated in the variety of all the solutions of system (7) in the space $\bar{F}_3^s$. Hence, the linear form $u_0 + x_1 u_1 + \ldots + x_s u_s$ divides the polynomial $\psi_0(z^{(2)})$, where $\psi_0 \in \mathbb{Q}[\varepsilon_1, \varepsilon_3, Z_0, \ldots, Z_n, u_0, \ldots, u_s]$ (see systems (8), (8′) and lemma 8). Consider some irreducible (over $\mathbb{Q}$) factor $\Omega \in \mathbb{Q}[\varepsilon_1, \varepsilon_3, Z_0, \ldots, Z_n, u_0, \ldots, u_s]$ of the polynomial ψ_0 for which $(u_0 + x_1 u_1 + \ldots + x_s u_s) | \Omega(z^{(2)})$.

Let $W^{(\varepsilon)}$ be the unique element of the partition $\mathscr{U}(\{p_e\}_e)$ such that $z^{(0)} \in W^{(\varepsilon)}$. Lemma 5(b) entails the inclusion $W \subset W^{(\varepsilon)}$. For a point $z \in W^{(\varepsilon)}$ and an element $\theta \in F_3$ we denote

$$(\theta, z)_{\gamma, \Omega, i_1, \ldots, i_n} = \left(z + \varepsilon_2 v_{i_1, \ldots, i_n}, \sum_j \frac{a_j^{(1)}(z + \varepsilon_2 v_{i_1, \ldots, i_n})}{b(z + \varepsilon_2 v_{i_1, \ldots, i_n})} \theta^j, \ldots, \right.$$

$$\left. \sum_j \frac{a_j^{(s)}(z + \varepsilon_2 v_{i_1, \ldots, i_n})}{\beta(z + \varepsilon_2 v_{i_1, \ldots, i_n})} \theta^j \right) \in F_3^{s+n+1}$$

(see lemma 10). According to lemma 9 and to the construction of a primitive element in section 3, there is a root $\theta^{(0)} \in F_3$ of polynomial $\Phi(z^{(2)})(T)$, for which $(z^{(2)}, x) = (\theta^{(0)}, z^{(0)})_{\gamma, \Omega, i_1, \ldots, i_n}$, furthermore,

$$\theta^{(0)} = \sum_{1 \leqslant i \leqslant s} \lambda_i x_i$$

for suitable natural numbers $1 \leqslant \lambda_i \leqslant \deg_{u_0, \ldots, u_s}(\Omega)$. Since point x is a solution of system (7), the following equality in particular is fulfilled:

$$g((\theta^{(0)}, z^{(0)})_{\gamma, \Omega, i_1, \ldots, i_n}) = 0. \tag{10}$$

Moreover, taking into account that point $(z^{(2)}, x) \in V_4^{(\varepsilon)} \subset V_1^{(\varepsilon)} \subset V^{(\varepsilon)}$ belongs to $V^{(\varepsilon)}$ the following inequalities are true:

$$f_i((\theta^{(0)}, z^{(0)})_{\gamma, \Omega, i_1, \ldots, i_n}) + \varepsilon_1 > 0; \quad 0 \leqslant i \leqslant k. \tag{11}$$

Consider the unique element $G_1 \subset F_3^{n+2}$ of the partition $\mathscr{U}(\{g_j\}_j)$ which contains point $(\theta^{(0)}, z^{(0)}) \in G_1$. In view of the definition of partition $\mathscr{U}(\{g_j\}_j)$ and of lemma 10 for an arbitrary point $(\theta, z) \in G_1$ the value of polynomial $b(z + \varepsilon_2 v_{i_1, \ldots, i_n}) \neq 0$ and, moreover, the conditions obtained from (10), (11) by replacing the point $(\theta^{(0)}, z^{(0)})$ by the point (θ, z), respectively, are valid for the given $\gamma, \Omega, i_1, \ldots, i_n$. Furthermore, $y \in F_3^s$, where

$$(z + \varepsilon_2 v_{i_1, \ldots, i_n}, y) = (\theta, z)_{\gamma, \Omega, i_1, \ldots, i_n}$$

satisfies system (7) by virtue of lemmas 8, 9.

One can observe that the intersection $G_1 \cap (F_3 \times W^{(\varepsilon)})$ coincides with a union of some sets of the sort $\{(q_{m_1}(z), z)\}_{z \in W^{(\varepsilon)}}$, where z runs over all points of the set $W^{(\varepsilon)}$ for appropriate indices $1 \leqslant m_1 \leqslant t$ (see above the beginning of the present section). Otherwise, suppose that

$$G_1 \cap (F_3 \times W^{(\varepsilon)}) \supset \{(\theta, z) : q_{m_1}(z) < \theta < q_{m_1 + 1}(z)\}$$

for a certain m_1, then for any point $z \in W$ (for instance, one can take $z = z^{(0)}$), there are infinitely many points of the kind $(\theta^{(1)}, z) \in G_1$, but on the other hand, every such $\theta^{(1)}$ is a root of polynomial $\Phi(z + \varepsilon_2 v_{i_1, \ldots, i_n})(T)$ (see lemma 10), that leads to contradiction. Consider such unique index $1 \leqslant m \leqslant t$ that

$$G_1 \cap (F_3 \times W^{(\varepsilon)}) \supset \{(q_m(z), z)\}_{z \in W^{(\varepsilon)}}$$

and point $(\theta^{(0)}, z^{(0)}) \in \{(q_m(z), z)\}_{z \in W^{(\varepsilon)}}$. Let us define $q = q_m$ for brevity.

For any point $z^{(1)} \in W$ the standard part $st((q(z^{(1)}), z^{(1)})_{\gamma, \Omega, i_1, \ldots, i_n}) \in \tilde{\mathbb{Q}}^{s+n+1}$ is defined, since $f_0((q(z^{(1)}), z^{(1)})_{\gamma,\Omega,i_1,\ldots,i_n}) + \varepsilon_1 > 0$ (cf. (4)), taking into account that for a point $z \in W^{(\varepsilon)}$ a point $(q(z), z) \in G_1$ satisfies the inequalities obtained from (11) by replacing point $(\theta^{(0)}, z^{(0)})$ by point $(q(z), z)$. Therefore,

$$0 \leqslant st(f_i((q(z), z)_{\gamma,\Omega,i_1,\ldots,i_n})) = f_i(st((q(z), z)_{\gamma,\Omega,i_1,\ldots,i_n})), \quad 0 \leqslant i \leqslant k,$$

provided that $st((q(z), z)_{\gamma,\Omega,i_1,\ldots,i_n})$ is defined. This means that point $st((q(z), z)_{\gamma,\Omega,i_1,\ldots,i_n}) \in V$. For the completion of the proof of the lemma it suffices to show that for any point $z^{(1)} \in W$ point $st((q(z^{(1)}), z^{(1)})_{\gamma,\Omega,i_1,\ldots,i_n})$ belongs to V_1, taking into account that

$$\pi_1(st((q(z^{(1)}), z^{(1)})_{\gamma,\Omega,i_1,\ldots,i_n})) = z^{(1)}.$$

Let us fix an arbitrary point $z^{(1)} \in W$. By means of lemma 3(b) one can join the points $z^{(0)}, z^{(1)} \in W$ by a closed connected semialgebraic curve $z^{(0)}, z^{(1)} \in C \subset W \cap \mathscr{D}_0(R)$, where $R \in \mathbb{Q}$. In view of lemmas 4(a), 5(a) there exists such a closed connected semialgebraic curve $C^{(\varepsilon)} \subset W^{(\varepsilon)} \cap \mathscr{D}_0(R)$ that $C^{(\varepsilon)} \supset st(C^{(\varepsilon)}) = C$. Consider the image $\rho(C^{(\varepsilon)})$ of the curve $C^{(\varepsilon)}$ under the action of the continuous on $W^{(\varepsilon)}$ semialgebraic mapping

$$\rho : z \to (q(z), z)_{\gamma,\Omega,i_1,\ldots,i_n} \in F_3^{s+n+1}.$$

Lemma 3(a) entails that $\rho(C^{(\varepsilon)}) \subset F_3^{s+n+1}$ is a connected semialgebraic curve.

We claim that $\rho(C^{(\varepsilon)}) \subset \mathscr{D}_0(R+1)$. Indeed, let us denote

$$\rho(z) = \rho(z_0, \ldots, z_n) = (z + \varepsilon_2 v_{i_1,\ldots,i_n}, y)$$

for an arbitrary point $z \in C^{(\varepsilon)}$. Then the euclidean norm $\|z\| \leqslant R$, furthermore,

$$\|\rho(z)\|^2 = \|z + \varepsilon_2 v_{i_1,\ldots,i_n}\|^2 + \|y\|^2 < \|z\|^2 + z_0 + 2\varepsilon_1 < R^2 + R + 1$$

by virtue of inequality (11) in the case $i = 0$ for the point $(q(z), z) \in G_1$. This proves the claim. Hence, the standard part $st(\rho(C^{(\varepsilon)})) \subset \tilde{\mathbb{Q}}^{s+n+1}$ is defined and $st(\rho(C^{(\varepsilon)})) \subset V$, according to inequalities (11) (cf. above).

The projection $\pi_1(st(\rho(C^{(\varepsilon)}))) \subset \tilde{\mathbb{Q}}^{n+1}$ contains both points $z^{(0)}, z^{(1)}$, since $C^{(\varepsilon)} \supset C$ and $\pi_1(st(\rho(z))) = z$ for any point $z \in C \subset W$. To complete the proof of the lemma it remains to show the inclusion $st(\rho(C^{(\varepsilon)})) \subset V_1$. In view of lemma 4(b) it suffices to check that $st(\rho(C^{(\varepsilon)})) \cap V_1 \neq \phi$, taking into account that the curve $\rho(C^{(\varepsilon)}) \subset \mathscr{D}_0(R+1)$ is connected and $st(\rho(C^{(\varepsilon)})) \subset V$. With the help of lemma 3(b) one can find a closed connected semialgebraic curve $C_1 \subset V_4^{(\varepsilon)}$ joining the points $(z^{(2)}, x), (z^{(2)}, \tilde{x}) \in C_1$. Then $C_1 \subset V_4^{(\varepsilon)} \subset \mathscr{D}_{(z^{(2)}, 0)}((z_0^{(0)} + 2\varepsilon_1)^{1/2})$ (see above). Lemma 6(a) implies the inclusion $st(C_1) \subset V$ since $C_1 \subset V_4^{(\varepsilon)} \subset V_3^{(\varepsilon)} \subset V_1^{(\varepsilon)} \subset V^{(\varepsilon)}$. On the other hand, point $st(z^{(2)}, \tilde{x}) \in V_1 \cap st(C_1)$, hence $st(C_1) \subset V_1$ by virtue of lemma 4(b), in particular, point $st(z^{(2)}, x) \in V_1$. Finally, the equality $(z^{(2)}, x) = \rho(z^{(0)})$ entails

$$st(z^{(2)}, x) = st(\rho(z^{(0)})) \in st(\rho(C^{(\varepsilon)})),$$

i.e.

$$st(z^{(2)}, x) \in V_1 \cap st(\rho(C^{(\varepsilon)})) \neq \phi.$$

The lemma is proved.

Now we proceed to the description of the algorithm which verifies formula (1). Let us first design by recursion on $0 \leqslant \alpha \leqslant a - 1$ (we remind the reader that a is the number of quantifier alternations in formula (1)), a certain procedure consisting of $a - 1$ stages. Before the implementation of the first stage the algorithm sets polynomials $g_i^{(0)} = f_i$, $0 \leqslant i \leqslant k$ (see formula (1)). As a result of the implementation of α stages of the procedure (we assume here that $\alpha \leqslant a - 2$) a family of polynomials $\{g_j^{(\alpha)}\}_j$ is produced, where

$$g_j^{(\alpha)} \in \mathbb{Q}[X_{1,1}, \ldots, X_{1,s_1}, \ldots, X_{a-\alpha,s_{a-\alpha}}].$$

The algorithm enumerates all $(\{g_j^{(\alpha)}\}_j)$-cells K_1, K_2, ... based on lemma 1 from section 1.

Let us fix for the time being a certain $(\{g_j^{(\alpha)}\}_j)$-cell K_t and let $K_t = \{\Pi_t\}$ for a suitable quantifier-free formula of Tarski algebra

$$\Pi_t = \Big(\underset{j \in J}{\&} (g_j^{(a)} = 0) \,\&\, \underset{j \in J_1}{\&} (g_{j_1}^{(\alpha)} > 0) \,\&\, \underset{j_2 \in J_2}{\&} (g_{j_2} < 0) \Big)$$

(see section 1). Consider the following formula of Tarski algebra

$$\exists \, X_{a-\alpha,1} \ldots \exists \, X_{a-\alpha,s_{a-\alpha}} \, \Pi_t \tag{12}$$

and let us apply to it the construction from section 3, taking (12) as the input formula (i.e. (12) plays the role of the formula (2) in the construction from section 3). As a result, the algorithm produces a formula of the form $\exists \, T(P_t)$ (see lemma 10). Here the quantifier-free formula P_t has atomic subformulas of the form $(g_{m,t}^{(\alpha+1)} \geqslant 0)$, where the polynomials

$$g_{m,t}^{(\alpha+1)} \in \mathbb{Q}[\varepsilon_1, \varepsilon_2, \varepsilon_3][T, Z_0, X_{1,1}, \ldots, X_{1,s_1}, \ldots, X_{a-\alpha-1,1}, \ldots, X_{a-\alpha-1,s_{a-\alpha-1}}].$$

We recall that (see (4), (5)) formula (12) is equivalent (over the field $\tilde{\mathbb{Q}}$) to formula $\exists \, Z_0 \in \tilde{\mathbb{Q}} \, \exists \, T \in F_3(P_t)$ (being not a formula of the first-order theory). Observe, that we do not use the latter claim.

As in the beginning of the present section the procedure yields polynomials

$$p_{l,t}^{(\alpha+1)} \in \mathbb{Q}[\varepsilon_1, \varepsilon_2, \varepsilon_3][Z_0, X_{1,1}, \ldots, X_{a-\alpha-1,s_{a-\alpha-1}}]$$

by means of the construction from section 2 in Wüthrich (1976) (see also Collins, 1975), which is applied here to the family of polynomials $\{g_{m,t}^{(\alpha+1)}\}_m$ and to the variable T. Next, we write

$$p_{l,t}^{(\alpha+1)} = \sum_{j_1, j_2, j_3} p_{l,t}^{(\alpha+1)(j_1, j_2, j_3)} \varepsilon_1^{j_1} \varepsilon_2^{j_2} \varepsilon_3^{j_3},$$

as above, where the polynomials

$$p_{l,t}^{(\alpha+1)(j_1, j_2, j_3)} \in \mathbb{Q}[Z_0, X_{1,1}, \ldots, X_{a-\alpha-1,s_{a-\alpha-1}}].$$

Finally, the procedure applies again the construction from section 2 in Wüthrich (1976) to the family $\{p_{l,t}^{(\alpha+1)(j_1, j_2, j_3)}\}_{j_1, j_2, j_3, l, t}$ consisting of polynomials corresponding to the totality of $(\{g_j^{(\alpha)}\}_j)$-cells K_t (cf. formula (12)), and to variable Z_0. As a result, the procedure obtains a family $\{g_i^{(\alpha+1)}\}_i$ of polynomials

$$g_i^{(\alpha+1)} \in \mathbb{Q}[X_{1,1}, \ldots, X_{1,s_1}, \ldots, X_{a-\alpha-1,1}, \ldots, X_{a-\alpha-1,s_{a-\alpha-1}}].$$

This completes the description of the recursive procedure which produces the polynomials $\{g_j^{(\alpha)}\}_{j,\alpha}$. Let us denote by $\{K_t^{(\alpha)}\}_t$ the family of all $(\{g_j^{(\alpha)}\}_j)$-cells. Note that the procedure produces also this family. Let us adopt the convention that for $\alpha = a$, the families $\{K_t^{(a)}\}_t$ and $\{g_j^{(a)}\}_j$ are empty.

LEMMA 12. *For any element W of partition $\mathcal{U}(\{g_j^{(\alpha)}\}_j)$ of the space $\tilde{\mathbb{Q}}^{s_1 + \cdots + s_{a-\alpha}}$ its natural projection onto the space $\tilde{\mathbb{Q}}^{s_1 + \cdots + s_{a-\alpha-1}}$ coincides with a union of a suitable collection of elements of partition $\mathcal{U}(\{g_i^{(\alpha+1)}\}_i)$.*

PROOF. Let us denote $n_1 = s_1 + \ldots + s_{a-\alpha-1}$ and consider a commutative diagram

$$\begin{array}{ccc}
\tilde{\mathbb{Q}}^{n_1+s_{a-\alpha}+2} & \stackrel{\pi_1}{\to} & \tilde{\mathbb{Q}}^{n_1+1} \\
\sigma_2 \downarrow & & \downarrow \sigma_1 \\
\tilde{\mathbb{Q}}^{n_1+s_{a-\alpha}} & \stackrel{\pi_2}{\to} & \tilde{\mathbb{Q}}^{n_1},
\end{array}$$

where all four mapping are linear projections. The space $\tilde{\mathbb{Q}}^{n_1}$ has coordinates $X_{1,1}, \ldots, X_{a-\alpha-1, s_{a-\alpha-1}}$; the space $\tilde{\mathbb{Q}}^{n_1+1}$ has the additional coordinate Z_0, the space $\tilde{\mathbb{Q}}^{n_1+s_{a-\alpha}}$ has coordinates $X_{1,1}, \ldots, X_{a-\alpha, s_{a-\alpha}}$ and at last the space $\tilde{\mathbb{Q}}^{n_1+s_{a-\alpha}+2}$ has coordinates $X_{1,1}, \ldots, X_{a-\alpha, s_{a-\alpha}}, X_{n_1+s_{a-\alpha}+1}, Z_0$.

The set $W \subset \tilde{\mathbb{Q}}^{n_1+s_{a-\alpha}}$ is a component of connectivity of a certain $(\{g_j^{(\alpha)}\}_j)$-cell K_t, being given by a quantifier-free formula

$$\Pi_t = \left(\underset{j \in J}{\&} (g_j^{(\alpha)} = 0) \& \underset{j_1 \in J_1}{\&} (g_{j_1}^{(\alpha)} > 0) \& \underset{j_2 \in J_2}{\&} (g_{j_2}^{(a)} < 0) \right)$$

(cf. (12) above). Let us introduce a semialgebraic set $U_t \subset \tilde{\mathbb{Q}}^{n_1+s_{a-\alpha}+2}$, which is given by a formula

$$\Pi_t \& \left(X_{n_1+s_{a-\alpha}+1} \left(\prod_{j_1 \in J_1} g_{j_1}^{(\alpha)} \right)\left(\prod_{j_2 \in J_2} (-g_{j_2}^{(\alpha)}) \right) \geqslant 1 \right) \&$$

$$\left(Z_0 - X_{n_1+s_{a-\alpha}+1}^2 - X_{a-\alpha, 1}^2 - \ldots - X_{a-\alpha, s_{a-\alpha}}^2 \geqslant 0 \right)$$

(cf. formulas (3), (4), herein variable $X_{n_1+s_{a-\alpha}+1}$ plays a role similar to the role of variable X_s in formula (3)). Then $\sigma_2(U_t) = K_t$.

For any point $x \in K_t$ the intersection $\sigma_2^{-1}(x) \cap U_t \subset \sigma_2^{-1}(x) \simeq \tilde{\mathbb{Q}}^2$ of its inverse image with set U_t is a connected semialgebraic set. We show that semialgebraic set $\sigma_2^{-1}(W) \cap U_t \subset \tilde{\mathbb{Q}}^{n_1+s_{a-\alpha}+2}$ is connected. Indeed, let us pick out an arbitrary pair of points $x^{(1)}, x^{(2)} \in W$. According to lemma 3(b) one can find a closed connected semialgebraic curve $C \subset W \cap \mathcal{D}_0(R)$ joining the points $x^{(1)}, x^{(2)} \in C$ for a certain $R \in \mathbb{Q}$. Then set

$$\left\{ \left(c = (x_{1,1}, \ldots, x_{a-\alpha, s_{a-\alpha}}), \left(\prod_{j_1 \in J_1} (g_{j_1}^{(\alpha)}(c)) \prod_{j_2 \in J_2} (-g_{j_2}^{(\alpha)}(c)) \right)^{-1}, \right.\right.$$

$$\left.\left. \left(\prod_{j_1 \in J_1} (g_{j_1}^{(\alpha)}(c)) \prod_{j_2 \in J_2} (-g_{j_2}^{(\alpha)}(c)) \right)^{-2} + x_{a-\alpha, 1}^2 + \ldots + x_{a-\alpha, s_{a-\alpha}}^2 \right) \right\}_{c \in C} \subset U_t$$

is a connected semialgebraic curve by virtue of lemma 3(a) and, furthermore, the projection of this curve under the projection σ_2 contains points $x^{(1)}, x^{(2)}$. This implies, in view of the arbitrariness of the choice of points $x^{(1)}, x^{(2)}$, that there is a unique component of connectivity $\mathcal{W}$ of the set U_t such that $\sigma_2(\mathcal{W}) = W$ and, moreover, $\sigma_2^{-1}(W) \cap U_t = \mathcal{W}$.

Lemma 11 entails that the projection

$$\pi_1(\mathcal{W}) = \bigcup_{\beta} \mathcal{V}_{\beta}^{(1)} \subset \tilde{\mathbb{Q}}^{n_1+1}$$

coincides with a union of an appropriate collection of elements $\mathcal{V}_{\beta}^{(1)}$ of the partition $\mathcal{U}(\{p_{l,t}^{(\alpha+1)(j_1, j_2, j_3)}\}_{j_1, j_2, j_3, l})$ for the given t. A fortiori

$$\pi_1(\mathcal{W}) = \bigcup_{v} \mathcal{V}_v$$

coincides with a union of a suitable collection of elements $\mathcal{V}_v$ of a finer partition $\mathcal{U}(\{p_{l,t}^{(\alpha+1)(j_1, j_2, j_3)}\}_{j_1, j_2, j_3, l, t})$. According to theorem 1 from Wüthrich (1976) (see also Collins,

1975), the projection

$$\sigma_1(\mathcal{V}_v) = \bigcup_{\varkappa} \mathcal{V}_{v,\varkappa} \subset \tilde{\mathbb{Q}}^{n_1}$$

coincides with a union of a certain collection of elements $\mathcal{V}_{v,\varkappa}$ of the partition $\mathcal{U}(\{g_i^{(\alpha+1)}\}_i)$. Thus,

$$\pi_2(W) = \pi_2\sigma_2(\mathcal{W}) = \sigma_1\pi_1(\mathcal{W}) = \bigcup_{v,\varkappa} \mathcal{V}_{v,\varkappa},$$

which completes the proof of the lemma.

Now we describe one more recursive process, consisting of a stages. At the first stage it applies lemma 2 from section 1 to a family of polynomials $\{g_i^{(a-1)}\}_i$. As a result, the process produces a representative set $\{w_{m_1}^{(a-1)}\}_{m_1} \subset \tilde{\mathbb{Q}}^{s_1}$ for the partition $\mathcal{U}(\{g_i^{(a-1)}\}_i)$ (see section 1).

Assume that for a certain $1 \leqslant \beta \leqslant a-1$ a finite set of points

$$\{w_{m_1,\ldots,m_\beta}^{(a-\beta)}\}_{m_1,\ldots,m_\beta} \subset \tilde{\mathbb{Q}}^{s_1+\ldots+s_\beta}$$

is already produced by recursion. For each point $w_{m_1,\ldots,m_\beta}^{(a-\beta)}$ the process applies lemma 2 to the intersection of partition $\mathcal{U}(\{g_i^{(a-\beta-1)}\}_i)$ with the $s_{\beta+1}$-dimensional plane

$$\Xi_{m_1,\ldots,m_\beta} = \{(w_{m_1,\ldots,m_\beta}^{(a-\beta)}, x) : x \in \tilde{\mathbb{Q}}^{s_{\beta+1}}\} \subset \tilde{\mathbb{Q}}^{s_1+\ldots+s_{\beta+1}},$$

in other words, lemma 2 is applied to a family of polynomials (in the notations of lemma 2)

$$\{\hat{g}_i\}_i = \{g_i^{(a-\beta-1)}(w_{m_1,\ldots,m_\beta}^{(a-\beta)}, X)\}_i,$$

where polynomial

$$g_i^{(a-\beta-1)}(w_{m_1,\ldots,m_\beta}^{(a-\beta)}, X) \in \tilde{\mathbb{Q}}[X_{\beta+1,1}, \ldots, X_{\beta+1,s_{\beta+1}}].$$

As a result, the process gets a representative set

$$\{(w_{m_1,\ldots,m_\beta}^{(a-\beta)}, w_{m_{\beta+1}})\}_{m_{\beta+1}} = \{w_{m_1,\ldots,m_\beta,m_{\beta+1}}^{(a-\beta-1)}\}_{m_{\beta+1}} \subset \tilde{\mathbb{Q}}^{s_1+\ldots+s_{\beta+1}}$$

for the partition formed by the components of connectivity of the intersections of all elements of partition $\mathcal{U}(\{g_i^{(a-\beta-1)}\}_i)$ with the plane $\Xi_{m_1,\ldots,m_\beta}$.

By means of the following lemma one can easily complete the decision algorithm for Tarski algebra (cf. theorem 3 in Wüthrich, 1976).

LEMMA 13. *Formula* (1) *is equivalent to quantifier-free formula*

$$\bigvee_{m_1} \underset{m_2}{\&} \ldots \bigvee_{m_a} P(w_{m_1,m_2,\ldots,m_a}^{(0)}). \tag{13}$$

PROOF. We shall prove by induction on $0 \leqslant \alpha \leqslant a$ that for all points $x, w_{m_1,\ldots,m_{a-\alpha}}^{(\alpha)} \in K_t^{(\alpha)}$, which belong to the same element of partition $\mathcal{U}(g_j^{(\alpha)}\}_j)$, the following formula of Tarski algebra (depending on point x)

$$\exists X_{a-\alpha+1,1} \ldots \exists X_{a-\alpha+1,s_{a-\alpha+1}} \exists X_{a-\alpha+2,1} \ldots \exists X_{a-\alpha+2,s_{a-\alpha+2}} \cdots$$

$$\exists X_{a,1} \ldots \exists X_{a,s_a}(P(x, X_{a-\alpha+1,1}, \ldots, X_{a-\alpha+1,s_{a-\alpha+1}}, \ldots, X_{a,1}, \ldots, X_{a,s_a})) \tag{$1_x^{(\alpha)}$}$$

is equivalent to the quantifier-free formula (depending on indices $m_1, \ldots, m_{a-\alpha}$)

$$\bigvee_{m_{a-\alpha+1}} \neg \bigvee_{m_{a-\alpha+2}} \neg \ldots \bigvee_{m_a} (P(w_{m_1,\ldots,m_a}^{(0)})) \tag{$13_{m_1,\ldots,m_{a-\alpha}}^{(\alpha)}$}$$

Note that formula $(1^{(a)})$ (respectively $(13^{(a)})$) is identical with formula (1) (respectively with (13)).

The induction basis for $\alpha = 0$ can be deduced from the fact that the truth value of formula $P(x)$ for a point $x \in \tilde{\mathbb{Q}}^{s_1 + \cdots + s_a}$ is determined uniquely by signs

$$\operatorname{sgn} f_i(x) = \operatorname{sgn} g_i^{(0)}(x) = \operatorname{sgn} g_i^{(0)}(w_{m_1, \ldots, m_a}^{(0)}) = \operatorname{sgn} f_i(w_{m_1, \ldots, m_a}^{(0)})$$

for all $1 \leqslant i \leqslant k$. Let us observe that here we did not exploit the statement that in every element of partition $\mathscr{U}(\{g_i^{(0)}\}_i)$ one can find at least one point of the sort $w_{m_1, \ldots, m_a}^{(0)}$ (a proof of this statement can be extracted from the further proof of the present lemma).

Suppose that the equivalence of formulas $(1_y^{(\alpha)})$ and $(13_{m_1, \ldots, m_{a-\alpha}}^{(\alpha)})$ is already proved for arbitrary points y and $w_{m_1, \ldots, m_{a-\alpha}}^{(\alpha)}$ which belong to the same element of the partition $\mathscr{U}(\{g_j^{(\alpha)}\}_j)$. Consider points $x, w_{m_1, \ldots, m_{a-\alpha-1}}^{(\alpha+1)} \in \mathscr{W}$, where $\mathscr{W}$ is a certain element of partition $\mathscr{U}(\{g_i^{(\alpha+1)}\}_i)$. Assume that the plane $\Xi_{m_1, \ldots, m_{a-\alpha-1}}$ has a nonempty intersection with some element W of partition $\mathscr{U}(\{\beta_j^{(\alpha)}\}_j)$. The image $\pi_2(W)$ under projection

$$\pi_2 : \tilde{\mathbb{Q}}^{s_1 + \cdots + s_{a-\alpha}} \to \tilde{\mathbb{Q}}^{s_1 + \cdots + a_{a-\alpha-1}}$$

coincides with a union

$$\pi_2(W) = \bigcup_{v, \varkappa} \mathscr{V}_{v, \varkappa}$$

of a suitable collection of elements $\mathscr{V}_{v, \varkappa}$ of partition $\mathscr{U}(\{g_i^{(\alpha+1)}\}_i)$ by virtue of lemma 12.

According to the process of construction of the points $\{w_{m_1, \ldots, m_{a-\alpha-1}, m_{a-\alpha}}^{(\alpha)}\}_{m_{a-\alpha}}$ there exists such an index m that

$$w_{m_1, \ldots, m_{a-\alpha-1}, m}^{(\alpha)} \in W \cap \Xi_{m_1, \ldots, m_{a-\alpha-1}}.$$

Since

$$\pi_2(w_{m_1, \ldots, m_{a-\alpha-1}, m}^{(\alpha)}) = w_{m_1, \ldots, m_{a-\alpha-1}}^{(\alpha+1)},$$

the intersection $\pi_2(W) \cap \mathscr{W} \neq \phi$. Hence, $\mathscr{W} = \mathscr{V}_{v, \varkappa} \subset \pi_2(W)$ for appropriate indices $v, \varkappa$. Therefore, one can find a point $x_W \in W$ with the projection $\pi_2(x_W) = x$.

So assume that formula $(13_{m_1, \ldots, m_{a-\alpha-1}}^{(\alpha+1)})$ is valid, then for a suitable index $m_{a-\alpha}$ the formula $(13_{m_1, \ldots, m_{a-\alpha-1}, m_{a-\alpha}}^{(\alpha)})$ is false. Let point $w_{m_1, \ldots, m_{a-\alpha-1}, m_{a-\alpha}}^{(\alpha)} \in W_1$ for a relevant element W_1 of partition $\mathscr{U}(g_j^{(\alpha)}\}_j)$. Then $\pi_2(x_{W_1}) = x$ for a certaint $x_{W_1} \in W_1$ in view of the fact proved above. The inductive hypothesis implies that formula $(1_{x_{W_1}}^{(\alpha)})$ is false. This entails the truth of formula $(1_x^{(\alpha+1)})$, which was to be shown.

Conversely, suppose that formula $(1_x^{(\alpha+1)})$ is valid, then there exist $x_1, \ldots, x_{s_{a-\alpha}} \in \tilde{\mathbb{Q}}$ such that formula $(1_{\tilde{x}}^{(\alpha)})$ is false, where we denote a point

$$\tilde{x} = (x, (x_1, \ldots, x_{s_{a-\alpha}})) \in \tilde{\mathbb{Q}}^{s_1 + \cdots + s_{a-\alpha}}.$$

Let the point $\tilde{x} \in W_1$ for a suitable element W_1 of the partition $\mathscr{U}(\{g_j^{(\alpha)}\}_j)$. Taking into account that $\pi_2(\tilde{x}) = x$, one concludes that $\pi_2(W_1) \cap \mathscr{W} \neq \phi$, whence $\mathscr{W} \subset \pi_2(W_1)$ by virtue of lemma 12 (cf. above), in particular point $w_{m_1, \ldots, m_{a-\alpha-1}}^{(\alpha+1)} \in \pi_2(W_1)$. Thus, there is an index $m_{a-\alpha}$ for which point $w_{m_1, \ldots, m_{a-\alpha-1}, m_{a-\alpha}}^{(\alpha)} \in W_1$ (see above). Then formula $(13_{m_1, \ldots, m_{a-\alpha-1}, m_{a-\alpha}}^{(\alpha)})$ is false according to the inductive hypothesis. Therefore, formula $(13_{m_1, \ldots, m_{a-\alpha-1}}^{(\alpha+1)})$ is true, which completes the proof of the lemma.

5. Time Analysis of the Decision Algorithm

First of all we estimate the time required for the algorithm in section 3 constructing the formula $\exists\, T(P_1)$ (see lemma 10) and the size of this formula. Let formula (2) be given. In the beginning, the algorithm yields the family of vectors $\Gamma \subset \mathbb{Z}^{s-1}$. According to Grigor'ev & Vorobjov (1987) (see also Vorobjov & Grigor'ev, 1985) $\operatorname{card}(\Gamma) \leqslant \mathscr{P}((kd)^{s^2})$.

Furthermore, for each vector $(\gamma_2, \ldots, \gamma_s) \in \Gamma$ the inequalities $1 \leqslant \gamma_i \leqslant \mathscr{P}((kd)^s)$, $2 \leqslant i \leqslant s$ are correct; finally, the algorithm yields the family Γ within time $\mathscr{P}((kd)^{s^2})$.

Next, the algorithm constructs the matrix A with entries in the ring $\mathbb{Q}[\varepsilon_1, \varepsilon_3][Y, Z_0, \ldots, Z_n, u_0, \ldots, u_s]$ corresponding to system (8') (see proposition 7) and applies to it the Gaussian algorithm. As a result, the polynomials

$$\psi_1, P_1 \in \mathbb{Q}[\varepsilon_1, \varepsilon_3][Y, Z_0, \ldots, Z_n, u_0, \ldots, u_s]$$

are produced. The number τ of rows of the matrix $A = (a_{ij})$ can be estimated by $\tau \leqslant \mathscr{P}((kd)^s)$, each entry a_{ij} of A is of degree

$$\deg_{\varepsilon_1, \varepsilon_3, Y, Z_0, \ldots, Z_n, u_0, \ldots, u_s}(a_{ij}) \leqslant 0(kd)$$

and the size $l(a_{ij}) \leqslant 0(M + n \log(kd))$. Taking into account that both ψ_1, P_1 are the products of not more than τ minors of the matrix A we obtain the bounds

$$\deg_{\varepsilon_1, \varepsilon_3, Y, Z_0, \ldots, Z_n, u_0, \ldots, u_s}(\psi_1), \deg_{\varepsilon_1, \varepsilon_3, Y, Z_0, \ldots, Z_n, u_0, \ldots, u_s}(P_1) \leqslant \mathscr{P}((kd)^s);$$

$$l(\psi_1), l(P_1) \leqslant (M + n)\mathscr{P}((kd)^s).$$

Similar bounds are valid for the polynomials $\psi_1^{(j_0)}, \psi_1^{(j_0, m_0)}$. Since executing the Gaussian algorithm with the matrix A requires $\mathscr{P}((kd)^s)$ arithmetical operations with entries $a_{ij}^{(\rho)}$ of intermediate matrices $A^{(\rho)}$, and taking into account that the bit sizes of rational functions

$$a_{ij}^{(\rho)} = \det(\Delta_{ij}^{(\rho)})/\det(\Delta_{i_{\rho-1}, j_{\rho-1}}^{(\rho-1)}) \in \mathbb{Q}(\varepsilon_1, \varepsilon_3)(Z_0, \ldots, Z_n, u_0, \ldots, u_s)$$

do not exceed $\mathscr{P}(M, (kd)^{s(n+s)})$ according to the bounds obtained on degrees and on sizes l, one concludes that the time necessary to construct the matrix A and the polynomials $\psi_1, P_1, \psi_1^{(j_0)}, \psi_1^{(j_0, m_0)}$ can be estimated by $\mathscr{P}(M, (kd)^{s(n+s)})$.

Then the algorithm calculates the quotient $\psi_1^{(j_0)}/\psi_1^{(j_0, m_0)} = \psi_0/P_2$, where polynomials $P_2 \in \mathbb{Q}[\varepsilon_1, \varepsilon_3][Z_0, \ldots, Z_n]$ and $\psi_0 \in \mathbb{Q}[\varepsilon_1, \varepsilon_3][Z_0, \ldots, Z_n, u_0, \ldots, u_s]$ with the help of proposition 1 factoring polynomials $\psi_1^{(j_0)}$ and $\psi_1^{(j_0, m_0)}$. Then the algorithm computes the polynomial $P = \delta_{K(1)}^{(1)} \delta_{K(2)}^{(2)} P_2$, where $\delta_{K(1)}^{(1)}, \delta_{K(2)}^{(2)} \in \mathbb{Q}[\varepsilon_1, \varepsilon_3][Z_0, \ldots, Z_n]$ are some nonzero coefficients of the polynomials $P_1, \psi_1^{(j_0, m_0)}$ respectively. Proposition 1 implies the bounds

$$\deg_{\varepsilon_1, \varepsilon_3, Z_0, \ldots, Z_n}(P), \deg_{\varepsilon_1, \varepsilon_3, Z_0, \ldots, Z_n, u_0, \ldots, u_s}(\psi_0) \leqslant \mathscr{P}((kd)^s);$$

$$l(P), l(\psi_0) \leqslant (M + n)\mathscr{P}((kd)^s)$$

and the time required to produce P, ψ_0 does not exceed $\mathscr{P}(M, (kd)^{s(n+s)})$.

After that the algorithm factorises the polynomial ψ_0 over the field $\mathbb{Q}$, picks out a factor $\Omega | \psi_0$ of the polynomial ψ_0 irreducible over $\mathbb{Q}$. By virtue of proposition 1

$$\deg_{\varepsilon_1, \varepsilon_3, Z_0, Z_n}(\Omega) \leqslant \mathscr{P}((kd)^s); \qquad l(\Omega) \leqslant (M + n)\mathscr{P}((kd)^s)$$

and the algorithm yields Ω within time $\mathscr{P}(M, (kd)^{s(n+s)})$. Then for each s-tuple $(\lambda_1^{(0)}, \ldots, \lambda_s^{(0)})$, where

$$1 \leqslant \lambda_i^{(0)} \leqslant D_1 = \deg_{u_0, \ldots, u_s}(\Omega) \leqslant \tau; \quad 1 \leqslant i \leqslant s$$

the algorithm tests irreducibility over the field H_2 of the polynomial

$$\Omega_0(u_0) = \Omega(u_0, -\lambda_1^{(0)}, \ldots, -\lambda_s^{(0)})$$

within time $\mathscr{P}(M, (kd)^{sn})$ again by proposition 1 and the bounds

$$\deg_{\varepsilon_1, \varepsilon_3, Z_0, \ldots, Z_n, u_0}(\Omega_0) \leqslant \mathscr{P}((kd)^s), \qquad l(\Omega_0) \leqslant (M + n)\mathscr{P}((kd)^s).$$

Since the whole number of s-tuples $(\lambda_1^{(0)}, \ldots, \lambda_s^{(0)})$ is less than $\mathscr{P}((kd)^{s^2})$, one concludes that the algorithm produces a primitive element

$$\theta^{(v)} = \sum_{1 \leqslant i \leqslant s} \lambda_i(\zeta_i^{(v)}/\zeta_0^{(v)})$$

and its minimal polynomial

$$\Phi = \sum_{0 \leqslant j \leqslant D_1} (\alpha_j/\beta)T^j \in H_2[T]$$

within time $\mathscr{P}(M, (kd)^{s(n+s)})$, herewith

$$\deg_{\varepsilon_1, \varepsilon_3, Z_0, \ldots, Z_n, T}(\phi) \leqslant \mathscr{P}(kd)^s), \quad l(\Phi) \leqslant (M+n)\mathscr{P}((kd)^s).$$

Next, the algorithm finds the expressions $(\zeta_i^{(v)}/\zeta_0^{(v)})(\theta^{(v)})$ again by proposition 1 factoring the polynomial

$$\Omega_i = \Omega(-\lambda_1^{(0)}, \ldots, -\lambda_{i-1}^{(0)}, -\lambda_i^{(0)} + u_i, -\lambda_{i+1}^{(0)}, \ldots, -\lambda_s^{(0)}).$$

Hence,

$$\deg_{\varepsilon_1, \varepsilon_3, Z_0, \ldots, Z_n}((\zeta_i^{(v)}/\zeta_0^{(v)})(\theta^{(v)})) \leqslant \mathscr{P}((kd)^s), \quad l((\zeta_i^{(v)}/\zeta_0^{(v)})(\theta^{(v)})) \leqslant (M+n)\mathscr{P}((kd)^s)$$

and the required time for finding $(\zeta_i^{(v)}/\zeta_0^{(v)})(\theta^{(v)})$ does not exceed $\mathscr{P}(M, (kd)^{s(s+n)})$.

After that the algorithm yields a polynomial $\chi^{(\theta_0)} = \Omega/L^{(\theta_0)}$ factoring the polynomial Ω over the field $H_2[\theta_0]$ based on proposition 1. Proposition 1 entails the bounds

$$\deg_{\varepsilon_1, \varepsilon_3, Z_0, \ldots, Z_n, u_0, \ldots, u_s, T}(\chi^{(T)}) \leqslant \mathscr{P}((kd)^s), \quad l(\chi^{(T)}) \leqslant (M+n)\mathscr{P}((kd)^s)$$

and that the time required to produce $\chi^{(T)}$ can be estimated by $\mathscr{P}(M, (kd)^{s(n+s)})$.

Thereupon, the algorithm finds the polynomials

$$\tau_1 = (\Omega - L^{(T)}\chi^{(T)})/\Phi,$$

$$\tau_2 = \left(T - \sum_{1 \leqslant i \leqslant s} \lambda_i \sum_j (a_j^{(i)}/b)T^j\right)\bigg/ \Phi \in H_2[T][u_0, \ldots, u_s]$$

dividing each coefficient at a monomial in the variables $u_0, \ldots, u_s$ on the polynomial Φ in the ring $H_2[T]$. It takes time $\mathscr{P}(M, (kd)^{s(n+s)})$ by virtue of, e.g. proposition 1 (though this is a rather strong tool to use it for dividing polynomials), and the following bounds are true:

$$\deg_{\varepsilon_1, \varepsilon_3, Z_0, \ldots, Z_n, u_0, \ldots, u_s, T}(\tau_1), \deg_{\varepsilon_1, \varepsilon_3, Z_0, \ldots, Z_n, u_0, \ldots, u_s, T}(\tau_2) \leqslant \mathscr{P}((kd)^s);$$

$$l(\tau_1), l(\tau_2) \leqslant (M+n)\mathscr{P}((kd)^s).$$

After that the algorithm finds the hypersurface $\mathscr{L} \subset \bar{F}_3^{n+1}$ (see (9)). It is given by not more than $(\text{card } \Gamma)(\deg \psi_0) \leqslant \mathscr{P}((kd)^{s^2})$ polynomials, each of degree less or equal to $\mathscr{P}((kd)^s)$. Hence, $N_1 \leqslant \mathscr{P}((kd)^{s^2})$, $N \leqslant n\mathscr{P}((kd)^{s^2})$. The algorithm finds the hypersurface $\mathscr{L}$ within time $\mathscr{P}(M, (kd)^{s(n+s)})$. At last, the algorithm produces the family of linear forms $Y_1, \ldots, Y_N$ and the family of vectors $\{v_{i_1, \ldots, i_n}\}_{1 \leqslant i_1 < \ldots < i_n \leqslant N}$ over the rationals. Then $l(v_{i_1, \ldots, i_n}) \leqslant \mathscr{P}(n, s, \log(kd))$; in addition, the whole number of vectors $\{v_{i_1, \ldots, i_n}\}$ and the time for producing them do not exceed

$$\mathscr{P}\left(\binom{N}{n}\right) \leqslant \mathscr{P}((kd)^{s^2 n})$$

(cf. section 2 in Chistov & Grigor'ev, 1983a).

Thus, we have proved the following

LEMMA 14. *For a given formula* (2), *the algorithm, described in section 3, yields a formula of the form* $\exists T(P_1)$ *(see lemma 10) which is equivalent to formula* (5), *within time*

$\mathscr{P}(M, (kd)^{s^2 n})$. *Furthermore, the quantifier-free formula* P_1 *contains* $\mathscr{P}((kd)^{s^2 n})$ *atomic subformulas of the sort* $(g_j \geqslant 0)$ *where polynomials* $g_j \in \mathbb{Q}[\varepsilon_1, \varepsilon_2, \varepsilon_3][T, Z_0, Z_1, \ldots, Z_n]$, *with*

$$\deg(g_j) \leqslant \mathscr{P}((kd)^s); \quad l(g_j) \leqslant M\mathscr{P}(n, (kd)^s).$$

We recall that the decision algorithm for Tarski algebra described in section 4, after implementation of α stages of the procedure enumerates first all $(\{g_j^{(\alpha)}\}_j)$-cells. Let us introduce the notations: $k^{(\alpha)}$ is the number of polynomials $g_j^{(\alpha)}$, i.e. $1 \leqslant j \leqslant k^{(\alpha)}$; next

$$d^{(\alpha)} = \max_{1 \leqslant j \leqslant k(\alpha)} \deg(g_j^{(\alpha)}), \quad M^{(\alpha)} = \max_{1 \leqslant j \leqslant k(\alpha)} l(g_j^{(\alpha)}).$$

Based on lemma 1 the algorithm enumerates all $(\{g_j^{(\alpha)}\}_j)$-cells within time $\mathscr{P}(M^{(\alpha)}, (k^{(\alpha)}d^{(\alpha)})^{n^2}$, where $n = s_1 + \ldots + s_a$. Note that the whole number of all $(\{g_j^{(\alpha)}\}_j)$-cells does not exceed $\mathscr{P}((k^{(\alpha)}d^{(\alpha)})^n)$ in view of lemma 1.

Next, the algorithm applies the construction from section 3 to a formula of the form (12) and outputs a formula of the sort $\exists\, T(P_t)$ within time $\mathscr{P}(M^{(\alpha)}, (k^{(\alpha)}d^{(\alpha)})^{s_a^2 - \alpha n})$ according to lemma 14. The quantifier-free formula P_t contains atomic subformulas $(g_{m,t}^{(\alpha+1)} \geqslant 0)$. By virtue of lemma 14 the following bounds are true:

$$m \leqslant \mathscr{P}((k^{(\alpha)}d^{(\alpha)})^{s_a^2 - \alpha n}), \quad \deg(g_{m,t}^{(\alpha+1)}) \leqslant \mathscr{P}((k^{(\alpha)}d^{(\alpha)})^{s_a - \alpha}),$$

$$l(g_{m,t}^{(\alpha+1)}) \leqslant M^{(\alpha)}\mathscr{P}(n, (k^{(\alpha)}d^{(\alpha)})^{s_a - \alpha}), \tag{14}$$

moreover, the polynomials "hidden" in the notations $\mathscr{P}$ do not depend on α.

In the next step the algorithm produces polynomials $\{P_{l,t}^{(\alpha+1)}\}_l$ within the same time-bound $\mathscr{P}(M^{(\alpha)}, (k^{(\alpha)}d^{(\alpha)})^{s_a^2 - \alpha n})$, where bounds similar to (14) are correct for the polynomials $P_{l,t}^{(\alpha+1)}$ in view of the construction in theorem 1 in Wüthrich (1976). Hence, the same time-bound and the bounds similar to (14) are satisfied for polynomials $P_{l,t}^{(\alpha+1)(j_1, j_2, j_3)}$. Finally, the algorithm applies the construction from Wüthrich (1976) to the family of polynomials $\{P_{l,t}^{(\alpha+1)(j_1, j_2, j_3)}\}_{j_1, j_2, j_3, l, t}$ and obtains, as a result, polynomials $\{g_i^{(\alpha+1)}\}_i$ within the same time-bound. In addition, the polynomials $g_i^{(\alpha+1)}$ again satisfy bounds analogous to (14) and, as above, the polynomials "hidden" in the notations $\mathscr{P}$ do not depend on α.

Let us denote $N^{(\alpha)} = k^{(\alpha)}d^{(\alpha)}$. One can infer by induction on α (taking into account for the induction basis $\alpha = O$ the estimates on the parameters of formula (1)) inequalities

$$N^{(\alpha+1)} \leqslant \mathscr{P}((N^{(\alpha)})^{s_a^2 - \alpha n}) \leqslant (kd)^{(O(n))^{3(\alpha+1)}}$$

(see 14)), where the constant factor "hidden" in the notation $O(n)$ does not depend on α. In addition,

$$M^{(\alpha+1)} \leqslant M(kd)^{(O(n))^{3\alpha}s_a - \alpha} \leqslant M(kd)^{(O(n))^{3\alpha+1}}.$$

The algorithm produces polynomials $\{g_j^{(\alpha)}\}$ for all j, $0 \leqslant \alpha \leqslant a-1$ within time $\mathscr{P}(M, (kd)^{(O(n))^{3(a-1)}})$.

Thereafter, the algorithm produces by recursion on $1 \leqslant \beta \leqslant a$ the representative set of points $\{w_{m_1, \ldots, m_\beta}^{(a-\beta)}\}_{m_1, \ldots, m_\beta}$. Assume that for a certain $0 \leqslant \beta < a$ the algorithm has constructed for each β-tuple of indices $m_1, \ldots, m_\beta$ a polynomial

$$\Phi_\beta(Z) = \Phi_{m_1, \ldots, m_\beta}(Z) \in \mathbb{Q}[Z],$$

irreducible over $\mathbb{Q}$, and expressions

$$w_\beta^{(p)}(\theta_1) = \sum_{0 \leqslant j < \deg(\Phi_\beta)} \rho_j^{(p)}\theta_1^j$$

for the coordinates of the point

$$w^{(a-\beta)}_{m_1,\ldots,m_\beta} = (w^{(1)}_\beta(\theta_1), \ldots, w^{(s_1+\ldots+s_\beta)}_\beta(\theta_1)) \in (\mathbb{Q}[\theta_1])^{s_1+\ldots+s_\beta},$$

where

$$\rho^{(p)}_j \in \mathbb{Q}, \ 1 \leqslant p \leqslant s_1 + \ldots + s_\beta, \ 0 \leqslant j < \deg(\Phi_\beta) \quad \text{and} \quad \theta_1 \in \tilde{\mathbb{Q}}, \ \Phi_\beta(\theta_1) = 0.$$

Furthermore,

$$\theta_1 = \sum_{1 \leqslant p \leqslant s_1+\ldots+s_\beta} \lambda^{(1)}_p w^{(p)}_\beta(\theta_1)$$

for certain natural numbers $1 \leqslant \lambda^{(1)}_p \leqslant \deg(\Phi_\beta)$. Besides, we assume that the algorithm has found a pair of rational numbers $c_1, c_2 \in \mathbb{Q}$ such that in the interval $(c_1, c_2) \subset \mathbb{R}$, θ_1 is the only root of the polynomial Φ_β. Let us define

$$d_\beta = \deg(\Phi_\beta), \quad M_\beta = \max_p \{l(\Phi_\beta), l(w^{(p)}_\beta(\theta_1)), l(c_1), l(c_2)\}.$$

Applying lemma 2 to the family of polynomials

$$\{\hat{g}_i\}_i = \{g^{(a-\beta-1)}_i(w^{(a-\beta)}_{m_1,\ldots,m_\beta}, X)\}_i$$

the algorithm produces for given $m_1, \ldots, m_\beta$ the set of points

$$\{w^{(a-\beta-1)}_{m_1,\ldots,m_\beta,m_{\beta+1}}\}_{m_{\beta+1}} \subset \tilde{\mathbb{Q}}^{s_1+\ldots+s_\beta+s_{\beta+1}},$$

a corresponding polynomial

$$\Phi_{\beta+1} = \Phi_{m_1,\ldots,m_{\beta+1}} \in \mathbb{Q}[Z],$$

expressions

$$w^{(l)}_{\beta+1}(\theta) \in \mathbb{Q}[\theta], \ \Phi_{\beta+1}(\theta) = 0 \quad \text{for} \quad 1 \leqslant l \leqslant s_1 + \ldots + s_{\beta+1}$$

(cf. above) and, finally, a pair of rational numbers $b_1, b_2 \in \mathbb{Q}$ such that in the interval $(b_1, b_2) \subset \mathbb{R}$, θ is the unique root of the polynomial $\Phi_{\beta+1}$. Lemma 2 implies the following bounds (taking into account the inequalities for $N^{(\alpha)}$, $M^{(\alpha)}$ proved above):

$$d_{\beta+1} \leqslant \mathscr{P}((N^{(a-\beta-1)}d_\beta)^n); \quad M_{\beta+1} \leqslant (M^{(a-\beta-1)} + M_{\beta+n})\mathscr{P}((N^{(a-\beta-1)}d_\beta)^n).$$

From this one can deduce by induction on β the bounds

$$d_{\beta+1} \leqslant (kd)^{(O(n))^{3a+\beta-2}}; \quad M_{\beta+1} \leqslant M(kd)^{(O(n))^{3a+\beta-2}}.$$

Moreover, the constant factors "hidden" in the notations $O(n)$, do not depend on β (cf. above).

By virtue of lemma 2, for given $m_1, \ldots, m_\beta$, the number of points in the set $\{w^{(a-\beta-1)}_{m_1,\ldots,m_\beta,m_{\beta+1}}\}_{m_{\beta+1}}$ does not exceed $(kd)^{(O(n))^{3a+\beta-1}}$. The algorithm can produce the set $\{w^{(a-\beta-1)}_{m_1,\ldots,m_\beta,m_{\beta+1}}\}_{m_{\beta+1}}$ within time $\mathscr{P}(M, (kd)^{(O(n))^{3a+\beta-1}})$, again according to lemma 2 and to the bounds on d_β, M_β obtained above. Thus, the number of all points from the representative set $\{w^{(0)}_{m_1,\ldots,m_a}\}_{m_1,\ldots,m_a}$ is not greater than $(kd)^{(O(n))^{4a-2}}$, the time for the construction of this set can be estimated by $\mathscr{P}(M, (kd)^{(O(n))^{4a-2}})$.

At the end of its work the algorithm evaluates the signs $\text{sgn}(f_j(w^{(0)}_{m_1,\ldots,m_a}))$ for all $1 \leqslant j \leqslant k$; $m_1, \ldots, m_a$, and thereby verifies formula (13) (see lemma 13). One can evaluate the sign $\text{sgn}(f_j(w^{(0)}_{m_1,\ldots,m_a}))$ for given $j, m_1, \ldots, m_a$ in the following way. First, the algorithm replaces the coordinates of the point $w^{(0)}_{m_1,\ldots,m_a} \in (\mathbb{Q}[\theta_0])^n$ by their expressions via appropriate real root $\theta_0 \in \tilde{\mathbb{Q}}$ of polynomial Φ_a (see above the representation of points $w^{(a-\beta)}_{m_1,\ldots,m_\beta}$). So, the algorithm gets the expression $f(w^{(0)}_{m_1,\ldots,m_a}) = h_j(\theta_0)$ for a suitable polynomial $h_j(Z) \in \mathbb{Q}[Z]$. Obviously, $f(w^{(0)}_{m_1,\ldots,m_a}) = 0$ is valid iff polynomial $\Phi_a | h_j$. If the latter relation is false, the algorithm finds a suitable rational approximation $\omega_0 \in \mathbb{Q}$ to the root θ_0 in order to guarantee the equality $\operatorname{sgn} h_j(\omega_0) = \operatorname{sgn} h_j(\theta_0) \neq 0$ (cf. section 3 in Grigor'ev & Vorobjov, 1987). The rational approximation can be found, e.g. by means of Heindel (1971) within time $\mathscr{P}(M, (kd)^{(O(n))^{4a-5}})$ in view of the bounds on d_a, M_a proved above. This completes the proof of the theorem (see the introduction).

6. Computing the Dimension of a Semialgebraic Set

Let $V = \{\Pi\} \subset \mathbb{R}^n$ be a semialgebraic set, given by a quantifier-free formula Π of Tarski algebra. We shall prove the following statement: If the dimension $\dim V = m < n$, then there is an index $1 \leqslant j \leqslant n$ such that $\dim(\pi^{(j)}(V)) = m$, where π^j denotes the projection onto the coordinates $X_1, \ldots, X_{j-1}, X_{j+1}, \ldots, X_n$. We conduct the proof by induction on m. When $m = 0$, the statement is trivial, so we assume further that $m \geqslant 1$. For any number $x \in \mathbb{R}$ consider a semialgebraic set $V^{(x)} = V \cap \{X_1 = x\}$. If for a certain $x_0 \in \mathbb{R}$ the equality $\dim V^{(x_0)} = m$ is valid, then one can take $j = 1$. Suppose that, on the contrary, $\dim V^{(x)} \leqslant m - 1$ for all $x \in \mathbb{R}$. Let us denote by $W \subset \mathbb{R}$ the semialgebraic set, which consists of all numbers $x \in \mathbb{R}$ such that $\dim(V^{(x)}) = m - 1$ (the set W is semialgebraic since the condition $x \in W$ can be written as a suitable formula of Tarski algebra). Then W contains some interval, taking into account that $\dim V = m$. For each $2 \leqslant i \leqslant n$ we introduce the semialgebraic set $W^{(i)} \subset W$ consisting of all points $x \in W$ for which $\dim(\pi^{(i)}(V^{(x)})) = m - 1$. The inductive hypothesis entails the equality

$$W = \bigcup_{2 \leqslant i \leqslant n} W^{(i)}.$$

Therefore, for a suitable $2 \leqslant i_0 \leqslant n$, the set $W^{(i_0)}$ contains a certain interval. Setting $j = i_0$ completes the proof of the statement.

One concludes from the statement that $\dim V = m$ iff m is the greatest natural number such that there exist indices $1 \leqslant i_1 < \ldots < i_m \leqslant n$, for which $\dim(\pi_{i_1,\ldots,i_m}(V)) = m$ where $\pi_{i_1,\ldots,i_m}: \mathbb{R}^n \to \mathbb{R}^m$ denotes the projection onto the coordinates $X_{i_1}, \ldots, X_{i_m}$. On the other hand, for given $i_1, \ldots, i_m$, the condition $\dim(\pi_{i_1,\ldots,i_m}(V)) = m$ is equivalent to the requirement that the semialgebraic set $\pi_{i_1,\ldots,i_m}(V) \subset \mathbb{R}^m$ contains some m-dimensional ball. The latter requirement is equivalent in turn to the following formula of Tarski algebra

$$\exists\, w \in \mathbb{R}^m\ \exists\, \tau > 0\ \forall\, w_1 \in \mathbb{R}^m\ \exists\, v_1 \in \mathbb{R}^{n-m}((\|w_1 - w\| \leqslant \tau) \Rightarrow (\langle w_1, v_1 \rangle \in V))$$

in $n + m + 1$ variables with the number of quantifier alternations $a = 3$. Here, $\langle w_1, v_1 \rangle = (u^{(1)}, \ldots, u^{(n)}) \in \mathbb{R}^n$ denotes the unique vector such that $\pi_{i_1,\ldots,i_m}(\langle w_1, v_1 \rangle) = w_1$ and $u^{(j_l)} = v_1^{(l)}$ for $1 \leqslant l \leqslant n - m$, where

$$\{i_1, \ldots, i_m\} \cup \{j_1, \ldots, j_{n-m}\} = \{1, \ldots, n\},$$

indices $j_1 < \ldots < j_{n-m}$, point $v_1 = (v_1^{(1)}, \ldots, v_1^{(n-m)})$. The algorithm verifies for every m and set of indices $i_1, \ldots, i_m$ the designed formula of Tarski algebra based on the theorem (see the introduction) and thereby determines $\dim V$. Thus, the following corollary is true.

COROLLARY. *Let Π be a quantifier-free formula of Tarski algebra, which contains k atomic subformulas $(f_i \geqslant 0)$, where f_i are polynomials in $\mathbb{Q}[X_1, \ldots, X_n]$ with $\deg(f_i) < d$, $l(f_i) \leqslant M$, $1 \leqslant i \leqslant k$. Then one can compute the dimension $\dim\{\Pi\}$ within time $\mathscr{P}(M, (kd)^{(O(n))^{10}})$.*

References

Ben-Or, M., Kozen, D., Reif, J. (1984). The complexity of elementary algebra and geometry. *Proc. 16 ACM Symp. Th. Comput.* 457–464.

Chistov, A. L., Grigor'ev, D. Yu. (1982). *Polynomial-time Factoring of Multivariable Polynomials over a Global Field.* Preprint LOMI E–5–82, Leningrad.

Chistov, A. L., Grigor'ev, D. Yu. (1983a). *Subexponential-time Solving Systems of Algebraic Equations. I.* Preprint LOMI E–9–83, Leningrad.

Chistov, A. L., Grigor'ev, D. Yu. (1983b). *Subexponential-time Solving Systems of Algebraic Equations. II.* Preprint LOMI E–10–83, Leningrad.

Chistov, A. L., Grigor'ev, D. Yu. (1984). Complexity of quantifier elimination in the theory of algebraically closed fields. *Springer Lec. Notes Comp. Sci.* **176**, 17–31.

Chistov, A. L. (1984). Polynomial-time factoring of polynomials and finding compounds of a variety within the subexponential time. *Notes of Sci. Seminars of Leningrad Department of Math. Steklov Inst.* **137**, 124–188 (in Russian). (English transl. to appear in J. Soviet Math.)

Cohen, P. (1969). Decision procedures for real and p-adic fields. *Commun. Pure Appl. Math.* **22**, 131–153.

Collins, G. E. (1975). Quantifier elimination for real closed fields by cylindrical algebraic decomposition. Automata Theory and Formal Languages 2nd GI Conf., Kaiserslautern (Brakhage, H., ed.). *Springer Lec. Notes Comp. Sci.* **33**, 134–183.

Fischer, M., Rabin, M. (1974). Super-exponential complexity of Presburger arithmetic. In: *Complexity of Computations* (SIAM–AMS Proc., 7), pp. 27–41.

Grigor'ev, D. Yu. (1984). Factoring multivariable polynomials over a finite field and solving systems of algebraic equations. *Notes of Sci. Seminars of Leningrad Department of Math. Steklov Inst.* **137**, 20–79 (in Russian). (English transl. to appear in J. Soviet Math.)

Grigor'ev, D. Yu. (1985). Complexity of deciding first-order theory of real closed fields. *Proc. All-Union Conf. Appl. Logic, Novosibirsk,* 64–66 (in Russian).

Grigor'ev, D. Yu. (1987). Computational Complexity in Polynomial Algebra. Proc. International Congress of Mathematicians, Berkeley, California.

Grigor'ev, D. Yu., Vorobjov, N. N., Jr (1987). Solving systems of polynomial inequalities in subexponential time. (Submitted.)

Heindel, L. E. (1971). Integer arithmetic algorithms for polynomial real zero determination. *J. Assoc. Comp. Mach.* **18**(4), 533–548.

Heintz, J. (1983). Definability and fast quantifier elimination in algebraically closed fields. *Theor. Comp. Sci.* **24**, 239–278.

Lang, S. (1965). *Algebra.* New York: Addison-Wesley.

Lazard, D. (1981). Résolution des systèmes d'équation algébriques. *Theor. Comp. Sci.* **15**, 77–110.

Lenstra, A. K. (1984). Factoring multivariable polynomials over algebraic number fields. *Springer Lec. Notes Comp. Sci.* **176**, 389–396.

Mayr, E. W., Meyer, A. R. (1982). The complexity of the word problem for commutative semigroups and polynomial ideals. *Adv. Math.* **46**, 305–329.

Milnor, J. (1964). On a Betti numbers of real varieties. *Proc. Amer. Math. Soc.* **15**(2), 275–280.

Monk, L. (1974). *An Elementary-recursive Decision Procedure for Th(R, +,·).* Ph.D. Thesis, Berkeley.

Seidenberg, A. (1954). A new decision method for elementary algebra and geometry. *Ann. Math.* **60**, 365–374.

Shafarevich, I. R. (1974). *Basic Algebraic Geometry.* Berlin: Springer-Verlag.

Tarski, A. (1951). *A Decision Method for Elementary Algebra and Geometry.* University of California Press.

Vorobjov, N. N. Jr, Grigor'ev, D. Yu. (1985). Finding real solutions of systems of algebraic inequalities in subexponential time. *Soviet Math. Dokl.* **32**(1), 316–320.

Wüthrich, H. (1976). Ein Entscheidungsverfahren für die Theorie der reel-abgeschlossenen Körper. In Komplexität von Entscheidungs-Problemen. (Specker, E., Strassen, V., Eds). *Springer Lec. Notes Comp. Sci.* **43**, 138–162.

J. Symbolic Computation (1988) **5**, 109–119

Some Aspects of Complexity in Real Algebraic Geometry

JEAN-JACQUES RISLER

*U.E.R. de Mathématiques, Université Paris 7,
2 place Jussieu, 75251 Paris Cedex 05, France*

(*Received* 15 *October* 1985)

This paper gives a description of some aspects of complexity in real algebraic geometry: explicit bounds for the topology of real algebraic and semi-algebraic sets are given in term of the degree of the equations and inequations (results by Milnor-Thom), and then in term of the additive complexity of the equations and inequations (i.e. in term of the minimal number of addition-substractions needed to evaluate the equations) (results by Grigoriev-Risler).

Those results are then used to give lower bounds for the complexity of various algorithms (results of Ben-Or).

The work of Collins and its school about C.A.D. of semi-algebraic sets is mentioned, and the paper ends with a description of a class of real analytic sets to which it is possible to generalise some of the previous methods (results by Hovansky).

Introduction

Real algebraic and semialgebraic sets are of great importance in several aspects of algorithmic problems, because they are the geometric objects that are the most calculable, and, to a first approximation, every "geometric" set in R^n can be assumed semialgebraic.

For instance, the "piano mover's problem" (which is studied in "theoretical robotics" (Sharir & Schwartz, 1983), and several parts of "graphics" research (Barsky, 1982), use semialgebraic sets, intersection of algebraic and semialgebraic sets, etc.

I will, in this partial survey, give some results and ideas which seem important to me and are perhaps not well known by some mathematicians working in the algorithmic aspect of the question. Those ideas are not about algorithms themselves, but on what comes "before" the production of algorithms, namely, what can be the "complexity" of a (semialgebraic) set, how to compute it, or to bound it (in terms of the data) etc.

The second step (which is up to now not well developed) would be to relate those "complexities" to complexity of algorithms (for instance for a decision procedure or for the projection of semialgebraic sets), to find specific algorithms for objects of low complexity, etc.

The paper is organised as follows: section 1 gives bounds for the topology of real algebraic and semialgebraic sets in relation to the degree of the equations and inequalities defining them; section 2 introduces the notion of additive complexity and gives bounds in terms of it for the topology of semialgebraic sets; section 3 gives an application to the complexity of some algorithms, section 4 gives references for some known methods and algorithms for the determination of the topology of a semialgebraic set given in R^n by polynomial equations and inequalities, and section 5 gives some results (due essentially to the Russian mathematician A. Hovansky) on a possible generalisation of semialgebraic objects to objects defined with transcendental functions.

1. Bounds for the Topology of Real Semialgebraic Sets in Terms of Degree

The most usual notion of complexity for a polynomial $P \in K[X_1, \ldots, X_n]$ (where K is any field) is its *degree* (total degree).

DEFINITION 1.1 (cf. Yomdin, 1985). Let $A \subset R^n$ be a semialgebraic set, finite union of sets A_α defined by

$$A_\alpha \begin{cases} P_{\alpha i} = 0 & (1 \leqslant i \leqslant r_\alpha) \\ P_{\alpha j} > 0 & (1 \leqslant j \leqslant s_\alpha) \\ P_{ak} \geqslant 0 & (1 \leqslant k \leqslant t_\alpha). \end{cases}$$

We will call the *diagram* of this representation of A (or simply diagram of A) the set $D(A)$ which is the union of the lists l_α, l_α being the set of degrees of all equations and inequalities $P_{\alpha i}$, $P_{\alpha j}$, and $P_{\alpha k}$.

Let us first recall the results of Milnor (1964) and Thom (1965) which bound the number of connected components of A.

Let $B_i(A)$ be the ith Betti number of A; $B_i(A)$ is the dimension (over $Z_2 = Z/2Z$) of the space $H_i(A, Z_2)$; B_0 is the number of connected components of A. We have then

THEOREM 1.2 (Milnor, 1964). *Let $A \subset R^n$ be defined by polynomial equations*

$$\begin{cases} f_1(X_1, \ldots, X_n) = 0 \\ \vdots \\ f_p(X_1, \ldots, X_n) = 0 \end{cases}$$

the degree of each f_i being $\leqslant d$; then $\sum_i B_i(A) \leqslant d(2d-1)^{n-1}$ (and so $B_0(A) \leqslant d(2d-1)^{n-1}$).

COROLLARY 1.3. *Let $A \subset R^n$ be a semialgebraic set of diagram $D(A)$; then there exists $C(D(A))$, depending only on $D(A)$, such that $B_0(A) \leqslant C(D(A))$.*

EXAMPLES.

(a) (Milnor, 1964) if $A \subset R^n$ is defined by $f_1 \geqslant 0, \ldots, f_p \geqslant 0$, where $f_i \in R[X_1, \ldots, X_n]$, and if $d = \deg f_1 + \cdots + \deg f_p$, then

$$\sum_i B_i(A) \leqslant 1/2(2+d)(1+d)^{n-1}$$

(and so $B_0(A) \leqslant 1/2(2+d)(1+d)^{n-1}$).

(b) (Ben-Or, 1983) if $A \subset R^n$ is defined by polynomials

$$\begin{cases} q_1 = \ldots = q_m = 0 \\ p_1 > 0 \ldots p_s > 0 \\ p_{s+1} \geqslant 0 \ldots p_h \geqslant 0 \end{cases}$$

with $d = \sup(2, \deg p_i, \deg q_j)$ and n the number of variables, then

$$B_0(A) \leqslant d(2d-1)^{n+h-1}.$$

The proof is as follows: first replace A by A_ε defined by

$$\begin{cases} q_1 = \ldots = q_m = 0 \\ p_1 \geqslant \varepsilon \ldots p_s \geqslant \varepsilon \\ p_{s+1} \geqslant 0 \ldots p_h \geqslant 0 \end{cases}$$

if ε is small enough, we have $B_0(A) \leqslant B_0(A_\varepsilon)$.

Let $Y_1, \ldots, Y_h$ be new unknowns; then A_ε is the projection of $B \subset R^{n+h}$ defined by

$$B \begin{cases} q_1(X_1, \ldots, X_n) = \ldots = q_m(X_1, \ldots, X_n) = 0 \\ p_1(X_1, \ldots, X_n) = Y_1^2 + \varepsilon, \ldots, p_s(X_1, \ldots, X_n) = Y_s^2 + \varepsilon \\ p_{s+1}(X_1, \ldots, X_n) = Y_{s+1}^2, \ldots, p_h(X_1, \ldots, X_n) = Y_h^2. \end{cases}$$

Then apply 1.2 to B.

(c) If A is defined by

$$A \begin{cases} q_1 = \ldots = q_m = 0 \\ p_1 > 0, \ldots, p_h > 0 \end{cases}$$

and if $d = \sup(\deg q_i, \deg p_i)$, one has the bound: $B_0(A) \leqslant (2hd+1)^{n+1}$ which is better than the bound in (b); the proof is again a variant of Milnor's paper: if H is defined in R^n by

$$\begin{cases} q_1 = \ldots = q_m = 0 \\ p_1 \ldots p_h \neq 0 \end{cases}$$

each connected component of A is a connected component of H, and if $\pi : R^{n+1} \to R^n$ is the projection defined by

$$\pi(X_1, \ldots, X_{n+1}) = (X_1, \ldots, X_n),$$

H is the projection $\pi(H')$, where H' is defined by

$$\begin{cases} q_1 = \ldots = q_m = 0 \\ X_{n+1} p_1 \ldots p_h = 1 \end{cases}$$

and one has $B_0(H) \leqslant B_0(H')$.

But $B_0(H') \leqslant (hd+1)(2hd+1)^n$ by Theorem 1.2 and so $B_0(H') \leqslant (2hd+1)^{n+1}$.

(d) If A is defined as in (b) by

$$\begin{cases} q_1 = \ldots = q_m = 0 \\ p_1 > 0, \ldots, p_s > 0 \\ p_{s+1} \geqslant 0, \ldots, p_h \geqslant 0 \end{cases}$$

the same method proves that $B_0(A) \leqslant (2hd+1)^{n+2}$.

In Thom (1965) there are similar results and another method is suggested to bound the topology of real algebraic sets based on "Smith theory".

PROPOSITION 1.4. *Let $X \subset RP^n$ be a real (closed) algebraic variety in real projective space of dimension n, $\tilde{X} \subset CP^n$ its complexification ($\tilde{X}$ is defined in CP^n by the same equations as X in RP^n; $\tilde{X}$ is then compact). Then*

$$\sum_i B_i(X) \leqslant \sum_i B_i(\tilde{X}).$$

It is then possible with this proposition to bound $B_0(X)$ in the projective case if one can compute $B_i(\tilde{X})$.

The results obtained with this method are then often more precise than with Theorem 1.2; for instance if $X \subset R^2$ is a smooth cubic curve, 1.4 gives $B_0(X) \leqslant 4$ (it is an easy exercise, looking at the projective closure of X), which is the best possible, and 1.2 gives $B_0(X) \leqslant 15$.

Yomdin (1985) shows similar results, but for metric properties of semialgebraic sets:

PROPOSITION 1.5. *Let $A \subset R^n$ be a semialgebraic set with diagram $D(A)$. There exist functions $C_i(D(A))$ such that if B_r is the ball of centre 0 and radius r, we have*

 (a) $B_0(A) \leqslant C_1(D(A))$.
 (b) *If k is the dimension of A and μ_k the k-dimensional Lebesgue measure on A,*
 $\mu_k(A \cap B_r) \leqslant C_2(D(A))r^k$.
 (c) *Any two points of a connected component of $A \cap B_r$ can be joined in A by a path of length $1 \leqslant C_3(D(A))r$.*

REMARK 1.6. For C_1, we gave explicit bounds above; for C_2 it is also possible to give explicit bounds (because (b) uses (a) and a "Crofton formula" of integral geometry). I do not know any explicit bound for C_3.

2. Additive Complexity

The additive complexity of a polynomial P is roughly speaking the minimal space needed to write P; the interesting fact is that in the real case, this complexity is related to the topology of $P^{-1}(0)$.

A polynomial $P \in R[X_1, \ldots, X_n]$ of high degree may need little space to be written, for instance if it has few terms; the following results will prove that semialgebraic sets defined by such polynomials P will have a "low complexity", for instance a few numbers of connected components, small Betti numbers, etc.

Let $P \in R[X_1, \ldots, X_n]$; the additive complexity $L^R_+(P)$ of P over R is by definition the minimum number of additions–subtractions required to evaluate P, beginning with the constants and the unknowns $X_1, \ldots, X_n$.

For instance, the polynomial $X^6 + 6X^5 + 9X^4 + 18X^3 + 9X^2 + 12X + 4$ can be written $((X+1)^3 + 1)^2$, so its additive complexity is $\leqslant 2$. In a more precise way we have:

DEFINITION 2.1. Let $P \in R[X_1, \ldots, X_n]$ and k a positive integer; then $L^R_+(P) \leqslant k$ if there exists a "program" where X^{α_i} (resp. X^{β_i}) represents a monomial like $X_1^{\alpha_{1i}} \ldots X_n^{\alpha_{ni}}$ (resp. $X_1^{\beta_{1i}} \ldots X_n^{\beta_{ni}}$):

$$\left\{ \begin{aligned}
S_1 &= c_1 X^{\alpha_1} + d_1 X^{\beta_1} \\
S_2 &= c_2 X^{\alpha_2} S_1^{a_{12}} + d_2 X^{\beta_2} S_1^{b_{12}} \\
&\vdots \\
S_k &= c_k X^{\alpha_k} \cdot \prod_{i=1}^{k-1} S_i^{a_{ik}} + d_k X^{\beta_k} \prod_{i=1}^{k-1} S_i^{b_{ik}} \\
S_{k+1} &= c_{k+1} X^{\alpha_{k+1}} \cdot \prod_{i=1}^{k} S_i^{a_{i,k+1}}
\end{aligned} \right.$$

with $c_i, d_i \in R$, $a_{ij}, b_{ij} \in Z$, the polynomial P being evaluated from S_{k+1} by successive elimination of the S_i $(1 \leqslant i \leqslant k)$.

We have then:

THEOREM 2.2 (Risler, 1985a). *There exists a constant $C > 0$ such that if $P \in R[X]$ satisfies $L_+^R(P) \leqslant k$, then P has at most C^{k^2} real roots.*

REMARKS.
  (a) C is a universal constant which can be shown to be $\leqslant 5$.
  (b) A similar result has been obtained by Gregoriev (1982).
  (c) One may conjecture that the bound C^{k^2} can be replaced by 3^k, which is the best known bound, reached by "Chebyshev polynomials".
  (d) The first result in this direction appeared in Borodin & Cook (1970).
  (e) For general results about complexity, cf. Strassen (1984).

PROPOSITION 2.3 (Risler, 1985a). *There exists a function $\varphi(n, k)$ such that for every polynomial $P \in R[X_1, \ldots, X_n]$ of additive complexity $\leqslant k$, $P^{-1}(0)$ has at most $\varphi(n, k)$ connected components.*

COROLLARY 2.4 (a *"real analogue of Bezout's theorem"*). *There exists a function $\psi(n, k)$ such that if $P_1, \ldots, P_s \in R[X_1, \ldots, X_n]$ are of additive complexity $\leqslant k$, the set of common solutions of $P_1 = \ldots = P_S = 0$ has at most $\psi(n, k)$ connected components. (To prove this corollary, just apply 2.3 to ΣP_i^2.)*

Those results are the consequence of a result of Hovansky (1980) which bounds the number of non-degenerate solutions of a system of n polynomial equations in n unknowns only in terms of the number of monomials involved; the proof of 2.3 is similar to the one of Milnor for 1.2, and uses the following lemma:

LEMMA 2.5 (Risler, 1985a). *Let $P(X_1, \ldots, X_n) \in R[X_i, \ldots, X_n]$ such that $L_+^R(P) \leqslant k$. Then*

$$L_+^R(\partial P/\partial X_i) \leqslant k(k+1) \quad (1 \leqslant i \leqslant n).$$

There are also results exactly similar to 1.6 (Risler, 1985b), replacing the diagram $D(A)$ by a similar notion $D'(A)$, where the degree is replaced by the additive complexity.

In particular, there exists a function $C(D'(A))$ such that if A is a smooth compact algebraic set of "complexity" $D'(A)$ contained in the ball B_r, for every connected component A_1 of A, and for any points x and y in A_1, there exists a path in A_1 from x to y of length $\leqslant C(D'(A))r$.

The proof is quite different from the one for the similar result of section 1, because it is not known if the additive complexity of a projection $\pi(A)$ can be bounded in terms of the additive complexity of A (cf. section 4).

This proof appears in Risler (1985b).

3. Applications to the Complexity of Algorithms

For general results about complexity of algorithms cf. Strassen, 1984. I will give here an outline of a paper by Ben-Or (1983), which uses results of section 1.

Let $W \subset R^n$ be any set. Consider the problem $P(W)$: given $(x_1, \ldots, x_n) \in R^n$, determine if $x = (x_1, \ldots, x_n) \in W$. An example of such a problem is the problem of distinctness: given $(x_1, \ldots, x_n) \in R^n$, is there a pair i, j, with $i \neq j$, and $x_i = x_j$?

$$\left(\text{here } W = \left\{ (x_1, \ldots, x_n) \, \middle| \, \prod_{i \neq j} (x_i - x_j) \neq 0 \right\} \subset \mathbb{R}^n \right).$$

We will obtain lower bounds on algorithms for solving $P(W)$ that allow both arithmetic operations and tests.

Formally an algebraic computation tree T is a binary tree with a function that assigns

— to any vertex v with exactly one son an operational instruction of the form $f_v := f_{v_1} \circ f_{v_2}$ or $f_v := c \circ f_{v_1}$ or $f_v := \sqrt{f_{v_1}}$, where v_i is an ancestor of v in T, or $f_{v_i} \in \{x_1, \ldots, x_n\}$, $\circ \in \{+, -, \times, /\}$, and $c \in R$;

— to any vertex v with two sons ("branching vertex") a test instruction of the form $f_{v'} > 0$ or $f_{v'} \geq 0$ or $f_{v'} = 0$, where v' is an ancestor of v, or $f_{v'} \in \{x_1, \ldots, x_n\}$;

— to any leaf an output YES or NO.

Given an input $x \in R^n$, the program traverses a path $P(x)$ in the tree T down from the root. At each simple vertex the arithmetical operation is performed and at each branching vertex a branch is made according to the test at the vertex. When a leaf is reached the answer YES or NO is returned. We say that "x passes through a vertex v" if v is on the path $P(x)$. We require that if an input x passes through a vertex v with a division instruction $f_v := f_{v_1}/f_{v_2}$ that $f_{v_2} \neq 0$, and if $f_v := \sqrt{f_{v_1}}$ that $f_{v_1} \geq 0$.

We say that the computation tree T solves the problem $P(W)$ if the answer returned is correct for every input $x \in R^n$. Let cost (x, T) denote the number of vertices that x passes through. The complexity of T, $C(T)$, is given by the maximum of cost (x, T) for any x. We have now:

THEOREM 3.1 (Ben-Or, 1983). *Let $W \subseteq R^n$ be any set, and T a computation tree which solves $P(W)$. If N is the number of connected components of W and $h = C(T)$, then $2^h 3^{n+h} \geq N$.*

COROLLARY 3.2. *If a tree T solves the problem of distinctness (cf. above), then $C(T)$ is at least $O(n \log n)$.*

For other examples and applications, cf. Ben-Or (1983).

PROOF OF 3.1 (Ben-Or, 1983). Let $\gamma = (v_1, \ldots, v_t)$, $t \leq h$, be a path from the root $r = v_1$ of T to a leaf $1 = v_t$ with the answer YES, and let V be the set of inputs $x \in R^n$ leading to 1. Traversing the tree down from the root v_1, we get a system of equations Γ according to the operations (or tests) on the vertices of the path γ, by the following rules:

Operation	Equation
$f_{v_i} := f_{v_j} \pm f_{v_k}$	$f_{v_i} = f_{v_j} \pm f_{v_k}$, etc.

and if v_i is a branching vertex with a test

$$f_{v_i} > 0 \quad \text{or} \quad f_{v_j} \geq 0 \quad \text{or} \quad f_{v_j} = 0,$$

then add this equation to Γ if it should be satisfied, and add the negated equation otherwise.

Let $f_{u_1}, \ldots, f_{u_r}$ be the set of new variables in Γ, and let s be the number of inequalities in Γ. Then $r + s \leqslant t$, since each step adds at most one new variable or one inequality. Let U be the set of solutions $(x_1, \ldots, x_n, f_{u_1}, \ldots, f_{u_r}) \in R^{n+r}$ to the system Γ. It is easily seen that the projection of U on the x coordinate is exactly V so we have $B_0(V) \leqslant B_0(U)$. Since the degree of Γ is $\leqslant 2$, we know by Corollary 1.3(b) that

$$B_0(V) \leqslant B_0(U) \leqslant 2 \cdot 3^{n+r+s-1} \leqslant 3^{n+h},$$

$$\begin{array}{ccc} U & \subset & R^{n+r} \\ \downarrow & & \downarrow \pi \\ V \subset W & \subset & R^n \end{array}$$

Since each leaf of T is correctly labelled, we have $V \subset W$, and so each connected component of V must be completely contained in some connected component of W.

Since the number of leaves of $T \leqslant 2^h$, and since W is the union of the V's corresponding to leaves with answer YES, we have $2^h 3^{n+h} \geqslant N$.

4. Algorithms for Semialgebraic Sets

The problem is to describe more or less precisely the set of solutions of a family of real polynomial equations and inequations, and the methods are versions of Tarski's famous principle of elimination of quantifiers for real closed fields.

Let $A \subset R^n$ be a semialgebraic set described by a given set of equations and inequalities; specific problems are: is A non-empty?, what is the dimension of A?, if $\pi : R^n \to R^{n-k}$ is a linear projection; find a description of $\pi(A)$, etc. The last problem is equivalent to the problem of elimination of quantifiers which is namely, given a formula F of the first order in the theory of real closed fields, find an equivalent formula F' quantifier free.

This last problem is quite difficult and in any case of great complexity (cf. the papers by Weispfenning and Davenport-Heintz in this issue).

The only algorithm really implemented up to now is based on the C.A.D. decomposition algorithm of Collins (cf. Collins, 1982a,b for references), with ameliorations by Arnon $et\ al.$ (1984).

If n is the number of variables involved, k the number of polynomials, and M a bound of the size of the polynomials (the size is the log of the absolute value of the largest coefficient), the algorithm goes in time polynomial in $(Mkd)^{2^n}$, that is doubly exponential in the number of variables.

For recent results about improvement of these algorithms, see Böge (1985).

Recently a new algorithm has been described (Grigoriev, 1985; Grigoriev & Vorobjov, 1985) but with no implementation as far as I know. This algorithm runs in time polynomial in $(Mkd)^{n^{4a}}$ where a is the number of quantifiers involved (in fact the number of alternations of quantifiers).

It would be very interesting to produce faster algorithms, and in particular algorithms which do not depend on the degrees for their complexity, but more on the additive complexity. It is natural to consider two problems:

(a) Is the additive complexity of the resultant of two polynomials P and P' bounded in terms of the additive complexity of P and P'? (I do not believe in this.)

(b) Let $A \subset R^n$ be a semialgebraic set, $\pi : R^n \to R^{n-1}$ a linear projection; is the additive complexity $D'(\pi(A))$ of $\pi(A)$ bounded in terms of the additive complexity of A? (This is perhaps true, may be with a weaker notion of complexity.)

5. Some Results on Sets Defined by Transcendental Equations

The results of this section are essentially due to the Russian mathematician Hovansky (for a survey see Risler, 1984).

(A) ONE DIMENSIONAL CASE

Let $I \subset \mathbb{R}$ be an open interval, maybe infinite. We denote by $\mathscr{A}$ the ring of all analytic functions defined in some subset $D_f \subset I$ whose complement $P_f = I \backslash D_f$ is finite.

Let $f \in \mathscr{A}$. Each point of P_f is called a *pole of* f and their number denoted by $p(f)$ (note that a pole in this sense is not necessarily a pole in the classical sense). The domain D_f of f is a union of $p(f) + 1$ open intervals. Over each of them, f either vanishes or has a discrete set of zeros. We denote by Z_f this set of discrete zeros and write $z(f) = \#Z_f$. Finally, put $n(f) = z(f) + p(f)$.

If a function $f \in \mathscr{A}$ has only discrete zeros and $z(f) < +\infty$, then $f^{-1} = 1/f$ is a well-defined function of $\mathscr{A}$. We have then

$$D_{f^{-1}} \supset D_f \backslash Z_f, \qquad Z_{f^{-1}} \subset P_f.$$

Let $f \in \mathscr{A}$. Then $f' \in \mathscr{A}$ and $D_{f'} \supset D_f$. Besides:

LEMMA 5.1.

$$z(f) \leqslant p(f) + z(f') + 1.$$

For, if $z(f) > p(f)$, by Rolle's theorem, two consecutive zeros of f, without poles between, determine a zero of f', and the remaining zeros and poles cancel pairwise.

In particular, if $z(f') < +\infty$, then $z(f) < +\infty$ too.

The first examples of functions in $\mathscr{A}$ are rational functions. But they all have the stronger property $z(f) < +\infty$. Our aim here is to study this condition for arbitrary functions in $\mathscr{A}$. So, we shall say that a set $A \subset \mathscr{A}$ has the *finiteness property* if $z(f) < +\infty$ for all $f \in A$. With this terminology we have the following:

PROPOSITION 5.2 (Gelfand & Hovansky, 1980) *Let A be a subring of $\mathscr{A}$ closed under derivation, which has the finiteness property. Let $g \in \mathscr{A}$ verify any of the conditions (i), (ii) or (iii) below*

$$(i) \ g^{-1} \in A; \quad (ii) \ g' \in \mathscr{A}; \quad (iii) \ g'g^{-1} \in \mathscr{A}.$$

Then $A[g] \subset \mathscr{A}$ is closed under derivation and has the finiteness property.

From 5.2. it follows immediately:

COROLLARY 5.3. *Let A be a subring of $\mathscr{A}$ closed under derivation which has the finiteness property. Let $g_1, \ldots, g_r$ be functions in $\mathscr{A}$, each verifying some condition amid (i), (ii) and (iii) in 5.2. Then the subring $B = A[g_1, \ldots, g_r] \subset \mathscr{A}$ is closed under derivation and has the finiteness property.*

The previous results give some motivation to define:

DEFINITION 5.4. Let A be a subring of $\mathscr{A}$. A subring B of $\mathscr{A}$ is a *Liouville extension* of A if it is a union of finite A-algebras $A[g_1, \ldots, g_r]$, where each $A_l = A_{l-1}[g_l]$ ($A_0 = A$) is as in 5.2. (This notion is interesting only if A is closed under derivation.) Furthermore:

DEFINITION 5.5. Let A be a subring of $\mathscr{A}$. An $f \in \mathscr{A}$ is a *Liouville function over* A if it lies in some Liouville extension of A.

A *Liouville function* is a Liouville function over $A = \mathbb{R}$.

Examples of Liouville functions (defined on some open interval of R) are rational functions in X, e^X, $\log X$, $\arcsin X$, etc., and compositions of such functions.

For Liouville functions, there exists an algorithm which gives the "cells" of R defined by the signs of a given finite number of functions, this algorithm is a generalisation of the one for polynomials.

DEFINITION 5.6. $f_1, \ldots, f_p \in \mathscr{A}$ form a *separating family* if for any choice of symbols $?_i$ among $>$, $=$, $<$, the set

$$\bigcap_{i=1}^{p} \{f_i \,?_i\, 0\}$$

is connected and its closure is given by relaxing inequalities, provided it is not empty.

This last sentence means that if a set B is non-empty and is defined, say by a formula

$$f_1 = \ldots = f_k = 0, \quad f_{k+1} > 0, \ldots, f_p > 0,$$

then $\bar{B}$ is defined by

$$f_1 = \ldots = f_k = 0, \quad f_{k+1} \geqslant 0, \ldots, f_p \geqslant 0;$$

this is not always true; for instance, if B is defined by $(X^2 - 1)(X - 2)^2 < 0$, then the set defined by $(X^2 - 1)(X - 2)^2 \leqslant 0$ contains strictly $\bar{B}$.

PROPOSITION 5.7. There exists an algorithm which extends any finite set $(f_1, \ldots, f_q)$ of Liouville functions to a separating family $(f_1, \ldots, f_q, f_{q+1}, \ldots, f_p)$.

For a proof, see Risler (1983).

(B) GENERAL CASE

One possible generalisation of Liouville functions in the case of n variables is termed, by Hovansky (1984), "Pfaffian functions".

For such functions, the analogue of the Tarski–Seidenberg principle is not yet proved; research in this direction is made by Lou van den Dries (1985).

Nevertheless, for such "Pfaffian functions" there are finiteness properties: it is possible to define a complexity for such a function (generalising the degree or the additive complexity of polynomials) and to give a result similar to Bezout's theorem.

I will just mention two results with explicit bounds:

THEOREM. *Let* $F_1, \ldots, F_n \in \mathbb{R}[X_1, \ldots, X_n, y_1, \ldots, y_k]$ *where* $X_1, \ldots, X_n$ *are indeterminates and* $y_i = e^{\langle a^i, X \rangle}$, $i = 1, \ldots, k$ *with* $\langle a^i, X \rangle = a_1^i X_1 + \ldots + a_n^i X_n$, $a_j^i \in \mathbb{R}$, *all* i, j. *Then the number of non-degenerate solutions of the system*

$$(1) \quad \begin{cases} F_1(X, y(X)) = 0 \\ \quad \vdots \\ F_n(X, y(X)) = 0 \end{cases}$$

is finite and less than or equal to $\left(\prod_{i=1}^{n} m_i \right)(1 + \Sigma m_i)^k \cdot 2^{k(k-1)/2}$, *where* m_i *is the degree of* F_i.

REMARK. A point $x = (x_1, \ldots, x_n) \in \mathbb{R}^n$ is a non-degenerate solution of (1) if the Jacobian of $F_1(X; y(X)), \ldots, F_n(X, y(X))$ is not zero at x.

PROPOSITION 5.10. *Let $F_1, \ldots, F_n$ be n functions of n real unknowns $X_1, \ldots, X_n$, where*

$$F_i \in R[X_1, \ldots, X_n, Y_1, \ldots, Y_k, U_1, \ldots, U_1, V_1, \ldots, V_1]$$

with $Y_i = e^{\langle a^i, X \rangle}$ like above $(1 \leqslant i \leqslant k)$,

$$U_i = \sin \langle b_i, X \rangle, \quad V_i = \cos \langle b_i, X \rangle \quad (1 \leqslant i \leqslant 1, \, b_i \in R^n).$$

Then the number of non-degenerate solutions of the system

$$\begin{cases} F_1 = 0 \\ \quad \vdots \\ F_n = 0 \end{cases}$$

in the open set of R^n defined by $\langle b_i, X \rangle < \pi/2$ $(1 \leqslant i \leqslant 1)$ is bounded by

$$\left(\prod_{i=1}^{n} m_i \right) (\Sigma(m_i + l + 1)^{l+k} \cdot 2^{l + ((l+k)(l+k+1)/2)}.$$

For proofs, see Hovansky (1980, 1984) and Risler (1983).

References

Arnon, D., Collins, G. E., MacCallum, S. (1984). Cylindrical algebraic decomposition, 1 and 2. *SIAM J. Comp.* **13**, 865–877 and 878–888.

Barsky, Brian A. (1981). Computer aided geometric design: a bibliography. *IEEE Comp. Graph. Appl.*, **1**, (3), 87–109.

Ben-Or, M. (1983). Lower bounds for algebraic computation trees. *Proceedings of the 1983 ACM Symposium on Symbolic and Algebraic Computation.* Pp. 80–86.

Böge, W. (1985). Quantifier elimination for real closed fields. *Proceedings of AAECC-3, Grenoble, 1985.* Berlin: Springer-Verlag.

Borodin, A., Cook, S. (1970). On the number of additions to compute specific polynomials. *SIAM J. Comp.* **5**, 146–157.

Collins, G. E. (1982*a*). Quantifier elimination for Real closed fields by cylindrical algebraic decomposition. *Springer Lec. Notes Comp. Sci.* **33**, 79–81.

Collins, G. E. (1982*b*). Quantifier elimination for real closed fields: a guide to the literature. *Comp. Supplementum* **4**, 79–81.

Gelfand, O. A., Hovansky, A. G. (1980). Real Liouville functions. *Funct. Analysis* **14**, (2), 52–53.

Grigoriev, D. Yu. (1982). Lower bounds in algebraic complexity. *Notes Sci. Sem. of Leningrad Steklov Institute,* **118**, 25–82. (Engl. Translation in *J. Soviet Math.* 1985, pp. 1388–1425).

Grigoriev, D. Yu. (1985). Complexity of deciding Tarsky algebra. Preprint of Leningrad Steklov Institute; also, see this issue.

Grigoriev, D. Yu., Vorobjov, N. N. (1985). Solving systems of polynomial inequalities in subexponential time. Preprint, Leningrad Steklov Institute; also, see this issue.

Hovansky, A. G. (1980). On a class of systems of transcendental equations. *Soviet Math. Dokl.* **22**, (3), 762–765.

Hovansky, A. G. (1984). Real analytic varieties with the finiteness property and complex abelian integrals. *Funct. Analysis* **18**, (2), 119–127.

Milnor, J. (1964). On the Betti numbers of real algebraic varieties. *Proc. AMS* **15**, 275–280.

Risler, J. J. (1983). On Bezout's theorem in the real case. Publ. del inst. "Jorge Juan" de matematicas, Madrid.

Risler, J. J. (1984). Complexité et géométrie réelle (d'après Hovansky). *Séminaire Bourbaki* **637**, Nov. 1984.

Risler, J. J. (1985*a*). Additive complexity and zeros of real polynomials. *SIAM J. Comp.* **14**, (1), 178–183.

Risler, J. J. (1985*b*). Propriétiés métriques des ensembles semi-algébriques et complexité additive. Preprint, Université Paris 7, France. To appear in: publications de l'université Paris 7, no. 23.

Sharir, M., Schwartz, J. (1983). On the piano mover's problem 2: General techniques for computing topological properties of real algebraic manifolds. *Adv. Appl. Math.* **4**, 298–351.

Strassen, V. (1984). *Algebraic complexity* (Algebraische Berechnungkomplexität). Birkhäuser.

Thom, R. (1965). Sur l'homologie des variétés algébriques réelles. In: (Cairn, S. S. ed.) *Diff. and Comb. Topology.* Pp. 255–265. Princeton University Press.

Van Den Dries, L. (1985). Tarski's problem and Pfaffian functions. Preprint, Stanford University.

Yomdin, Y. (1985). Metric properties of semi-algebraic sets and mappings and their applications in smooth analysis. *Proceedings of Larabida Conference, Spain, 1985.*

(*) *Note added in proof* (1987): in a forthcoming book by Benedetti-Risler-Tognoli, an amelioration of Theorem 1.2 is given in case of one equation, namely $B_0(A) \leqslant d(d+2)^{n-1}$.

J. Symbolic Computation (1988) **5**, 121–129

Thom's Lemma, the Coding of Real Algebraic Numbers and the Computation of the Topology of Semi-algebraic Sets

M. COSTE AND M. F. ROY

IRMAR, Université de Rennes I, Rennes, France

(*Received 7 July* 1986)

Thom's lemma, a very simple and basic result in real algebraic geometry, and explained in section 1, has a lot of interesting computational consequences.

We shall outline two of these.

The first one is the fact that a real root ξ of a polynomial P of degree n with real coefficients may be distinguished from the other real roots of P by the signs of the derivatives $P^{(i)}$ of P at ξ, $i = 1, \ldots, n-1$. This offers a new possibility for the coding of real algebraic numbers and for computation with these numbers (see section 2).

The second is based on a generalisation of Thom's lemma to the case of several variables. It gives, after a linear change of coordinates, a cylindric algebraic decomposition of a semi-algebraic set where the incidence relation between the cells is easily obtained (see section 3).

1. Thom's Lemma

DEFINITIONS AND NOTATIONS 1.1: A *sign condition* is any one of the symbols >0, <0 or $=0$. A *generalised sign condition* is any one of the symbols >0, <0, $=0$, $\geqslant 0$, $\leqslant 0$.

If $\varepsilon = (\varepsilon(i))_{i=0,\ldots,n-1}$ is an n-uple of generalised sign conditions, one denotes by $\underline{\varepsilon}$ the n-uple obtained by relaxing the strict inequalities of ε, that is by replacing >0 (resp. <0) by $\geqslant 0$ (resp. $\leqslant 0$).

THOM'S LEMMA

PROPOSITION 1.2. *Let P be a polynomial of degree n with real coefficients, $P', \ldots, P^{(i)}$ its derivatives and $\varepsilon = (\varepsilon(i))_{i=0,\ldots,n-1}$ a n-uple of generalised sign conditions. Let*

$$A(\varepsilon) = \{x \in \mathbb{R}/P^{(i)}(x)\varepsilon(i), i = 1, \ldots, n-1\}.$$

Then (a) $A(\varepsilon)$ is either empty or connected, (b) if $A(\varepsilon)$ is non empty the closure of $A(\varepsilon)$ is $A(\underline{\varepsilon})$.

PROOF. Easy, by induction on the degree of P.

REMARK 1.3. (a) In the usual formulation of Thom's lemma (Coste, 1982; Bochnak *et al.*, 1987) ε consists of sign conditions. The form given here is useful for the topological study of a real algebraic curve.

(b) One does not obtain in general the closure of a set defined by strict inequalities by relaxing these inequalities, as is shown by the example of $\{x \in \mathbb{R}|x^3 - x^2 > 0\}$.

0747–7171/88/010121+09 \$03.00/0　　　　　　　　

2. Coding Real Algebraic Numbers

Algorithms proposed up to now to work on real algebraic numbers are semi-numerical: one characterises a real algebraic number ξ by means of a square-free defining polynomial P with integer coefficients and an interval that isolates ξ from all other roots of P. The approach we propose here is purely formal and relies on Thom's lemma.

2.a. CHARACTERISATION OF A REAL ALGEBRAIC NUMBER

PROPOSITION 2.1. *Let P be a polynomial of degree n with integer coefficients. Let ξ and ξ' be two real roots of P. Suppose the signs $\varepsilon(i)$ and $\varepsilon'(i)$ of $P^{(i)}(\xi)$ and $P^{(i)}(\xi')$, $i = 1, \ldots, n-1$, are given. Then:*

(i)   *If $\varepsilon(i) = \varepsilon'(i)$ for all $i = 1, \ldots, n-1$ then $\xi = \xi'$.*
(ii) *In the other case one can decide whether $\xi < \xi'$ or $\xi > \xi'$: let k be the smallest integer such that $\varepsilon(n-k)$ and $\varepsilon'(n-k)$ are different;*
(ii.a) *$\varepsilon(n-k+1) = \varepsilon'(n-k+1)$ is different from $= 0$,*
(ii.b) *if $\varepsilon(n-k+1) = \varepsilon'(n-k+1)$ is >0 one has $\xi > \xi'$ if and only if $P^{(n-k)}(\xi)$ is bigger than $P^{(n-k)}(\xi')$ [which is known by looking at $\varepsilon(n-k)$ and $\varepsilon'(n-k)$],*
(ii.c) *if $\varepsilon(n-k+1) = \varepsilon'(n-k+1)$ is <0 one has $\xi < \xi'$ if and only if $P^{(n-k)}(\xi)$ is bigger than $P^{(n-k)}(\xi')$ [which is known by looking at $\varepsilon(n-k)$ and $\varepsilon'(n-k)$].*

PROOF. (i) is a consequence of Thom's lemma. (ii) is also a consequence of Thom's lemma: (ii.a) because of point (i) in Thom's lemma applied to $P^{(n-k+1)}$, (ii.b) and (ii.c) because the set $\{x \in \mathbb{R} / P^{(i)}(x)\varepsilon(i), \ i = n-k+1, \ldots, n-1\}$ is connected by Thom's lemma applied to $P^{(n-k+1)}$, and that on a connected set, the sign of the derivative of a polynomial gives its behaviour (increasing or decreasing).

2.b. ALGORITHMS FOR COMPUTING ON REAL ALGEBRAIC NUMBERS

In section 2b let P be a polynomial of degree n with integer coefficients having only simple roots and d real roots. It is clear that Proposition 2.1 gives a way of coding a real algebraic number root of P (by the sign it gives to the derivatives of P) and that, given such a code, the sign of any polynomial Q with integer coefficients at a given real algebraic number is fixed. We propose in the following an algorithm for computing the codes of the roots of P and more generally algorithms for computing sign conditions given by roots of P to other polynomials.

All algorithms are based on a generalisation of Sturm theorem due to Sylvester (1853) (algorithm b_1), revisited by Ben-Or, Kozen & Reif (1986) (algorithm b_3).

Let us introduce some notation. Let $R, Q_1, \ldots, Q_k$ be polynomials with integer coefficients and let $\varepsilon = (\varepsilon(1), \ldots, \varepsilon(k))$ be a sequence of sign conditions ($>0, <0, =0$). One denotes by $c_\varepsilon(R; Q_1, \ldots, Q_k)$ the number of real roots of R giving to $Q_1, \ldots, Q_k$ the signs $\varepsilon(1), \ldots, \varepsilon(k)$.

Let P_0 be R, P_1 be Q, P_{i+1} be the opposite of the remainder of the division of P_{i-1} by P_i. One calls the sequence of polynomials so obtained the *generalised Sturm sequence associated with R and Q* and one denotes by $v_{R,Q}(-\infty)$ and $v_{R,Q}(+\infty)$ the number of sign changes in the sequences of the signs of the P_i at $-\infty$ and $+\infty$. The *Sturm sequence of P* is the generalised Sturm sequence associated to P and P'.

PROPOSITION 2.2. *With the above definitions and notations, if R and Q are coprime, one has*
$$c_{>0}(R; Q) - c_{<0}(R; Q) = v_{R, P'Q}(-\infty) - v_{R, P'Q}(+\infty)).$$

PROOF. Similar to the proof of Sturm theorem (see [S]).

2.b.1. *Generalised Sturm algorithm for an inequality (algorithm b_1)*

Let Q be a polynomial with integer coefficients, prime to P. The aim of the algorithm is to compute $c_{>0}(P; Q)$ and $c_{<0}(P; Q)$. Let $v(1) = v_{P, P'}(-\infty) - v_{P, P'}(+\infty)$ and $v(2) = v_{P, P'Q}(-\infty) - v_{P, P'Q}(+\infty)$. Using Proposition 2.2 we have

$$\begin{bmatrix} 1 & 1 \\ 1 & -1 \end{bmatrix} \begin{bmatrix} c_{>0}(P; Q) \\ c_{<0}(P; Q) \end{bmatrix} = \begin{bmatrix} v(1) \\ v(2) \end{bmatrix}$$

We can easily deduce $c_{>0}(P; Q)$ and $c_{<0}(P; Q)$ from $v(1)$ and $v(2)$.

2.b.2. *Algorithm for computing the number of real roots of P giving a fixed sign condition ($>0, <0, =0$) to a polynomial Q (algorithm b_2)*

Let Q be a polynomial with integer coefficients. The aim of the algorithm is to compute $c_{>0}(P; Q)$, $c_{<0}(P; Q)$ and $c_{=0}(P; Q)$.

One computes the GCD R of P and Q. If R is of degree 0 one applies b_1. If R is of strictly positive degree, one defines $S = P/R$. The number $c_{=0}(P; Q)$ is equal to the number of real roots of R (computed by Sturm), $c_{>0}(P; Q)$ is equal to $c_{>0}(S; Q)$ and $c_{<0}(P; Q)$ is equal to $c_{<0}(S; Q)$. One computes then $c_{>0}(S; Q)$ and $c_{<0}(S; Q)$ b_1.

2.b.3. *Generalised Sturm algorithm for several inequalities (algorithm b_3)*

Let $Q_1, \ldots, Q_k$ be polynomials in one variable with integer coefficients. The *sign conditions realised by $Q_1, \ldots, Q_k$ at the real roots of P* are the k-uples of signs conditions ε such that $\{x \in \mathbb{R} | P(x) = 0$ and $Q_j(x) \in (j), j = 1, \ldots, k\}$ is *non empty*.

Let us suppose now that the Q_j are prime to P. The aim of the algorithm is to determine the number of real roots of P giving to $Q_1, \ldots, Q_k$ fixed signs. The idea is to compute the generalised Sturm sequences associated to the products of the subsets of $\{Q_1, \ldots, Q_k\}$ and use the sign variations of these generalised Sturm sequences and some linear algebra (generalising b_1) to deduce the values of the $c_\varepsilon(P; Q_1, \ldots, Q_k)$. If done without care, this leads to an exponential algorithm, since we have 2^k subsets of $\{Q_1, \ldots, Q_k\}$, hence 2^k generalised Sturm sequences to compute, to get the values c_ε for the 2^k-sign conditions ε. To avoid this exponential growth the idea (Ben-Or *et al.*, 1986) is to remark that the number $d(k)$ of distinct sign conditions realised by $Q_1, \ldots, Q_k$ at the real roots of P is, for any k, smaller than the number d of real roots of P.

More precisely one wants to determine the sign conditions $\varepsilon_{k,1}, \ldots, \varepsilon_{k, d(k)}$, with $d(k) \leqslant d$, realised by $Q_1, \ldots, Q_k$ at the real roots of P and the number $c(k, j) = c_{\varepsilon_{k, j}}(P; Q_1, \ldots, Q_k)$ of elements of the non empty set

$$\{x \in \mathbb{R} | P(x) = 0 \text{ and } Q_j(x) \varepsilon_{k, j}(i), i = 1, \ldots, k\}.$$

We are going to compute by induction on k

(1_k) the list $\varepsilon_k = (\varepsilon_{k, 1}, \ldots, \varepsilon_{k, d(k)})$, $d(k) \leqslant d$, of the k-uples of sign conditions realised by $Q_1, \ldots, Q_k$ at the real roots of P,

(2_k) the numbers $c(k, 1), \ldots, c(k, d(k))$

(3_k) a list of $S_k = (S_{k,1}, \ldots, S_{k,d(k)})$ of $d(k)$ subsets of $\{1, \ldots, k\}$,

(4_k) for each $j = 1, \ldots, k$ the number $v(k,j) = v_{P,Q_{k,j}}(-\infty) - v_{P,Q_{k,j}}(+\infty)$, with $Q_{k,j}$ the product of the Q_l for $l \in S_{k,j}$, and the list of the $Q_{k,j}$,

(5_k) an invertible matrix $\mathbf{A}_k$ of dimension $d(k)$ such that $\mathbf{A}_k \cdot c = v$, where c is the vector $(c(k,1), \ldots, c(k,d(k)))$ and v the vector $(v(k,1), \ldots, v(k,d(k)))$.

The case $k = 0$ is given by $d(0) = 0$.

Suppose that we are given $(1_k), \ldots, (5_k)$ and consider the case $k+1$.

We compute $d(k)$ generalised Sturm sequences associated to the $Q_{k,d(k)+j} = Q_{k+1}Q_{k,j}$, $j = 1, \ldots, d(k)$ and $v(k,l) = v_{PP',Q_{k,1}}(-\infty) - v_{PP',Q_{k,1}}(+\infty)$, $l = d(k)+1, \ldots, 2d(k)$. One considers the $2d(k)$ dimension matrix

$$\mathbf{A}' = \begin{bmatrix} A_k & A_k \\ A_k & -A_k \end{bmatrix},$$

and the equality $\mathbf{A}' \cdot c' = v'$, obtained from ($5_k$) and 2.2 where c' is the vector

$$(c_{\varepsilon_{k,j}, >0}(P; Q_1, \ldots, Q_{k+1}), \ j = 1, \ldots, d(k), \quad c_{\varepsilon_{k,j}, <0}(P; Q_1, \ldots, Q_{k+1}), \ j = 1, \ldots, d(k))$$

and V' the vector $(v(k,1), \ldots, v(k,2d(k)))$.

The matrix $\mathbf{A}'$ is invertible. One computes c' by inverting $\mathbf{A}'$.

Let us define $d(k+1)$ as the number ($\leqslant d$) of non zero elements in c' and let

$$m_1, \ldots, m_{d(k+1)}(m_1 < \ldots < m_{d(k+1)} \leqslant 2d(k))$$

be the numbers j such that the jth coordinate of c' is different from 0. The sign conditions $\varepsilon_{k+1,1}, \ldots, \varepsilon_{k+1,d(k+1)}$ realised by $Q_1, \ldots, Q_{k+1}$ at the real roots of P are the m_jth elements, $j = 1, \ldots, d(k+1)$, of the list

$$((\varepsilon_{k,1}, >0), \ldots, (\varepsilon_{k,d(k)}, >0), (\varepsilon_{k,1}, <0), \ldots, (\varepsilon_{k,d(k)}, <0)).$$

One can then extract from $\mathbf{A}'$ the columns of number $m_1, \ldots, m_{d(k+1)}$, which gives a $2d(k), d(k+1)$ matrix $\mathbf{A}''$. An invertible matrix $\mathbf{A}_{k+1}$, of dimension $d(k+1)$ can be extracted from $\mathbf{A}''$. Let us denote $l_1, \ldots, l_{d(k+1)}$, $l_1 < \ldots < l_{d(k+1)}$, the elements of $\{1, \ldots, 2d(k+1)\}$ such that the l_ith line of $\mathbf{A}''$ appears in $\mathbf{A}_{k+1}$.

The list $(S_{k+1,1}, \ldots, S_{k+1,d(k+1)})$ is obtained taking in the list

$$(S_{k,1}, \ldots, S_{k,d(k)}, S_{k,1} \cup \{k+1\}, \ldots, S_{k,d(k)} \cup \{k+1\})$$

the elements of number $l_1, \ldots, l_{d(k+1)}$.

It is easy to verify $(1_{k+1}), \ldots, (5_{k+1})$.

2.b.4. *Algorithm for determining the number of real roots of P giving fixed signs to polynomials $Q_1, \ldots, Q_k$ (algorithm b_4)*

Let $Q_1, \ldots, Q_k$ be polynomials with integer coefficients. The aim of the algorithm is to determine the number of real roots of P giving to $Q_1, \ldots, Q_k$ fixed signs. More precisely, we want to determine

($1'_k$) k-uples of signs $\varepsilon_{k,1}, \ldots, \varepsilon_{k,m}$ (with $m \leqslant d$) such that the $c_{\varepsilon_{k,j}}(P; Q_1, \ldots, Q_k)$ are different from zero;

($2'_k$) for each $\varepsilon_{k,j}$ (if one denotes $Q_{j,1}, \ldots, Q_{j,k'}$ the k' polynomials among $(Q_i)_{i=1,\ldots,k}$ such that $\varepsilon_{k,j}(i)$ is not zero and $\varepsilon'_{k,j}$ the k'-uple of strict signs on the $Q_{j,l}$ taken from $\varepsilon_{k,j}$), a polynomial S_j dividing P, prime with all the $Q_{j,l}$s and such that $c_{\varepsilon_{k,j}}(S_j; Q_{1,j}, \ldots, Q_{k,j})$ is equal to $c_{\varepsilon_{k,j}}(P; Q_1, \ldots, Q_k)$.

The case $k = 1$ *is* b_2.

Let us suppose the problem solved for $P, Q_1, \ldots, Q_k$ and let us add the polynomial Q_{k+1}. One considers for each k-uple of signs $(\varepsilon_{k,j})_{j=1,\ldots,m}$ given by (1_k) for $P, Q_1, \ldots, Q_k$ the GCD R of Q_{k+1} and S_j. One applies b_3 to $S_j/R, Q_{j,1}, \ldots, Q_{j,k'}, Q_{k+1}$, which gives the result for strict signs ($>0, <0$) on Q_{k+1}. For the case where the sign of Q_{k+1} is 0, one applies again b_3 to $R, Q_{j,1}, \ldots, Q_{j,k'}$.

2.b.5. *Algorithm for characterising the real roots of a polynomial (algorithm b_5)*

One wants to compute the signs given to the derivatives of P by the real roots of P. One applies b_4 to $P, P', \ldots, P^{(n-1)}$.

As seen in Proposition 2.1 a real root ξ of P can be coded by the sequence of signs it gives to the derivatives of P.

2.b.6. *Algorithm for deciding the sign of a polynomial with integer coefficients at a real algebraic number*

One supposes the real algebraic number ξ, root of P, coded as in b_5 and one wants to compute the sign of Q in ξ.

One applies b_4 to $P, P', \ldots, P^{(n-1)}, Q$.

2.c. COMPLEXITY OF THE ALGORITHMS

The precise complexity study of the algorithm is not included in the present paper and is in preparation (Roy and Szpirglas, in prep.).

Let us indicate quickly why algorithm b_3 is a polynomial-time algorithm (then clearly $b_4, \ldots, b_7$ will also be polynomial-time algorithms). If R is a polynomial in one variable with integer coefficients, let $C(R)$ be the maximum of the degree of R and of the length of the coefficients of R, written in base 2. Let $P, Q_1, \ldots, Q_k$, be as in b_3, let $N = \sup(k, C(P), C(Q_i), i = 1, \ldots, k)$, let (step i) be the passage from $Q_1, \ldots, Q_i$ to $Q_1, \ldots, Q_{i+1}$ in b_3. The number of generalised Sturm sequences to compute in (step i) is bounded by d, where d is the number of real roots of P, hence by N. All the computations in (step i) are polynomial in N if one uses the theory of subresultants (for example, see Loos, 1982) for the computation of generalised Sturm sequences.

2.d. GENERALISATION TO SEVERAL VARIABLES

It is not always the case that a real algebraic number is given as a root of a polynomial with integer coefficients. In many natural geometric situations it happens that it is given as a root of a polynomial having itself real algebraic coefficients. Of course, it is in principle possible, using the theorem of the primitive element, to compute a polynomial with integer coefficients having the real algebraic number as a root. It we want to avoid these primitive element computations, we can proceed as follows.

The coding of a real algebraic number given as a root of a polynomial P with real algebraic coefficients by the signs of the derivatives of P will be a particular case of the following general decision problem (*): let us consider

$$P_1 \in \mathbb{Q}[T_1], P_2 \in \mathbb{Q}[T_1, T_2], \ldots, P_n \in \mathbb{Q}[T_1, \ldots, T_n], Q_1, \ldots, Q_k \in \mathbb{Q}[T_1, \ldots, T_n];$$

compute the sign conditions $\varepsilon = (\varepsilon(1), \ldots, \varepsilon(k))$ such that there exists a real root ξ_1 of P_1, there exists a real root ξ_2 of $P_2[\xi_1, T_2], \ldots,$ there exists a real root ξ_n of $P[\xi_1, \xi_2, \ldots, \xi_{n-1}, T_n]$ such that $Q_i(\xi_1, \xi_2, \ldots, \xi_{n-1}, \xi_n)\varepsilon(i)$, and for each such ε the number of roots ξ_1 of P_1 such that there exists a fixed number m_2 of roots of $P(\xi_1, T_2) \ldots$ such that there exists a fixed number of roots m_n of $P[\xi_1, \xi_2, \ldots, \xi_{n-1}, T_n]$ such that $Q_i(\xi_1, \xi_2, \ldots, \xi_{n-1}, \xi_n)\varepsilon(i)$ for all $i = 1, \ldots, k$.

The case $n = 1$ corresponds obviously to b_4. The case n is treated recursively from case $n-1$, by considering the polynomials $P_n, Q_1, \ldots, Q_k$ as polynomials in the variable T_n. Computing the generalised Sturm sequences corresponding to the situation, we get, by considering signs conditions on leading terms on generalised Sturm sequences, a new problem (*) in the variables $T_1, \ldots, T_{n-1}$.

2.e REMARK

The algorithms proposed can be used in any computable real closed fields, for example in the field of real Puiseux series which are algebraic over $\mathbb{Q}(X)$, which appears in natural geometric situations. Since these fields are not archimedean, the technique of dichotomy for the isolation of real roots can no more be applied here.

2.f. APPLICATION TO THE COMPUTATION OF THE TOPOLOGY OF A CURVE

This subject will be developed in another paper (Roy, 1987). Let us simply say that the computation of the topology of a curve given by the equation $Q(x, y) = 0$ (number of connected components, isotopy type of the embedding of the curve in the real plane, number of multiple points and of branches passing through them, number of real isolated points) can be made if one knows the answer to the following questions

(1) determine above the real roots ξ of the discriminant D of $Q(X, Y)$ with respect to Y the number of real roots of $Q(\xi, Y)$,

(2) determine just to the left and to the right of the real roots of the discriminant D the number of real half-branches and how they glue over ξ.

Using generalised Sturm theorem and Thom's lemma one can show that answering these questions is equivalent to characterising the real roots of D (in the sense of b_5) and, using 2.d, evaluating the signs of other polynomials at these real roots (by b_6, which is a special case of b_3).

3. Computing the Topology of Semi-algebraic Sets

3.a. CYLINDRIC ALGEBRAIC DECOMPOSITION

We recall that a semi-algebraic subset of $\mathbb{R}^n$ is a subset defined by a boolean combination of equations and inequalities involving a finite number of polynomials in $\mathbb{R}[X_1, \ldots, X_n]$. Any systematic study of semi-algebraic sets is based on the following result (Theorem 3.2). We introduce first some terminology from Schwartz & Sharir (1983).

DEFINITION 3.1. Let $T \subset \mathbb{R}[X_1, \ldots, X_n]$ be a family of polynomials and $A \subset \mathbb{R}^k$ any subset. We say that A is *T-invariant* when any polynomial $f \in T$ is either strictly positive, or strictly negative, or identically zero on the whole of A.

THEOREM 3.2. *For any finite family $S \subset \mathbb{R}[X_1, \ldots, X_n]$ one can produce a finite family $T \subset \mathbb{R}[X_1, \ldots, X_{n-1}]$ such that for any connected subset $A \subset \mathbb{R}^{n-1}$ which is T-invariant, there are an integer $q \in \mathbb{N}$ and continuous functions*

$$\xi_1 < \ldots < \xi_q : A \to \mathbb{R}$$

which give, for any $a \in A$, all the real roots of all the polynomials $f(a, X_n), f \in S$. Moreover, if A is semi-algebraic, the graphs of the ξ_is and the slices of the cylinder $A \times \mathbb{R}$ cut out by these graphs are semi-algebraic.

PROOF. See, for example, Bochnak *et al.* (1987), ch. 2, th. 2.3.1.

DEFINITION 3.3. In the situation of the preceding theorem, we say that the family T *cylindrifies* the family S.

Theorem 3.2 above goes back to Lojasiewicz (1965, pp. 105–110), which contains the first systematic study of semi-algebraic sets. Now, given a finite family $S \subset \mathbb{R}[X_1, \ldots, X_n]$, this result yields by induction on n a finite decomposition of $\mathbb{R}^n$ into semi-algebraic cells C_i, $i = 1, \ldots, p$, such that each C_i is homeomorphic to $]0, 1[^{d_i}$ for some d_i, and is S-invariant. Collins has given the name of *cylindric algebraic decomposition* (c.a.d.) to such a decomposition of $\mathbb{R}^n$, and he has described an algorithm for the production of c.a.d. (cf. Collins, 1975).

3.b. INCIDENCE RELATION AND STRATIFYING FAMILIES

Schwartz & Sharir (1983) have shown that, up to a linear change of coordinates, the c.a.d. has the property that the closure of any cell is a union of cells. Then an algorithm for determining the incidence relation between the cells would give the possibility to calculate all the topological information needed on semi-algebraic sets. For instance, the number and the description of connected components, a triangulation, the homology groups, etc. Schwartz and Sharir propose an algorithm for the incidence relation, but this algorithm is rather complicated except in the case when the dimension of the cells are n and $n-1$, respectively, which is all they need for their purpose.

Here we want to sketch how a result known to real algebraic geometers and which is a generalisation of Thom's lemma to the case of several variables provides an alternative algorithm for the incidence relation. This generalisation is essentially due to Efroymson (1976). A proof may be found in Coste (1982, th. 3.2), and a more detailed version is in Bocknak *et al.* (1987, ch. 9, sect. 1). We start with a definition.

DEFINITION 3.4. A *stratifying family* of polynomials in $\mathbb{R}[X_1, \ldots, X_n]$ is a family $(f_{i,j})$, $i = 1, \ldots, n, j = 1, \ldots, l_i$, such that:

  (i) For any (i, j) the polynomial $f_{i,j}$ is in $\mathbb{R}[X_1, \ldots, X_i]$ and its leading coefficient (as a polynomial in X_i) is a non zero constant.
  (ii) For any fixed i, the family $(f_{i,j}), j = 1, \ldots, l_i$, is stable under derivation with respect to X_i (that is, the derivative of $f_{i,j}$ with respect to X_i is either zero or one of the $f_{i,j'}, j' \neq j$).
  (iii) For any $i < n$, the family $(f_{i,j})$, $j = 1, \ldots, l_i$, cylindrifies the family $(f_{i+1,j})$, $j = 1, \ldots, l_{i+1}$.

A stratifying family yields a c.a.d. which is very satisfying.

THEOREM 3.5 (GENERALISED THOM'S LEMMA). *Let $(f_{i,j})$, $i = 1, \ldots, n$, $j = 1, \ldots, l(i)$, be a stratifying family of polynomials. Consider the finite decomposition of $\mathbb{R}^n$ given by all the non-empty semi-algebraic sets of the form*

$$C_\varepsilon = \bigcap_{i=1}^{p} \bigcap_{j=1}^{l(i)} \{x \in \mathbb{R}^n | \operatorname{sign}(f_{i,j}(x))\varepsilon(i,j)\}$$

where $\varepsilon = (\varepsilon(i,j))$ is a family of sign conditions <0, $=0$ or >0.
  (i) *Any such C is homeomorphic to $]0, 1[^d$ for some d.*
  (ii) *The closure of any such C is obtained by relaxing the strict inequalities which appear in its definition.*

PROOF. See Bocknak *et al.* (1987, ch. 9, th. 9.1.4).

So for the c.a.d. associated to a stratifying family the formulas describing the cells and the incidence relation between the cells are given free. A cell C is contained in the closure of a cell C' if and only if any sign condition on the $f_{i,j}$s which appears in the description of C also appears in the description of C'. The only problem left is to determine which collections of signs $(\varepsilon(i,j))$ give non empty cells. One has to use here a going-up process by induction on the dimension for which the algorithms for computing with real algebraic numbers sketched in 2 can be useful.

Finally, after a linear change of coordinates, any semi-algebraic set can be decomposed via a c.a.d. given by a stratifying family.

PROPOSITION 3.6. *Let $g_1, \ldots, g_k$ be polynomials in $\mathbb{R}[X_1, \ldots, X_n]$. Then one can produce a linear change of coordinates $u: \mathbb{R}^n \to \mathbb{R}^n$ and a stratifying family of polynomials $(f_{i,j})$ in $\mathbb{R}[X_1, \ldots, X_n]$ such that $g_j(u(X_1, \ldots, X_n)) = f_{n,j}(X_1, \ldots, X_n)$ for $j = 1, \ldots, l_n$.*

PROOF. The properties (i) and (ii) in the definition of the stratifying families (3.4) are easily obtained (Bocknak *et al.*, ch. 9, prop. 9.1.2). For the property (iii) of 3.4 one may use the algorithm of "augmented projection" in Collins (1975).

REMARK 3.7. The introduction of derivatives multiplies the number of cells of the c.a.d. But these cells may be grouped together, in the going-up process of the induction on the dimension, in semi-algebraic connected subsets of $\mathbb{R}^i$ over which the polynomials $(f_{i+1,j})$, $j = 1, \ldots, l_{i+1}$, have a constant number of roots, since the incidence relation in the c.a.d. of $\mathbb{R}^i$ is known. This remark is due to D. Lazard. We do not know if this remark improves the complexity of the cylindrical algebraic decomposition.

Let us remark also that, in some situations, when we want to get a description of the projection of a semi-algebraic set in a given direction, a linear change of coordinates is forbidden and the stratifying families cannot be used.

References

Ben-Or, M., Kozen, D., Reif, J. (1986). The complexity of elementary algebra and geometry. *J. Computation Systems Sci.* **32**, 251–264.

Bochnak, J., Coste, M., Roy, M.-F. (1987). *Géométrie algébrique réelle. Ergebnisse der Mathematik*, Springer-Verlag, Berlin.

Collins, G. (1975). Quantifier elimination for real closed fields by cylindric algebraic decomposition. In: *Second GI Conference on Automata Theory and Formal Languages. Lecture Notes in Computer Science, 33.* Springer-Verlag, Berlin, pp. 134–183.

Coste, M. (1982). Ensembles semi-algébrique. In: *Géométrie algébrique réelle et formes quadratiques. Lecture Notes in Mathematics, 959*. Springer-Verlag, Berlin, pp. 109–138.

Efroymson, G. (1976). Substitution in Nash functions. *Pac. J. Math.* **54,** 109–138.

Lojasiewicz, S. (1965). Ensembles semi-analytiques. *Lecture Notes, I.H.E.S.*

Loos, R. (1982). Generalized polynomial remainder sequences. In: *Computer Algebra, Symbolic and Algebraic Computation*. Springer-Verlag, Berlin.

Roy, M.-F. (1987) Computation of the topology of a real algebraic curve. Congress on "Computational geometry and topology and computation in teaching mathematics", Seville (to appear).

Roy, M.-F., Szpirglas, A. (In prep.). Complexity of computation on real algebraic numbers.

Schwartz, J. T., Sharir, M. (1983). On the "Piano mover's" problem. II. *Advan. Applied Math.* **4,** 298–351.

Sylvester, J. J. (1853). On a theory of syzygetic relations of two rational integral functions, comprising an application to the theory of Sturm's function. *Trans. Roy. Soc. London.*

J. Symbolic Computation (1988) **5**, 131–140

Algebraic Decomposition of Regular Curves.

STEFAN ARNBORG AND HUICHUN FENG

*Department of Numerical Analysis and Computing Science
The Royal Institute of Technology
S-100 44 Stockholm, Sweden*

(Received 7 July 1986)

The cylindrical algebraic decomposition method decomposes E^r into regions over which a given polynomial has constant sign by extension of one complicated decomposition of E^{r-1}. We investigate a method which decomposes E^r into sign-invariant regions by combining several but simpler decompositions of E^{r-1}. We can obtain a sign-invariant decomposition of E^2 defined by a bivariate polynomial of total degree n and coefficient size d in time $O(n^{12}(d+\log n)^2 \log n)$. Preliminary experiments suggest that the method is useful in practice.

Introduction

The cylindrical algebraic decomposition (henceforth CAD) algorithm of Collins [Arnon *et al.*, 1984] is a fascinating tool for solving, among other things, first and second order problems in geometry with applications in design and robotics. Despite numerous improvements which gave drastic performance gains on important classes of problems, [Arnon *et al.*, 1984], [Arnon, 1985], [McCallum, 1985], one still feels that some relatively simple problem instances take more computing resources than one would expect. The asymptotic worst case bound for CAD is doubly exponential in the number of variables, and there are good reasons to believe that this is the best that can be achieved [Ben-Or *et al.*, 1986]. It is clearly desirable to have a method which works better on 'simple' problem instances. What we mean by a simple problem instance could be inferred from what types of instances arise 'in practice' or we could mean simply instances on which our method works well. A study of the application in motion–planning convinced us that a useful notion of simplicity may be regularity. The results presented here are first steps that might lead to a method that is fast when applied to a set of regular polynomials and whose cost increases with the number of and complexity of the singularities in the polynomials. Specifically, we show how to obtain a decomposition of E^2, sign-invariant *w.r.t.* a single regular bivariate polynomial.

1. Definitions and outline

The following definitions are standard, see e.g. [Buchberger *et al.*, 1982]. We consider problems involving sets of polynomials in r (we soon specialize to the case $r = 2$)

0747–7171/88/010131 + 10 $03.00/0

variables with coefficients in Q or Z. A *decomposition* of E^r, Euclidean r–dimensional space, *sign–invariant with respect to S*, a set of r–variate polynomials, is a partitioning of E^r into connected regions such that the sign $(+,-$ or $0)$ of each polynomial in S is constant over each region. Two regions are *adjacent* if they have a connected union. The *size* of an integer k is $\log k$ and the size of a rational number p/q is $\log p + \log q$ when p and q are relatively prime. A hypersurface is *regular iff* an open neighborhood of each of its points is homeomorphic to the unit ball and it has a well defined and continuously varying tangent normal vector everywhere. A polynomial is *regular iff* there is no common real zero of the polynomial and its first order partial derivatives. A regular polynomial defines a regular surface but not vice versa. A hypersurface or polynomial which is not regular is *singular*, and the points where the condition for regularity is not fulfilled are its *singular points*.

We propose a method with the following characteristics, in E^r and for a regular hypersurface: For each coordinate x_i obtain a set of rational values $t_j^{(i)}$ which separate the stationary points of the surface *w.r.t.* x_i (probably more points are required than one in each interval between stationary points). The intersections of the hypersurface with the hyperplanes $x_i = t_j^{(i)}$ form regular hypersurfaces in E^{r-1}. Then we recursively obtain a decomposition of these intersections (which are problems in E^{r-1}). For every pair of adjacent parallel 'hyperslices', and for all pairs of intersecting hyperslices, find a number of pairs of regions that correspond to the same region in E^r, and form the transitive closure of the corresponding relation. The equivalence classes (of E^{r-1} regions) define the regions in E^r. The adjacency relation in E^r then follows from those in the slices. The major problem is to find a set of rational values and pairs such that the computation actually gives maximal sign-invariant regions and correct adjacencies, without going through all the machinery of the standard CAD algorithm. We have a solution to this problem in two variables. The reason why this method is interesting is that the degrees of the involved polynomials (in the slices) do not increase, so we can use many slices without approaching the cost of the extension phase of CAD.

Recall [Collins and Loos, 1982], [Davenport, 1985] that, given a bivariate polynomial $f(x,y)$ of total degree n and maximum coefficient size d, f_x and f_y have degree $n-1$ and coefficient size $O(\log n + d)$. If ρ is a rational number of size d', $f(\rho, y)$ and $f(x, \rho)$ are polynomials of degree n and coefficient size $nd' + d$. Resultants of f, f_y and f_x *w.r.t.* x and y have degree $O(n^2)$ and coefficient size $O(n(d + \log n))$. Evaluating a univariate polynomial of degree n and coefficient size d at a point of size d' takes time $O(n(d' + d))$. Evaluation of its Sturm sequence (using fast arithmetic) takes time $O(n^2(d + \log n + d'))$ and its roots can be isolated in time $O(n^4(\log n + d)^2 \log n\, \epsilon)$, where ϵ stands for a number of $\log \log$ terms that we will ignore in the following. The total size of the separating and bounding rational points for the roots is $O(n(\log n + d))$. As a by-product we also obtain the squarefree part of the polynomial.

2. Two-dimensional regular curves

The algorithm for finding a coarsest decomposition of E^2, sign–invariant *w.r.t.* a single polynomial defining a regular curve, finding its adjacency relation and solving the point location problem will now be described. Assume $f(x,y)$ is the polynomial, of total degree n and maximum coefficient size d, defining the regular curve. It is well known that a regular curve consists of a set of disjoint (maximal connected) components, each being a closed loop (Jordan curve) or infinite arch extending to infinity in both directions. In other words, there are no crossings, cusps or isolated points on a regular curve.

Let a *turning point* of a curve be a point where it has horizontal or vertical tangent. We will assume that the curve has no vertical or horizontal straight lines, and thus that the number of turning points is finite. This can be assumed without loss of generality since in case we had such lines we could either factor them out (they correspond to univariate factors of the polynomial defining the curve) or tilt the coordinate system as is done for obtaining a well-based decomposition in [Schwartz and Sharir, 1983]. Similarly, we can assume that no asymptotes to the curve are vertical or horizontal (this simplifies the analysis slightly but should probably not be assumed in practice). The first step of our algorithm is to place a *grid* of horizontal and vertical lines in the plane which isolates the turning points, *i.e.*, each *rectangle* whose boundary is contained in two pairs of adjacent lines, one pair vertical and one horizontal, contains at most one turning point of the curve. Obviously, at a turning point with vertical tangent, f and f_y are both zero and at one with horizontal tangent f and f_x are both zero. Thus, the x–coordinates of the turning points are among the zeros of $v(x) = \mathrm{res}_y(f, f_x)\mathrm{res}_y(f, f_y)$ and the y–coordinates of the turning points are among the zeros of $h(y) = \mathrm{res}_x(f, f_x)\mathrm{res}_x(f, f_y)$. Call the zeros (in increasing order) of $v(x)$ $\hat{x}_i$, $0 < i < O(n^2)$, those of $h(y)$ are called $\hat{y}_i$, $0 < i < O(n^2)$. The multiplicities of these zeros give upper bounds on the number of turning points on the corresponding vertical or horizontal line. A common zero of $\mathrm{res}_x(f, f_x)$ and $\mathrm{res}_x(f, f_y)$ indicates that there may be a singular point or that a vertical turning point has the same y-coordinate as a horizontal one. Since our only requirement is to isolate the turning points, it is possible to omit the vertical or horizontal lines in cases with no multiple real zeros. We do not describe this improvement since it is straightforward and does not change the worst-case behaviour of the method (as an example, if the curve has circular components, pairs of turning points can not be separated by tilting).

LEMMA 1.   *A grid isolating the turning points of the curve can be obtained in time* $O(n^{10}(d+\log n)^2 \log n)$. *The total size of the coordinates of these lines is* $O(n^3(d+\log n))$.

PROOF.   For the x-coordinates of the vertical lines we can take a sequence of rational numbers $(x_i)_{0<i<O(n^2)}$ bounding and separating the zeros of $v(x)$. $v(x)$ is of degree $O(n^2)$ and has coefficient size $O(n(\log n + d))$. Substituting these asymptotic bounds into the root separation cost and size of section 1 gives the asymptotic bounds of the Lemma. The y-coordinates of the horizontal lines are similarly obtained from $h(y)$. In the grid so obtained there is obviously not more than one point $(\hat{x}_i, \hat{y}_j)$ in any rectangle, and this point set contains all turning points of the curve. ■

Having obtained the grid lines $x = x_i$ and $y = y_i$, we proceed by intersecting the curve and the curves defined by the first–order derivatives f_x and f_y with these lines. Specifically, mutually isolate the zeros of $f(x_i, y)$, $f_x(x_i, y)$ and $f_y(x_i, y)$ from each other and from the y_j, for each i. Similarly, isolate the zeros of $f(x, y_i)$, $f_x(x, y_i)$ and $f_y(x, y_i)$ from each other and from the x_j. This operation gives us, for each intersection of the curve with a grid line, the sign of slope ('up' if the product of the first–order derivatives is negative, 'down' otherwise) and the rectangles whose boundaries are intersected. There are only $O(n^2)$ real turning points for a curve of degree n (since each turning point contributes one zero to one of the two discriminants of the polynomial, and these are polynomials of degree $O(n^2)$). The rectangles which contain a turning point are exactly those which have an odd number of 'up' (or 'down') intersections along their boundaries (where we do not count intersections unless all neighborhoods of the intersection on the curve intersect the interior of the rectangle, *e.g.*, a 'down' intersection with the lower lefthand corner of a rectangle is not counted).

For every intersection (except possibly one) with a side of a rectangle containing a turning point we also locate the intersection between an x_i and $\hat{x}_i$ or between an $\hat{x}_i$ and x_{i+1} (the intersection can also coincide with an x_i or an $\hat{x}_i$). Similarly for vertical lines.

LEMMA 2.   *The labelling and classification of the intersections of the curve with the grid can be computed in time $O(n^{12}(d + \log n)^2 \log n)$.*

PROOF.   Let d_i be the size of x_i and d'_j the size of y_j. By Lemma 1 we know that $\sum_i d_i$ and $\sum_j d'_j$ are both $O(n^3(d + \log n))$. The roots of $h(y)$ are already isolated, and the isolating and bounding points are the y_j. $f(x_i, y)$, $f_x(x_i, y)$ and $f_y(x_i, y)$ are all of degree $O(n)$ and have maximum coefficient sizes $O(nd_i + d)$, $O(nd_i + d + \log n)$ and $O(nd_i + d + \log n)$, respectively. The roots of each of these polynomials can therefore be isolated in time $O(n^4(\log n + nd_i + d)^2 \log n)$ or $O(n^{12}(\log n + d)^2 \log n)$ when summed over i. The isolation of zeros between these polynomials, for each i, can be done within the same time bound (consider isolation of the zeros of the product of the three polynomials). Location of the zeros among the y_j, for each i, requires $O(n^2)$ evaluations at points of sizes d'_j, $O(\sum_j n(d'_j + nd_i + d))$ or, when summed also over i, $O(n^7(d + \log n))$. Intersections along horizontal lines are treated in the same way. After identification of the $O(n^2)$ rectangles with a turning point inside, it remains to locate the intersections around the boundary of each such rectangle relative to the coordinates $(\hat{x}_i, \hat{y}_j)$ of its turning point. For a vertical side of such a rectangle we have a number of intersections with f which are isolated from each other. These correspond to zeros of multiplicity one of $f(x_i, y)$ (since the grid contains by construction no turning point). We shall now also isolate these zeros, except possibly one, from the zero along the side of $h(y)$, or rather its squarefree part which was obtained when we isolated its zeroes. This is done by evaluating $h(y)$ at the isolating points for the zeros of $f(x_i, y)$. Since $h(y)$ has degree $O(n^2)$ and coefficient size $O(n(\log n + d))$, and $f(x_i, y)$ has degree n and coefficient size $O(n(d_i + d)$ with total size of isolating points $O(n^2(d_i + d))$, the total evaluation cost is $O(n^7(d + \log n))$, when summed over i. ∎

Now consider the rectangle with corners (x_i, y_j), (x_i, y_{j+1}), (x_{i+1}, y_{j+1}) and (x_{i+1}, y_j), containing the potential turning point $(\hat{x}_i, \hat{y}_j)$. Let the *quadrants* of the rectangle boundary be SW, SE, NE, and NW, where SW is the line segments (x_i, y_j) to $(x_i, \hat{y}_j)$ and (x_i, y_j) to $(\hat{x}_i, y_j)$, *etc.*. We will construct a coarsest sign–invariant decomposition of the intersection of the curve with each such open rectangle and then show how these decompositions can be merged into one for the whole curve. Any component of the curve intersecting the open rectangle will intersect its boundary, otherwise there would be at least four turning points in the rectangle. Moreover, the intersection of the curve and the closed rectangle consists of a number of arches between two intersections with the boundary and no two such arches intersect or touch each other. We can thus describe the decomposition with a number of pairs of intersections of the rectangle boundary and the curve. We now construct the appropriate pairs. Label the intersections of the curve and the rectangle boundary as follows: An intersection of the curve with the SW or NE quadrant is labelled 'in' if the slope is 'up', otherwise it is labelled 'out'; and vice versa for the SE and NW quadrants. An intersection with the SE or NW corner is labelled 'in' if it is 'down', otherwise it is not considered; vice versa for intersections with the SW and NE corners of the rectangle. Intersections belonging to two quadrants (with one coordinate equal to that of the potential turning point) are always labelled 'out'. Note that an 'out'–labelled intersection cannot belong to an arch that passes through the turning point of the rectangle, so it belongs to an arch that has constant sign of slope throughout the rectangle. Note also that at most one intersection on each side of the rectangle is still unclassified, since we have not isolated it from the coordinate of the turning point. Doing so would lead (at least using our present tools) to a higher cost bound for the whole algorithm. However, we will prove that the correct decomposition is obtained if we classify these intersections as 'in' (and thus also assign them to a specific quadrant). This is because any unclassified intersection is furthest away from the corner of its quadrant and does not separate any other intersection from the corner of its quadrant. We will refer to these intersections as possibly misclassified and those who got the 'wrong' label are referred to as misclassified.

LEMMA 3.   *If there is an 'out'-labelled intersection on the boundary, then there is also an 'out'-labelled intersection I in a quadrant Q that is closest to the corner of Q, i.e., there is no other intersection between I and the corner of Q. I belongs to the same arch as the first intersection, with the same sign of slope as I, encountered when following the boundary from that corner, away from I.*

PROOF.   Assume *wlog* that there is an 'out'-labelled intersection I on the horizontal part of the SW quadrant, with slope sign 'down'. The other intersection of its arch is either in the same quadrant, also labelled 'out', or in the NW quadrant and labelled 'in'. In the first case all intersections between the SW corner and I are labelled 'out' and the first statement of the Lemma is satisfied. In the other case, if I is separated from the SW corner by an 'in'-labelled intersection, then this intersection belongs to an arch which also intersects the horizontal side of the NW quadrant, labelled 'out' and closer to the NW corner than I is to the SW corner.

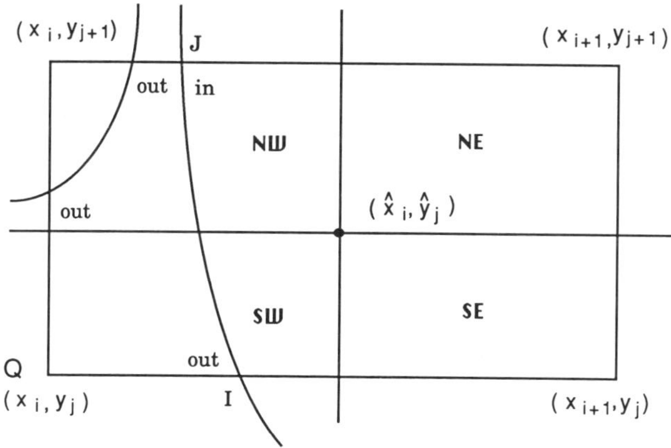

FIGURE 1.   Matching intersections by Lemma 3.

Since there are only finitely many intersections, an 'out'-labelled intersection is closest, on the horizontal sides of the rectangle, either to the SW corner or to the NW corner. Assume that I was in fact closest to the SW corner, and that intersection J is on the other end of the arch of I. The area enclosed by the arch and the part of the rectangle boundary with the SW corner contains no turning point, so no arch in this area can have negative slope (it would then have to intersect the boundary between the SW corner and I). But J has negative slope and is therefore the first intersection with negative slope encountered when following the boundary clockwise from the SW corner (see Figure 1). ∎

Lemma 3 allows us to pair off all 'out'-labelled intersections (some of which may be paired to an 'in'-labelled one). Clearly, the possibly misclassified intersections will not prevent us from correctly matching all 'out'-labelled intersections. We can think of the paired intersections as being 'removed'. The obvious algorithm for doing this uses $O(n)$ time for each of the matched, at most $O(n^3)$, intersections.

LEMMA 4.   *If there are no 'out'-labelled intersections, then at least one quadrant has no 'in'-labelled intersection. The first two intersections found along the boundary from this quadrant, one in each direction, belong to the same arch.*

PROOF.   An 'in'-labelled intersection belongs to an arch that either turns and intersects an adjacent quadrant or goes to an 'in'-labelled intersection in the opposite quadrant (assuming no 'out'-labelled intersections). If there is at least one intersection in each quadrant, then there are at least two arches, and one of them does not turn. Suppose that the latter intersects the SW and NE quadrants. All other arches that do not turn must obviously also intersect the SW and NE quadrants, otherwise arches would cross inside the rectangle. The arch that turns can only intersect one of the SE and NW quadrants, so one quadrant must be free from intersections. Assume *wlog* that the turning arch intersects SE and SW; then the NW quadrant is empty of intersections and the intersections nearest to the NW quadrant in both directions have positive slope and belong to the same arch because otherwise the two arches would have to cross or one of them would have to turn, which it does not, by assumption (Figure 2). ∎

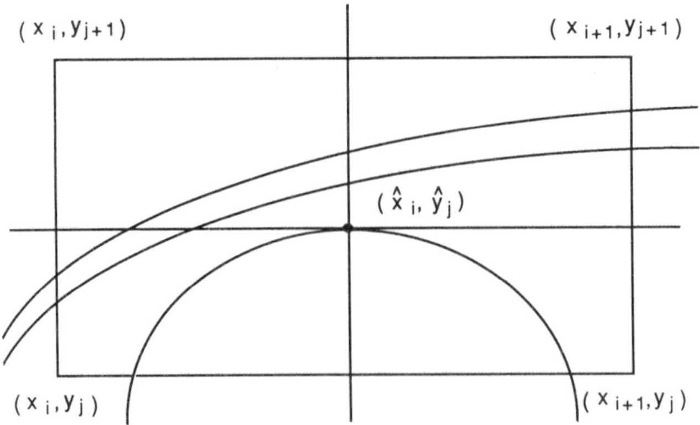

FIGURE 2.   Matching intersections by Lemma 4.

Lemma 4 allows us to pair off the remaining intersections by going around the boundary from the empty quadrant in both directions. It is clear that labelling and pairing of intersections around one rectangle can be done in time proportional to the number of intersections, $O(n^3)$. This is also the time required to find the $O(n^3)$ rectangles with non-empty intersection with the curve among the, in total, $O(n^4)$ rectangles. If some intersection were misclassified ('in' instead of 'out') we still get the right pairing, as is easily seen by considering the perturbed problem where misclassified intersections have been moved over the quadrant boundary (and get a correct classification). This can be done because the misclassified intersections are closest to the quadrant boundary.

Matching intersections around the rectangles without a turning arch, the boundaries of the semi-infinite 'strips' (bounded by three grid lines) and the four 'quadrants' of the plane (bounded by two grid lines) is simpler than for the rectangles with turning points. The simplest way to do this is to place the hypothetical turning point on a corner of a rectangle or sufficiently far away on an infinite boundary line for a semi-infinite region, classify the intersections 'in' or 'out' *w.r.t.* the hypothetical turning point and apply Lemma 3. If this does not match all intersections we place the hypothetical turning point at another corner or infinite boundary line until all four corners or two infinite boundary lines have been tried. It is easy to see that this will identify every branch which has two intersections, and the remaining (unpaired) intersections belong to branches that go to infinity in the region.

It should be clear that the equivalence classes of the reflexive, symmetric and transitive closure of the binary relation defined by the pairs are in one-to-one correspondence with the maximal connected components of the curve. This is because a path on the curve between two points on two arches exists *iff* for two intersections s and e incident to the two arches, there is a sequence $s = s_1, \ldots, s_n = e$ such that s_i and s_{i+1} are incident to a common arch, for $0 < i < n$. Each equivalence class consists of the intersection of one component of the curve with the grid.

The maximal connected regions of the curve's complement (in total one more than the number of components of the curve, since the latter is regular) can be defined by saying, for each component of the curve, whether it lies 'inside' or 'outside' the component (for infinite components of the curve we arbitrarily say that the side where y goes to $-\infty$ is 'outside'). We can find all non–empty such regions by traversing the intersection lists for all vertical lines in the grid, starting from low y–coordinates and remembering the parity of the number of intersections with each component along each line. This classifies all points on the vertical lines. To classify the points on the horizontal lines, follow the horizontal line segments from the grid points to the next grid point. Since the adjacency relation between regions is the same as the adjacency relation between their intersections with the grid, we now have the adjacencies between regions. To classify a point inside a rectangle (the infinite strips and quadrants are again simpler), find the rectangle in which it lies. If the curve does not intersect this rectangle, the given point has the same classification as any point on the boundary; otherwise the classification is obtained after counting (using Sturm sequences) the number of intersections with the curve along a straight line from the given point to $-\infty$; the line can be vertical if the curve has no vertical turning point in the rectangle and vice versa. So point location involves locating the point in the grid, evaluating one precomputed Sturm sequence, and a lookup in a map from number of alternations of the Sturm sequence to region.

Clearly, the step to which Lemma 2 relates is the most expensive one. The Sturm sequence computation is the most expensive step in the point location part, so we have:

THEOREM 1. *For a bivariate polynomial with total degree n and coefficient size d,* *whose zeros form a regular curve, we can find the minimum number of sign-invariant regions in time $O(n^{12}(d + \log n)^2 \log n)$. Then we can decide in which such region a given rational point of size d' lies in time $O(n^2(\log n + d + d'))$.*

3. Two examples

We will examine an example where we use an incremental method: Often, the intersections with the vertical grid lines are adequate for pairing off intersections into branches. Horizontal lines can be added successively when it appears that we must separate two potential turning points with the same x–coordinate. The curve $f(x, y) = 4y^4 + 17x^2y^2 - 20y^2 + 4x^4 - 20x^2 + 17$ is analyzed in [Arnon, 1983]. We compute the resultants of f and its partial derivatives f_x and f_y, $r_y = \text{res}_y(f, f_x)$ and $r_x = \text{res}_y(f, f_y)$. The zeros of these polynomials are the x-coordinates where the curve may have a horizontal or vertical tangent, respectively. Since their gcd is 1, the curve must be regular. We isolate the zeros of r_y from those of r_x, and find a sample point on each side of each zero of r_y. For each such sample point α we isolate the zeros of $f(\alpha, y)$ from those of $\text{res}_x(f, f_y)$, $f_y(\alpha, y)$ and $f_x(\alpha, y)$. This gives us the ordering of the branches through $x = \alpha$ and the zeros of $\text{res}_x(f, f_y)$, and also the slope (+ or -) of each branch. This is all information needed to pair the intersection points into branches. The computing time is 22 seconds, of which .6 are for computing resultants, 8 to isolate the roots of the discriminant, and the rest for isolating zeros along 10 vertical lines with rational x-coordinate. We can replace 5 of the latter by 6 vertical lines with algebraic x-value, but only computing the norm $\text{res}_x(f, \text{res}_y(f, f_y))$ and isolating its real zeros increases the total time by a factor of 5 on our system (we never used the obvious symmetries of the curve).

The 'random polynomial' $f(x, y, z) = (y - 1)z^4 + xz^3 + x(1 - y)z^2 + (y - x - 1)z + y$ is analyzed in [McCallum, 1985]. It is not possible to make a CAD of it in reasonable time. A Gröbner-basis of f, f_x, f_y and f_z has only a constant, so f is regular. Slicing the surface with a plane perpendicular to the z-axis gives us a hyperbola except where the coefficient of xy is zero, *i.e.*, at $z = 0$, and at saddle points. We find two real zeros (except 0) of

$$\text{res}_x(\text{res}_y(f_x, f), \text{res}_y(f_y, f)) = z(z^6 - z^4 + z^3 + z^2 - 1),$$

which are saddle points (since the intersection curve is never a closed loop). Since the saddle points are on each side of $z = 0$, the surface has only one maximal connected component and it divides its complement into two regions. At $z = 0$, $z = -\infty$ and $z = \infty$, one of the branches of the hyperbola goes to infinity. We can decide in which region a given point (x', y', z') lies by just looking at the sign of $f(x', y', z')$. Our total computing time for this analysis is 4 seconds as opposed to 700 hours estimated for a full CAD with adjacency computation (which, it should be observed, also gives a lot more information about the surface).

5. Conclusions

For the two-variable case, the decomposition described above has the same asymptotic cost as the CAD algorithm without extension of zero-dimensional regions in E^1, *i.e.*,

$O(n^{12}(d+\log n)^2 \log n)$ as opposed to $O(n^{16}(d+\log n)^2 \log n)$ for the full CAD algorithm [Davenport, 1985]. It should also be observed that we have not given an efficient way of deciding that a polynomial defines a regular curve, or even that a polynomial is regular. However, many conditions that imply regularity can be easily decided (and seem useful in practice). One such is the absence of real roots of $\gcd(\text{res}(f, f_x), \text{res}(f, f_y))$, where resultants are $w.r.t.$ x or y. Another is the absence of (real or) complex solutions to the system $\{f = 0, f_x = 0, f_y = 0\}$, which is equivalent to the Gröbner basis of $\{f, f_x, f_y\}$ being a constant. The Gröbner basis algorithm [Buchberger, 1985] is quite fast and provably polynomial [Buchberger, 1983] for two variables. The best known bound is exponential even for three variables [Winkler, 1984], but we have found it fast for our regular examples in three and four variables, at least with suitable term ordering. A probabilistic algorithm for deciding regularity of a polynomial is to make a 'random tilting' of the coordinate system and check that the polynomials defining the coordinates of the horizontal and vertical turning points have no common real roots.

We have some 3-dimensional and 4-dimensional applications where our method can be applied and gives very good performance. However, it is not obvious that a better worst-case bound than that of CAD is obtainable for higher dimensions. In the cases we have considered up to now, the main reason for the more spectacular performance improvements over CAD is that a coarsest sign-invariant decomposition has only a few regions while a cylindrical decomposition for the same surface may have thousands of regions. This may of course be a reflection of the general observation for combinatorial problems that unbounded quantifier alternation (for which cylindrical structure of the decomposition seems essential) usually makes a problem P-SPACE complete whereas the same class without quantifier alternations is (only) NP-complete.

As a continuation of the work reported here, it would of course be important to generalize the method to arbitrary dimension and to obtain a general characterization of the effect on problem complexity of singularities of different kinds as well as different types of intersections among a set of regular surfaces. It seems, however, as if our present intuitive geometrical reasoning does not extend easily.

James Davenport has given essential advice and made our work possible through his visit at our department. In the reported experiments we used REDUCE 3.0 on DEC 2060, with Davenport's Gröbner basis package and Joachim Hollman's real root isolator. A stimulating discussion of the manuscript with Scott McCallum is also acknowledged. The work is supported by the Swedish Board for Technical Development.

References

Arnon, D.S. (1983). Topologically reliable display of algebraic curves, *Computer Graphics*, **17**, 219–227.

Arnon, D.S. (1985). A cluster-based cylindrical algebraic decomposition algorithm, *EUROCAL 85*, LNCS **204**, 262–269.

Arnon, D.S., Collins, G.E. and McCallum,S. (1984). Cylindrical algebraic decomposition, *SIAM J. Comp.*, **13**, 865–877, 878–889.

Ben–Or, M., Kozen, D. and Reif, J. (1986). The complexity of elementary algebra and geometry, *JCSS*, **32**, 251–264.

Buchberger, B. (1983). A note on the complexity of constructing Gröbner bases, *EUROCAL 83*, LNCS **162**, 137–145.

Buchberger, B. (1985). *Gröbner–bases: An algorithmic method in polynomial ideal theory* . Multidimensional Systems Theory (Ed. N.K. Bose). D. Reidel Publishing Company 1985, 184–232.

Buchberger, B., Collins, G.E. and Loos, R.G.K. (1982). Computing Supplementum 4, *Springer-Verlag, Wien-New York, 1982*.

Collins, G.E. and Loos, R. (1982). Real zeros of polynomials, *Computing Supplementum 4*, 1982, 83-94.

Davenport, J.H. (1985). *Computer algebra for cylindrical algebraic decomposition*, TRITA–NA–8511, The Royal Institute of Technology.

McCallum, S. (1985). *An improved projection operation for cylindrical algebraic decomposition*, Computer Science Tech. Report 548, University of Wisconsin at Madison.

Schwartz, J.T. and Sharir, M. (1983). On the "piano movers" problem. II. general techniques for computing topological properties of real algebraic manifolds, *Advances in Applied Maths.*, 4, 298–351.

Winkler, F. (1984). On the complexity of the Gröbner basis algorithm over $K[x, y, z]$, *EUROSAM 84*, LNCS **174**, 184–194.

J. Symbolic Computation (1988) **5**, 141–161

An Improved Projection Operation
for Cylindrical Algebraic Decomposition
of Three-dimensional Space†

SCOTT McCALLUM‡

*Department of Computer Science,
University of Toronto, Toronto, Canada M5S 1A4*

(*Received* 21 *May* 1986)

A key component of the cylindrical algebraic decomposition (cad) algorithm of Collins (1975) is the projection operation: the *projection* of a set A of r-variate polynomials is defined to be a certain set of $(r-1)$-variate polynomials. The zeros of the polynomials in the projection comprise a "shadow" of the critical zeros of A. The cad algorithm proceeds by forming successive projections of the input set A, each projection resulting in the elimination of one variable. This paper is concerned with a refinement to the cad algorithm, and to its projection operation in particular. It is shown, using a theorem from complex analytic geometry, that the original projection set for trivariate polynomials that Collins used can be substantially reduced in size, without affecting its essential properties. Observations suggest that the reduction in the projection set size leads to a substantial decrease in the computing time of the cad algorithm.

1. Introduction

A fundamental procedure that pertains to the solution of polynomial equations in several variables is the *cylindrical algebraic decomposition* (*cad*) algorithm due to Collins (1975). This method was developed as part of a decision procedure for elementary algebra and geometry (formally speaking, the theory of real closed fields) that was shown to be more efficient than Tarski's (1951) original method and, indeed, any other subsequent method. The cad algorithm accepts as input a set of integral polynomials (that is, polynomials with integer coefficients) in some $r \geqslant 1$ variables, and produces as output a description of a certain cellular decomposition of r-dimensional Euclidean space $\mathbb{R}^r$. This cellular decomposition of $\mathbb{R}^r$ has the property that each polynomial in the input set is invariant in sign throughout every cell of the decomposition. The "solutions" of the polynomials occurring in the input are thus obtained by retaining those cells in which the sign of each input polynomial is zero.

A key component of the cad algorithm is the projection operation: the *projection* of a set A of r-variate integral polynomials is defined to be a certain set $PROJ(A)$ of $(r-1)$-variate integral polynomials. The zeros of the polynomials in $PROJ(A)$ comprise a "shadow" of the "critical" zeros of A. The set $PROJ(A)$ contains, amongst other elements, all principal subresultant coefficients of all pairs of reducta of elements of A (see section 3).

† This research was supported by NSF (Grant DCR-840817) and NSERC (Grant 3-640-126-30).
‡ Present address: Institut für Mathematik, Johannes Kepler Universität, A-4040 Linz, Austria.

The property of the map *PROJ* of particular relevance to the cad algorithm is that if S is any connected subset of $\mathbb{R}^{r-1}$ in which every element of *PROJ*(A) is invariant in sign and no element of A vanishes identically, then the portion of the zero set of A that lies in the cylinder $S \times \mathbb{R}$ over S consists of a number (possibly 0) of disjoint "layers" over S (that is, A is "delineable" on S). This property is stated as Theorem 5 by Collins (1975) and Theorem 3.4 by Arnon *et al.* (1984*a*). It follows from this property that any decomposition of $\mathbb{R}^{r-1}$ into connected regions such that every polynomial in *PROJ*(A) is invariant in sign throughout every region can be extended to a decomposition of $\mathbb{R}^r$ (consisting of the union of all of the above-mentioned layers and the regions in between successive layers, for each region of $\mathbb{R}^{r-1}$) such that every polynomial in A is invariant in sign throughout every region of $\mathbb{R}^r$.

This paper is concerned with a refinement to the projection operation in the cad algorithm. Collins (1975) observed that a smaller projection suffices for a set A of bivariate integral polynomials. Provided that the elements of A are squarefree and pairwise relatively prime, it suffices to define *PROJ*(A) to be the set of all leading coefficients, discriminants, and resultants (of pairs) of the elements of A. The reason is that the delineability property is readily seen to hold over any connected region of the real line in which just the leading coefficients, discriminants and resultants (of pairs) of the elements of A are invariant in sign. The main contribution of this paper is to show that a similar simplification can be made to the projection of a set of trivariate polynomials.

The main result underlying our refinement to the projection map is a theorem from complex analytic geometry which was stated in precise terms by Zariski (1965) (the essential idea used by us in this paper appears to have been known much earlier: Zariski, 1935). Zariski (1975) has, in fact, extended his result to higher dimensions. This extended result of Zariski is used in the author's (1984) PhD thesis to develop an improved projection operation for polynomials in an arbitrary number of variables. Another paper is planned to expose this work.

Section 2 of this paper provides background mathematical material that may be helpful to the reader. Section 3 defines the reduced projection map, states the relevant theorems on this map, and presents a cad construction algorithm that uses this map. Sections 4 and 5 contain the proofs of the theorems stated in section 3. Section 4 consists essentially of a derivation of the main theorem about the reduced projection from the theorem of Zariski (1965) mentioned above. Section 5 contains an exposition of Zariski's theorem. Section 6 comprises observations relating to the application of the cad algorithm from section 3 to two examples. Several details of the proofs from sections 4 and 5 are presented in the Appendix.

2. Background material

2.1. ANALYTIC FUNCTIONS OF SEVERAL VARIABLES

Let $\mathbb{R}$ denote the field of all real numbers, and let $\mathbb{C}$ denote the field of all complex numbers. Throughout this section K will denote either $\mathbb{R}$ or $\mathbb{C}$. A function $f: U \to K$ from an open subset U of K^n into K is said to be *analytic* (in U) if it has a multiple power series representation about each point of U. An analytic function is continuous and has continuous partial derivatives of all orders. A function defined as the sum of a convergent power series is analytic, and its partial derivatives can be obtained by differentiating the defining series term by term. Sums, products and quotients (where the denominator is non-zero) of analytic functions are analytic. The reader is referred to any of the texts

(Gunning & Rossi, 1965; Bochner & Martin, 1948, or Kaplan, 1966) for a more detailed discussion of the basic properties of analytic functions.

If $c \in K^n$, then a *neighbourhood* of c is an open subset W of K^n containing c. The *polydisc* in $\mathbb{C}^n$ about the point $c = (c_1, \ldots, c_n)$ of *polyradius* $(r_1, \ldots, r_n)$ is the set of points $(z_1, \ldots, z_n)$ in $\mathbb{C}^n$ satisfying $|z_1 - c_1| < r_1, \ldots, |z_n - c_n| < r_n$. Let Δ be a polydisc about 0 in $\mathbb{C}^{n-1}$, where $n \geqslant 2$, and let R be the ring of all analytic functions $f(z_1, \ldots, z_{n-1})$ in Δ. As Δ is connected, R is an integral domain (by the identity theorem, Theorem I–6, Gunning & Rossi, 1965). The units of R are the analytic functions which are non-zero throughout Δ. An element of the polynomial ring $R[z_n]$ is called a *pseudopolynomial* in Δ. Let z denote the $(n-1)$-tuple $(z_1, \ldots, z_{n-1})$. A monic pseudopolynomial

$$h(z, z_n) = z_n^m + a_1(z)z_n^{m-1} + \ldots + a_m(z)$$

of positive degree m, such that $a_i(0) = 0$ for each i, $1 \leqslant i \leqslant m$, is called a *Weierstrass polynomial* in Δ. The Weierstrass preparation theorem (Theorem 62, Chapter 9, Kaplan, 1966) states that every analytic function $f(z, z_n)$ defined in some neighbourhood of the origin in $\mathbb{C}^n$ either does not vanish at 0, or is associated to a Weierstrass polynomial in some polydisc about 0 (provided that $f(0, z_n)$ does not vanish identically).

Let f be an analytic function defined in some open domain U of K^n. Let p be a point of U. We say that f has *order* k at p, and write $\mathrm{ord}_p f = k$, provided that k is the least non-negative integer such that some partial derivative of f of order k does not vanish at p. If all partial derivatives of all orders vanish at p, then we say f has order ∞ at p, and write $\mathrm{ord}_p f = \infty$.

Let x denote $(x_1, \ldots, x_n)$. A mapping $G(x) = (g_1(x), \ldots, g_m(x))$ from the open subset U of K^n into the open subset V of K^m is said to be analytic if each of the component functions g_j is analytic. Where $f : V \to K$ is an analytic function and $G : U \to V$ is an analytic mapping, the composite function $f_o G$ of f and G is analytic, and its power series expansion about any point p of U can be obtained by formal substitution of the power series expansions about p of the component functions of G into the power series expansion about $G(p)$ of f (Bochner & Martin, 1948, p. 33). The following theorem is an immediate consequence of this fact:

THEOREM 2.1. *Let $U \subseteq K^n$ and $V \subseteq K^m$ be open sets, let $G : U \to V$ be an analytic mapping, and let $f : V \to K$ be an analytic function. Then, for every point p of U,*

$$\mathrm{ord}_{G(p)} f \leqslant \mathrm{ord}_p f_o G.$$

2.2. ANALYTIC SUBMANIFOLDS OF EUCLIDEAN SPACE

The original cad algorithm decomposes $\mathbb{R}^n$ into semi-algebraic subsets which Collins (1975) called *cells*. It was subsequently observed (Kahn, 1978) that the cells produced by this decomposition of n-space are actually bona fide cells in the sense of topology: that is, each cell is homeomorphic to an open unit ball in $\mathbb{R}^i$, for some i, $0 \leqslant i \leqslant n$. What is further true is that each cell is homeomorphic to an open unit ball via a mapping which is analytic: this smoothness property of the cells turns out to be quite important in developing an improved projection operation for the cad algorithm.

Before giving a precise definition of an analytic submanifold of $\mathbb{R}^n$ we define the notion of a regular point of an analytic mapping. Let $U \subseteq \mathbb{R}^n$ be open and let $F(x) = (F_1(x), \ldots, F_m(x))$ be an analytic mapping from U into $\mathbb{R}^m$. The point p of U is said to be a *regular* point of F if the rank of the Jacobian matrix $J_F(p) = (\partial F_i/\partial x_j(p))$ of F at p is equal to m. For example, let $F: \mathbb{R}^3 \to \mathbb{R}$ be defined by $F(x, y, z) = x^2 + y^2 + z^2 - 1$. Then $J_F = (2x, 2y, 2z)$, so every point of $\mathbb{R}^3$ other than the origin is a regular point of F. The non-empty subset S of $\mathbb{R}^n$ is an *analytic submanifold* of $\mathbb{R}^n$ of dimension s if for each point p of S there is a neighbourhood $W \subseteq \mathbb{R}^n$ of p and an analytic mapping $F: W \to \mathbb{R}^{n-s}$ which has p as a regular point, such that

$$S \cap W = \{x \in W : F(x) = 0\}.$$

The only kind of submanifold we shall consider in this paper is the analytic kind. Thus, we shall henceforth omit the term "analytic" when referring to submanifolds: all submanifolds will be understood to be analytic. For example, let

$$S^2 = \{(x, y, z) \in \mathbb{R}^3 : x^2 + y^2 + z^2 = 1\}$$

be the unit sphere in $\mathbb{R}^3$. For each point p in S^2 we may take $W = \mathbb{R}^3$ and $F: W \to \mathbb{R}$ to be the map $F(x, y, z) = x^2 + y^2 + z^2 - 1$. As noted above, F is regular at every point $p \neq 0$, and hence at every point p of S^2. Thus S^2 is a submanifold of $\mathbb{R}^3$ of dimension 2.

Let U and V be open subsets of $\mathbb{R}^n$. A homeomorphism $\Phi: U \to V$ such that both Φ and Φ^{-1} are analytic mappings is called an analytic isomorphism. Such a mapping Φ is also called a *coordinate system* in U; if $p \in U$, $0 \in V$, and $\Phi(p) = 0$, then Φ is called a coordinate system (in U) *about* p. The next theorem expresses the intuitive idea that a submanifold of $\mathbb{R}^n$ of dimension s is a set which "looks locally like Euclidean s-space".

THEOREM 2.2. *The non-empty subset S of $\mathbb{R}^n$ is a submanifold of $\mathbb{R}^n$ of dimension s, where $0 \leqslant s \leqslant n$, if and only if for every point p of S there is a neighbourhood $U \subseteq \mathbb{R}^n$ of p and a coordinate system $\Phi: U \to V$, $\Phi = (\phi_1, \ldots, \phi_n)$, about p such that*

$$S \cap U = \{x \in U : \phi_{s+1}(x) = 0, \ldots, \phi_n(x) = 0\}. \tag{2.1}$$

REMARK. If $\Phi = (\phi_1, \ldots, \phi_n)$ is a coordinate system about the point p, one often identifies $(y_1, \ldots, y_n)$ with $(\phi_1, \ldots, \phi_n)$ and speaks of the y-coordinates about p. Equation 2.1 can then be paraphrased "S is defined near p by the equations $y_{s+1} = 0, \ldots, y_n = 0$ in the y-coordinate system". Theorem 2.2 above is essentially Lemma 4F of Appendix II in Whitney (1972).

3. Cad construction using reduced projection map

Let A be a finite set of r-variate integral polynomials. An *A-invariant cylindrical algebraic decomposition* (*cad*) of $\mathbb{R}^r$ partitions $\mathbb{R}^r$ into a finite collection of cylindrically-arranged semialgebraic cells in each of which every polynomial in A is sign-invariant. A more precise definition of cad is given by Arnon *et al.* (1984*a*).

The cad algorithm (Arnon *et al.*, 1984*a*) accepts as input a finite set A of integral polynomials in r variables, and yields as output a description of an A-invariant cad D of $\mathbb{R}^r$. The description of D takes the form of a list of cell indices and sample points for the cells of D. The algorithm consists of three phases: projection (computing successive sets of polynomials in one fewer variables, the zeros of each set containing a "shadow" of the "critical" zeros in the next higher dimensional space), base (constructing a cad of $\mathbb{R}^1$), and extension (successive extension of the cad of $\mathbb{R}^i$ to a cad of $\mathbb{R}^{i+1}$, $i = 1, 2, \ldots, r-1$). Each of these phases is described by Arnon *et al.* (1984*a*).

The key component of the projection phase is the projection operation: the *projection*

PROJ(*A*) of a set *A* of *r*-variate integral polynomials is defined to be a certain set of $(r-1)$-variate integral polynomials.

In this section the map *PROJ* from Collins (1975) or Arnon *et al.* (1984*a*) is reviewed, and a new projection map *P* is defined. The map *P* is essentially just a reduced version of the original projection map *PROJ*. It is proved that, for an input set *A* of trivariate polynomials with integer coefficients, one can use the map *P* in place of its larger counterpart *PROJ* in constructing an *A*-invariant cad of $\mathbb{R}^3$.

In order to define the maps *PROJ* and *P* we first need to recall some definitions and notation from Collins (1975) or Arnon *et al.* (1984*a*). Let *R* be any commutative ring and let $f(x)$ be a polynomial over *R*. We denote by deg (f) the degree of $f(x)$, and take deg $(0) = -\infty$. We denote by *red* (f) the *reductum* of $f(x)$, that is, the difference of $f(x)$ and the leading term of $f(x)$ (*red* $(0) = 0$). We let red^k (f) denote the *k*th reductum of $f(x)$. Let $f(x)$ and $g(x)$ be non-zero polynomials over *R*, with deg $(f) = m$ and deg $(g) = n$. For $0 \leqslant j \leqslant \min(m, n)$, let $psc_j(f, g)$ denote the *j*th *principal subresultant coefficient* of *f* and *g*, that is, the coefficient of x^j in $S_j(f, g)$, the *j*th subresultant of *f* and *g*. Note that $psc_0(f, g)$ is the resultant of *f* and *g*, *res* (f, g).

Assume now that *R* is an integral domain, and let $f(x)$ be a polynomial over *R* of degree $m \geqslant 1$. Where *a* is the leading coefficient of $f(x)$, and $\alpha_1, \ldots, \alpha_m$ are the *m* roots of $f(x)$ in some algebraic closure of the quotient field of *R*, define the *discriminant* of $f(x)$, *discr*(f), as follows:

$$discr(f) = a^{2m-2} \prod_{i<j} (\alpha_i - \alpha_j)^2.$$

Let $f'(x)$ denote the derivative of $f(x)$. The following well-known theorem (Lang, 1984, Proposition V-10.5) relates *discr*(f) to *res*(f, f').

THEOREM 3.0. *Let R be an integral domain and let $f(x)$ be a polynomial of degree $m \geqslant 1$ over R with leading coefficient a. Assume that the characteristic of R, char R does not divide m. Then*

$$a \, discr(f) = (-1)^{m(m-1)/2} res(f, f').$$

COROLLARY. *With the hypotheses of the theorem, discr(f) is seen to be a polynomial in the coefficients of $f(x)$.*

Assume char *R* does not divide *m*. For $0 \leqslant j \leqslant \deg(f) - 1$, define the *j*th *principal subdiscriminant* of *f*, $psd_j(f)$, by the equation

$$a \, psd_j(f) = (-1)^{m(m-1)/2} psc_j(f, f').$$

Note that $psd_j(f)$ is a polynomial in the coefficients of $f(x)$ as *a* is a factor of $psc_j(f, f')$, and that $psd_0(f) = discr(f)$.

Let $\mathbb{Z}[x_1, \ldots, x_r]$ denote the ring of all integral polynomials in $x_1, \ldots, x_r$. We regard the elements of $\mathbb{Z}[x_1, \ldots, x_r]$ as polynomials in x_r over $\mathbb{Z}[x_1, \ldots, x_{r-1}]$. Thus, for example, the degree $\deg(f)$ of a polynomial *f* in $\mathbb{Z}[x_1, \ldots, x_r]$ means the degree of *f* in x_r. Let *A* be a set of polynomials in $\mathbb{Z}[x_1, \ldots, x_r]$. We now define several sets of $(r-1)$-variate polynomials. The *coefficient set of A*, *coeff*(*A*), is the set of all non-zero coefficients of all elements of *A*. The *reducta set of A*, *red*(*A*), is the set of all $red^k(f)$ such that *f* belongs to *A* and $0 \leqslant k \leqslant \deg(f)$ The *principal subdiscriminant set of A, psd*(*A*), is the set of all $psd_j(f)$ such that *f* belongs to *A* and $0 \leqslant j < \deg(f) - 1$. The *discriminant set of* A, *discr*(A), is the subset of *psd*(A) consisting of all discriminants of all elements f of A with $\deg(f) > 1$. The *principal subresultant coefficient set of* A, *psc*(A), is the set of all $psc_j(f, g)$, such that f and g are elements of A with $f \neq g$ and $0 \leqslant j < \min(\deg(f), \deg(g))$. The *resultant set of* A, *res*(A), is the subset of *psc*(A) consisting of all resultants of elements f and g of A, with $f \neq g$ and $\deg(f)$ and $\deg(g)$ both positive.

We can now define *PROJ*(A) and *P*(A):

$$PROJ(A) = coeff(A) \cup psd(red(A)) \cup psc(red(A));$$
$$P(A) = coeff(A) \cup discr(A) \cup res(A).$$

REMARKS.
(1) $P(A)$ is a subset of $PROJ(A)$.
(2) If A has m elements, with the degree of each polynomial in each variable at most n, then $PROJ(A)$ has $0(m^2 n^3)$ elements, whereas $P(A)$ has only $0(mn + m^2)$ elements.
(3) These definitions of the projection maps have been kept conceptually simple for ease of exposition. In practice, there may be elementary improvements that can be made to reduce the size of the sets $PROJ(A)$ and $P(A)$. Some of these will be discussed in section 6.

Recall a couple of basic concepts from Collins (1975). A set A of polynomials in $\mathbb{Z}[x_1, \ldots, x_r]$ is said to be a *squarefree basis* if the elements of A have positive degree, and are primitive, squarefree and pairwise relatively prime. Let x denote the $(r-1)$-tuple $(x_1, \ldots, x_{r-1})$. An r-variate polynomial $f(x, x_r)$ over the reals is said to be *delineable* on a subset S (usually connected) of $\mathbb{R}^{r-1}$ if

(1) the portion of the real variety of f that lies in the cylinder $S \times \mathbb{R}$ over S consists of the union of the graphs of some $k \geqslant 0$ continuous functions $\theta_1 < \ldots < \theta_k$ from S to $\mathbb{R}$: and
(2) there exist integers $m_1, \ldots, m_k \geqslant 1$ such that for every $a \in S$, the multiplicity of the root $\theta_i(a)$ of $f(a, x_r)$ (considered as a polynomial in x_r alone) is m_i.

(Remark that if f has no zeros in $S \times \mathbb{R}$, then f is delineable on S as we may take $k = 0$ in this definition.) In the above definition, the θ_i are sometimes called the *real root functions* of f on S, the graphs of the θ_i are called the *f-sections* over S, and the regions between successive *f*-sections are called *f-sectors*.

One more definition: let $K = \mathbb{R}$ or $\mathbb{C}$ and let U be an open subset of K^r; an analytic function $f : U \to K$ is said to be *order-invariant* in a subset S of U provided that the order of f (see section 2) is the same at every point of S.

Remark that if $K = \mathbb{R}$, and if the analytic function $f : U \to K$ is order-invariant in the connected subset S of U, then f is sign-invariant in S.

An example: let $\mathbb{R}^2 \to \mathbb{R}$ be given by $f(x, y) = x^2 - y^2$. Let C_f be the curve defined by $f(x, y) = 0$ and let $S = C_f - \{0\}$. Then f is order-invariant in S ($\text{ord}_{(x_0, y_0)} f = 1$ for every point (x_0, y_0) of S). However, f is not order-invariant in C_f ($\text{ord}_{(0, 0)} f = 2$) (but f is sign-invariant in C_f).

The main result pertaining to P follows.

THEOREM 3.1. *Let A be a finite squarefree basis of integral polynomials in r variables, where $r = 2$ or 3. Let S be a connected submanifold of $\mathbb{R}^{r-1}$ of positive dimension. Suppose that each element of $P(A)$ is order-invariant in S. Then each element of A is delineable on S, and* *the sections of the elements of A over S are pairwise disjoint. Moreover, if $r = 2$, then every such section is a submanifold of $\mathbb{R}^2$ and is order-invariant with respect to each element of A.*

REMARKS.
  (1) The counterpart of Theorem 3.1 for the map *PROJ* is Theorem 5 from Collins (1975) (also stated as Theorem 3.4 by Arnon *et al.*, 1984*a*). The proof of Theorem 5 from Collins (1975) makes essential use of the fundamental theorem of polynomial remainder sequences (Brown & Traub, 1971).
  (2) Theorem 3.1 has a generalisation to arbitrary r reported in McCallum (1984): this generalisation requires an additional hypothesis, namely, that each element of A does not vanish identically on S (an r-variate polynomial $f(x_1, \ldots, x_r)$ over $\mathbb{R}$ is said to *vanish identically* on a subset S of $\mathbb{R}^{r-1}$ if $f(p, x_r) = 0$ for every point p of S). The conclusions, however, are stronger: each element f of A is *analytic-delineable* on S in the sense that the sections of f over S are the graphs of analytic functions defined in S; moreover, each element of A is order-invariant in every such section.

Theorem 3.1 can be quite readily derived from the following theorem (an r-variate polynomial $f(x_1, \ldots, x_r)$ over $\mathbb{R}$ is said to be *degree-invariant* on the subset S of $\mathbb{R}^{r-1}$ if the degree of $f(p, x_r)$ (as a polynomial in x_r) is the same for every point p of S):

THEOREM 3.2. *Let x denote the $(r-1)$-tuple $(x_1, \ldots, x_{r-1})$. Regard elements of $\mathbb{R}[x, x_r]$ as polynomials in x_r over $\mathbb{R}[x]$. Let $r = 2$ or 3. Let $f(x, x_r)$ be a polynomial of positive degree in $\mathbb{R}[x, x_r]$, let $D(x)$ be the discriminant of $f(x, x_r)$, and suppose that $D(x) \neq 0$. Let S be a connected submanifold of $\mathbb{R}^{r-1}$ on which f is degree-invariant and not identically vanishing, and in which D is order-invariant. Then f is delineable on S. Moreover, if $r = 2$, then every f-section over S is a submanifold of $\mathbb{R}^2$ and is order-invariant with respect to f.*

Theorem 3.2. can, in turn, be derived in a straightforward manner from the following (recall the definition of Weierstrass polynomial given in section 2):

THEOREM 3.3 (Zariski). *Let $h(x, y, z)$ be a Weierstrass polynomial of degree $m \geqslant 1$ in the polydisc Δ_1 about 0 in $\mathbb{C}^2$, and assume that for every fixed (x, y) in Δ_1, every root of $h(x, y, z)$ (considered as a polynomial in z alone) is contained in the disc $|z| < \varepsilon$. Let $F(x, y)$ be the discriminant of $h(x, y, z)$, and assume that F does not vanish identically. Let $T_* = \{(x, 0) | x \in \mathbb{C}\}$ be the complex x-axis in $\mathbb{C}^2$, and assume that F is order-invariant in $T_* \cap \Delta_1$. Then there exists a polydisc $\Delta_2 \subseteq \Delta_1$ about 0 such that for every fixed $(x, 0)$ in $T_* \cap \Delta_2$, $h(x, 0, z)$ (as a polynomial in z) has exactly one root (necessarily of multiplicity m) in the disc $|z| < \varepsilon$.*

This theorem follows from Theorem 4.5 in Zariski (1965). The setting for Zariski's formulation of the theorem is more abstract than ours (he works over an arbitrary algebraically closed field of characteristic zero). In fact, a reader unfamiliar with abstract algebraic geometry may have difficulty discerning the relationship between our Theorem 3.3 and Zariski's Theorem 4.5. So that our presentation is as self-contained as possible, we present in Section 5 a proof of Theorem 3.3. Our exposition is different from Zariski's.

REMARK. The generalisation of Theorem 3.1 mentioned above takes quite a bit longer to prove than Theorem 3.1 itself. The proof is again based on work of Zariski (1975).

Another paper is planned to expose this generalisation and its proof. Parts of this present paper readily generalise to arbitrary r, and will be used in the forthcoming sequel.

We now present the

PROOF OF THEOREM 3.1. There is nothing to prove if A is empty, so assume A is non-empty. Let $A = \{f_1, \ldots, f_m\}$ and let f be the product of the f_i. Let x denote the $(r-1)$-tuple $(x_1, \ldots, x_{r-1})$, and let $D(x)$ be the discriminant of $f(x, x_r)$. By Lemma A.1 and Theorem 3 of Loos (1982),

$$D = \prod_{i=1}^{m} discr(f_i) \cdot \prod_{1 \leqslant i < j \leqslant m} res(f_i, f_j)^2.$$

It follows from this equation that $D(x) \neq 0$ (because each $discr(f_i)$ and each $res(f_i, f_j)$ are non-zero, as the f_i are squarefree and pairwise relatively prime). Now f is degree-invariant on S, as each element of $coeff(A)$ is order-invariant, hence sign-invariant, in S. Moreover, f is primitive (as the f_i are primitive) and hence, by Lemma A.2, f does not vanish identically on S (S has positive dimension and therefore comprises an infinite set of points). By hypothesis, each $discr(f_i)$ and each $res(f_i, f_j)$ are order-invariant in S. Hence, by Lemma A.3, D is order-invariant in S. Hence, by Theorem 3.2, f is delineable on S. Moreover, if $r = 2$, then every f-section over S is a submanifold of $\mathbb{R}^2$ and is order-invariant with respect to f. Therefore, by Lemma A.7, every f_i is delineable on S. It follows that the sections over S of all the f_i are pairwise disjoint. Moreover, if $r = 2$, then by Lemma A.3, each f_i is order-invariant in every such section. $\square$

The proofs of Theorems 3.2 and 3.3 occupy the next two sections respectively.

We now present a cad construction algorithm $CADR3$ which can be applied to any set A of polynomials in r variables, where $1 \leqslant r \leqslant 3$, yielding a list of cell indices and sample points for an A-invariant cad D of $\mathbb{R}^r$. The algorithm $CADR3$ is modelled on the algorithm CAD from Arnon *et al.* (1984a), which is in turn a summary of algorithm $DECOMP$ from Collins (1975).

Apart from the restriction on r in $CADR3$, there are two differences between $CADR3$ and CAD. The main difference is that in $CADR3$ the map P is used in place of the map $PROJ$. The other difference is that while squarefree basis computation is optional in CAD (Collins, 1975, p. 152), it is essential in $CADR3$ (because of the hypotheses of Theorem 3.1).

Some definitions first: let A be a subset of $\mathbb{Z}[x_1, \ldots, x_r]$, where $r \geqslant 1$. Define $cont(A)$ to be the set of non-zero non-unit contents of the elements of A. Define $prim(A)$ to be the set of those primitive parts of elements of A that have positive degree. Now suppose that A consists of primitive polynomials of positive degree. The *finest squarefree basis for A* is the set of all ample irreducible factors of the elements of A (see Collins, 1975, p. 146)

$$CADR3(r, A; I, S)$$

Inputs: r is an integer with $1 \leqslant r \leqslant 3$. A is a list of r-variate integral polynomials.
Outputs: I is a list of the indices of the cells comprising an A-invariant cad D of $\mathbb{R}^r$. S is a list of sample points for D.

(1) [Initialise.] Set $B \leftarrow$ the finest squarefree basis for $prim(A)$ (algorithms for polynomial factorization are given by Kaltofen, 1982). Set $I \leftarrow$ the empty list. Set $S \leftarrow$ the empty list.

(2) [$r = 1$.] If $r > 1$ then go to 3. Isolate the real roots of B. Construct the indices of the cells of D (as described by Arnon *et al.*, 1984a) and add them to I. Construct

sample points for the cells of D (as described by Arnon *et al.*, 1984*a*) and add them to S. Exit.

(3) [$r > 1$.] Set $P \leftarrow cont(A) \cup P(B)$. Call $CADR3$ recursively with inputs $r - 1$ and P to obtain outputs I' and S' which specify a P-invariant cad D' of $\mathbb{R}^{r-1}$. For each cell c of D', let i denote the index of c and let α denote the sample point for c; carry out the following sequence of steps: set $f*(x_r) \leftarrow$ the product of all the $f(\alpha, x_r)$ such that $f \in B$ and $f(\alpha, x_r) \neq 0$; ($f*(x_r)$ is constructed using exact arithmetic in $Q(\alpha)$, Loos, 1982); isolate the real roots of $f*(x_r)$ (Loos, 1982, section 2); use i, α and the isolating intervals for the roots of $f*$ to construct cell indices and sample points (as described by Arnon *et al.*, 1984*a*) for the sections and sectors over c of those elements of B that are not identically zero on c; add the new indices to I and the new sample points to S. Exit. $\square$

It is straightforward to prove the validity of algorithm $CADR3$ using Theorem 3.1.

Remarks.
   (1) Step 1 of $CADR3$ prescribes the computation of the *finest* squarefree basis for $prim(A)$. In fact, if $r > 1$, then *any* squarefree basis for $prim(A)$ can be computed in this step (see Collins, 1975, for the definition of a squarefree basis *for* a set of polynomials).
   (2) The generalisation of Theorem 3.1 mentioned above can be used to prove the validity of a cad construction algorithm (McCallum, 1984) for arbitrary r in which the map P is used in place of the map $PROJ$ provided that the input set of polynomials is assumed to be *well-oriented* (a set A of r-variate polynomials over $\mathbb{R}$ is said to be *well-oriented* if no element of $prim(A)$ vanishes identically on any submanifold of $\mathbb{R}^{r-1}$ of positive dimension and, moreover, this property holds recursively for the set $cont(A) \cup P(B)$, where B is the finest squarefree basis for $prim(A)$).

4. Proof of Theorem 3.2.

We assume that S has positive dimension. (The dimension 0 case is trivial.) By connectedness of S, it suffices to show that f is delineable on S near an arbitrary point p of S. That is, it is enough to show that for every point p of S, there exists a neighbourhood $N \subseteq \mathbb{R}^{r-1}$ of p such that f is delineable on $S \cap N$ (and that, if $r = 2$, then every f-section over $S \cap N$ is a submanifold of $\mathbb{R}^2$ and is order-invariant with respect to f). Let p be a point of S, and let the degree of $f(p, x_r)$ (considered as a polynomial in x_r alone) be l. Then $l \geqslant 0$ (that is, $f(p, x_r)$ is not the zero polynomial), as f is degree-invariant and not identically vanishing on S.

Let $\alpha_1 < \ldots < \alpha_k, k \geqslant 0$, be the real roots of $f(p, x_r)$, let $\alpha_{k+1}, \ldots, \alpha_t, k \leqslant t$, be the distinct non-real roots of $f(p, x_r)$, and let m_i be the multiplicity of the root α_i, for $1 \leqslant i \leqslant t$. Let

$$\kappa = \min\left(\{|\alpha_i - \alpha_j| : 1 \leqslant i < j \leqslant t\} \cup \{1\}\right).$$

Let $0 < \varepsilon < \kappa/2$, and let C_i be the circle of radius ε centred at α_i, $1 \leqslant i \leqslant t$. By root continuity (Theorem (1,4), Marden, 1966) and degree-invariance of f on S, there exists a neighbourhood $N_0 \subseteq \mathbb{R}^{r-1}$ of p such that for every fixed point x of $S \cap N_0$, the interior of each C_i contains exactly m_i roots (multiplicities counted) of $f(x, x_r)$ (considered as a polynomial in x_r alone).

To prove the delineability of f on S near p, it suffices to show that for each i, $1 \leqslant i \leqslant k$, there exists a neighbourhood $N_i \subseteq N_0$ of p such that for every fixed $x \in S \cap N_i$, the interior of C_i contains exactly one root, say $\theta_i(x)$ of $f(x, x_r)$ (considered as a polynomial in x_r alone), necessarily of multiplicity m_i and necessarily real. (For if this has been shown, then let

$$N = \bigcap_{i=0}^{i} N_i.$$

By root continuity, each θ_i is continuous in $S \cap N$. Let (p', α') be a point belonging to the cylinder $(S \cap N) \times \mathbb{R}$ over $S \cap N$, and assume $f(p', \alpha') = 0$. By degree-invariance of f on S, the degree of $f(p', x_r)$ is l. As $p' \in S \cap N \subseteq S \cap N_0$, the interior of each C_i, $1 \leqslant i \leqslant t$, contains exactly m_i roots (multiplicities counted) of $f(p', x_r)$. Since

$$\sum_{i=1}^{t} m_i = l,$$

every root of $f(p', x_r)$ is contained within one of the C_i. Each C_i with $k+1 \leqslant i \leqslant t$ contains no real points, however, as the non-real roots of $f(p, x_r)$ occur in conjugate pairs. Hence, as α' is a real root of $f(p', x_r)$, α' must lie inside a C_i with $1 \leqslant i \leqslant k$, so $\alpha' = \theta_i(p')$. This proves that f is delineable on $S \cap N$.)

We now proceed to prove that for each i, with $1 \leqslant i \leqslant k$, there exists a neighbourhood $N_i \subseteq N_0$ of p such that for every fixed $x \in S \cap N_i$, the interior of C_i contains exactly one root, say $\theta_i(x)$, of $f(x, x_r)$ (as a polynomial in x_r), necessarily of multiplicity m_i. (It will be shown that, informally speaking, each real root α_i of $f(p, x_r)$ does not "split" into many roots as p is perturbed a little within S.)

That the root α_i does not split into many roots as p is perturbed a little within S is quite easy to see in the case in which the dimension of S is equal to $r-1$. For in this case, S is an open subset of $\mathbb{R}^{r-1}$. Hence, as f is degree-invariant on S, the leading coefficient of f (with respect to x_r) vanishes nowhere in S. Also, as $D(x)$, a non-zero polynomial, is order-invariant in S, D vanishes nowhere in S. Therefore, for fixed $x \in S$, every root of $f(x, x_r)$ (as a polynomial in x_r) is simple; hence $m_i = 1$. It follows that the graph of each real root function $\theta_i : S \cap N_0 \to \mathbb{R}$ is a submanifold of $\mathbb{R}^r$ (of dimension $r-1$) and is order-invariant with respect to f (because $x_r = \theta_i(x)$ if and only if $f(x, x_r) = 0$, for all $(x, x_r) \in (S \cap N_0) \times C_i$, and $\partial f / \partial x_r \neq 0$ in the graph of θ_i).

The remaining case to consider is that in which $r = 3$ and the dimension of S is 1, that is, S is a smooth curve in the plane. For the remainder of the proof, let (x, y, z) denote the triple (x_1, x_2, x_3), and let the coordinates of the point p be (a, b). There is no loss of generality in assuming that $\alpha_i = 0$. By Theorem 2.2, we choose coordinates (u, v) about the point p such that S is defined locally by the equation $v = 0$ in the new coordinate system. Let $g(u, v, z)$ denote the function $f(x, y, z)$ transformed into the new coordinates (that is, if Φ is the coordinate system mapping from the (x, y)-plane to the (u, v)-plane, then $g(u, v, z) \sim f(\Phi^{-1}(u, v), z))$. Then $g(u, v, z)$ is a polynomial in z whose coefficients are (real) analytic functions of (the real variables) u and v, defined near the origin (the analyticity here comes from the analyticity of the coordinate system mapping Φ). The discriminant $E(u, v)$ of $g(u, v, z)$ is analytic near 0, and is order-invariant in the u-axis near 0, by Theorem 2.1.

Each coefficient of $g(u, v, z)$ can be expanded in a convergent double power series about 0 (by definition of analyticity). By the two-variable analogue (Theorems 54–56 of Kaplan, 1966) of a well-known result on convergence, each of these double power series is absolutely convergent in a polydisc $\Delta_1 : |u| < r_1, |v| < s_1$ about 0 in complex 2-space $\mathbb{C}^2$, and sums to a function that is analytic in Δ_1. In this way, each coefficient of $g(u, v, z)$, and hence also $g(u, v, z)$ itself, can be extended (uniquely) to a neighbourhood of 0 in $\mathbb{C}^2$. We do not use new notation for this extension of g: henceforth, $g(u, v, z)$ will denote the complex pseudopolynomial (section 2.1) that extends the real g. It is not difficult to show (Lemma A.4) that the discriminant $E(u, v)$ of $g(u, v, z)$ is order-invariant in the complex u-axis $T*$ near 0 (the complex u-axis is the subset $\{(u, 0)|u \in \mathbb{C}\}$). By refining Δ_1 to a smaller polydisc about 0 if necessary, let us assume that $E(u, v)$ is order-invariant in $T* \cap \Delta_1$.

It will be shown that the root $\alpha_i = 0$ of $g(0, 0, z)$ does not split into many roots as $(u, v) = (0, 0)$ is perturbed a little within $T*$. (This will imply, in the old coordinates, the desired result that the root α_i of $f(a, b, z)$ does not split into many roots as (a, b) is perturbed a little within S.) To do this, it will be convenient to focus attention on the zero set of $g(u, v, z)$ near the origin in $\mathbb{C}^3$, using the Weierstrass preparation theorem (Theorem 62 of Kaplan, 1966; see also section 2.1 of the present paper) from the theory of several complex variables. Recall that $\alpha_i = 0$ is a root of $g(0, 0, z)$ of multiplicity $m := m_i$, and that $g(0, 0, z) \neq 0$ for $0 < |z| \leqslant \varepsilon$. By the Weierstrass preparation theorem, there is a polydisc $\Delta_2 \subseteq \Delta_1$, a function $q(u, v, z)$ analytic and nowhere-vanishing in the polydisc $\Delta' : (u, v) \in \Delta_2, |z| < \varepsilon$, and a Weierstrass polynomial

$$h(u, v, z) = z^m + a_1(u, v)z^{m-1} + \ldots + a_m(u, v)$$

in Δ_2, such that

$$g(u, v, z) = q(u, v, z)h(u, v, z) \tag{3.1}$$

for all $(u, v, z) \in \Delta'$, and such that for each fixed $(u, v) \in \Delta_2$, all the m roots of $h(u, v, z)$ (as a polynomial in z) are contained in the disc $|z| < \varepsilon$. By (3.1), as $q(u, v, z) \neq 0$ for all $(u, v, z) \in \Delta'$, the zero set of g is the same as that of h, in Δ'. Thus, the non-splitting of the root $\alpha_i = 0$ of $g(0, 0, z)$ as $(u, v) = (0, 0)$ is perturbed a little within $T*$ will follow from the non-splitting of the root $\alpha_i = 0$ of $h(0, 0, z)$. But the non-splitting of the root $\alpha_i = 0$ of $h(0, 0, z)$ as $(u, v) = (0, 0)$ is perturbed a little within $T*$ is precisely the conclusion of Theorem 3.3. It remains to show that the hypotheses of Theorem 3.3 are satisfied.

Let $F(u, v)$ be the discriminant of $h(u, v, z)$: we shall prove that F does not vanish identically, and that F is order-invariant in $T* \cap \Delta_2$. To do this we first need to take a closer look at the function q: this is done in the proof of Lemma A.5, whose conclusion is that $q(u, v, z)$ is, in fact, a pseudopolynomial in Δ_2. We shall find a function $Q(u, v)$, analytic in Δ_2, such that

$$E(u, v) = Q(u, v)F(u, v) \tag{3.2}$$

for all $(u, v) \in \Delta_2$. Let d be the degree of $g(u, v, z)$. If $d > m$, in which case $q(u, v, z)$ has positive degree $d - m$, then set

$$Q(u, v) = G(u, v)R(u, v)^2,$$

where $G(u, v)$ is the discriminant of $q(u, v, z)$ and $R(u, v)$ is the resultant of $q(u, v, z)$ and $h(u, v, z)$. Equation (3.2) holds by Lemma A.1. If $d = m$, in which case $q(u, v, z) = q(u, v)$ has degree 0, then set

$$Q(u, v) = q(u, v)^{2m-2}.$$

Equation (3.2) holds by the definition of discriminant. Now it follows by (3.2) that F does not vanish identically (as $D(x, y)$, hence $E(u, v)$, does not vanish identically). Furthermore, by Lemma A.3, F is order-invariant in $T* \cap \Delta_2$. The hypotheses of

Theorem 3.3 are satisfied. Hence, by Theorem 3.3, there exists a polydisc $\Delta_3 \subseteq \Delta_2$ about 0 such that for every fixed $(u, v) \in T* \cap \Delta_3$, $h(u, v, z)$ (as a polynomial in z) has exactly one root (necessarily of multiplicity m) in the disc $|z| < \varepsilon$. Hence, by (3.1), the same holds true for $g(u, v, z)$, and the desired conclusion as to the non-splitting of the root α_i of $f(a, b, z)$ as (a, b) is perturbed a little within S now follows. Theorem 3.2 has been proved. $\square$

5. Proof of Theorem 3.3 (and Lemmas)

Let Δ be a polydisc about 0 in $\mathbb{C}^{n-1}$, where $n \geqslant 2$, and let R be the ring of all analytic functions $f(z_1, \ldots, z_{n-1})$ in Δ. We have noted previously that R is an integral domain, and have called an element of the polynomial ring $R[z_n]$ a *pseudopolynomial* in Δ. Now R is not a unique factorisation domain (Whitney, 1972, Appendix IV, Example 2C) and, hence, neither is $R[z_n]$. However, it follows by induction on the degree that every monic pseudopolynomial h of positive degree can be factored as a product of monic irreducible psuedopolynomials thus:

$$h = h_1 \ldots h_k.$$

It will follow from Lemma 5.2 below that the h_i are uniquely determined, provided that the discriminant of h (an element of R) does not vanish identically.

We use the notation $z = (z_1, \ldots, z_{n-1})$. Let U be an open subset of $\mathbb{C}^{n-1}$. A continuous function $\Gamma : [0, 1] \rightarrow U$, such that $\Gamma(0) = w_0$ and $\Gamma(1) = w_1$, is called a *path* in U from w_0 to w_1.

LEMMA 5.1. *Let Δ be a polydisc about 0 in $\mathbb{C}^{n-1}$ and let $h(z, z_n)$ be a monic pseudopolynomial of positive degree in Δ. Let U be an open subset of Δ in which the discriminant of h vanishes nowhere. Let p and q be points of U, not necessarily distinct, and let Γ be a path in U from p to q. Let α be a root of $h(p, z_n)$ (a polynomial in z_n). Then there exists a unique path ϕ in $\mathbb{C}^1$ such that $\phi(0) = \alpha$ and $h(\Gamma(t), \phi(t)) = 0$ for all $t \in [0, 1]$.*

PROOF. Let the degree of h be m, let $V = \{(z, z_n) \in U \times \mathbb{C} | h(z, z_n) = 0\}$, and let w be a point of U. As h is monic and the discriminant of h is non-zero at w, $h(w, z_n)$ (as a polynomial in z_n) has m distinct, simple roots. By m applications of the implicit function theorem (Gunning & Rossi, 1965, Theorem I-4) there exists a neighbourhood $W \subseteq U$ of w such that the portion of V contained in $W \times \mathbb{C}$ consists of the disjoint graphs of m analytic functions from W into $\mathbb{C}$. Hence (Munkres, 1975, Chapter 8), V is an m-fold covering of U, with covering map $\pi : V \rightarrow U$ given by $\pi(z, z_n) = z$. The existence and uniqueness of ϕ now follow by the path lifting property (Munkres, 1975, Chapter 8, Lemma 4.1.) $\square$

REMARK. With the notation of the above lemma, we set $\Gamma_h[\alpha] := \phi(1)$ and say Γ *carries* α *into* $\phi(1)$ *(via h)*. We also say α is continued along Γ to $\phi(1)$.

The following is an important lemma about monic irreducible pseudopolynomials:

LEMMA 5.2. *Let Δ be a polydisc about 0 in $\mathbb{C}^{n-1}$ and let $h(z, z_n)$ be a monic irreducible pseudopolynomial in Δ. Let $G(z)$ be the discriminant of $h(z, z_n)$, let $F(z)$ be an analytic function in Δ such that $F(z) \neq 0$ implies $G(z) \neq 0$ for all $z \in \Delta$, and let $U = \{z \in \Delta | F(z) \neq 0\}$. Let $w \in U$ and let α be a root of $h(w, z_n)$ (as a polynomial in z_n). Then for every $w' \in U$ and every root α' of $h(w', z_n)$ (a polynomial in z_n), there exists a path Γ in U from w to w' such that $\Gamma_h[\alpha] = \alpha'$.*

PROOF. A proof of this result, using slightly different notation and terminology, is given by Bochner & Martin (1948, Chapter 9, Sec. 3, pp. 194–198). $\square$

REMARK. The above lemma implies that the monic irreducible factors of a monic pseudopolynomial of positive degree whose discriminant does not vanish identically are uniquely determined.

IDEA OF PROOF OF THEOREM 3.3. The hypothesis as to the order-invariance of the discriminant of h in the complex x-axis near 0 amounts to assuming that the zero set of the discriminant of h is identical with the complex x-axis near 0 (or is empty). Our approach is to consider a monic irreducible factor h_i of h first. Now h_i satisfies the same hypotheses as h. Thus the complement U of the zero set of the discriminant of h_i is topologically the product of a punctured plane and a full plane (or, if the zero set is empty, simply $\mathbb{C} \times \mathbb{C}$). Lemma 5.2 implies that all roots of h_i over U can be obtained from a given root by continuation along some path in U. Furthermore, continuous deformation (or homotopy) of any such path within U yields the same root of h_i. But any such path in U from a point (a, b') to itself can be continuously deformed within U to a path with constant x-value $x = a$. We conclude that $h_i(a, y, z)$, as a pseudopolynomial in y and z, remains irreducible, and hence that there is just one root (necessarily of multiplicity equal to the degree of h_i) of $h_i(a, 0, z)$ (as a polynomial in z). By considering the resultant of the pair of monic irreducible factors h_i, h_j of h we then see that the root of $h_i(a, 0, z)$ must be equal to the root of $h_j(a, 0, z)$. $\square$

We now give the details of the

PROOF OF THEOREM 3.3. By Lemma A.6, there exists a polydisc $\Delta_2 \subseteq \Delta_1$ about 0 such that the zero set of F in Δ_2 is either empty or equal to $T_* \cap \Delta_2$. Let (r, s) be the polyradius of Δ_2. Factor h into irreducible Weierstrass polynomials:

$$h = h_1 \ldots h_k.$$

Let $1 \leqslant i \leqslant k$, and let $G(x, y)$ be the discriminant of $h_i(x, y, z)$. By Lemma A.3, G is order-invariant in $T_* \cap \Delta_2$ (as G is a factor of F, by Lemma A.1, and F is order-invariant in $T_* \cap \Delta_2$, by hypothesis). But the zero set of G in Δ_2 is contained in the zero set of F in Δ_2 (as G is a factor of F). Hence, the zero set of G in Δ_2 is either empty or equal to $T_* \cap \Delta_2$.

We shall show that for each point a with $|a| < r$, there exists exactly one distinct root of $h_i(a, 0, z)$ (considered as a polynomial in z). This is clearly true if $G \neq 0$ in Δ_2, as in this case the degree of h_i is 1. Suppose, on the other hand, that the zero set of G in Δ_2 is equal to $T_* \cap \Delta_2$, and that for some a with $|a| < r$, $h_i(a, 0, z)$ has $l \geqslant 2$ distinct roots, say $\alpha_1, \ldots, \alpha_l$, with multiplicities $m_1, \ldots, m_l$ respectively. Choose disjoint open discs $D_1, \ldots, D_l$ in the z-plane about $\alpha_1, \ldots, \alpha_l$ respectively. By root continuity of $h_i(a, y, z)$ (considered as a pseudopolynomial in y and z), there exists a disc $D' : |y| < s'$ in the y-plane such that $s' \leqslant s$, and for every fixed b in D', there are exactly m_j roots (counted according to multiplicity) of $h_i(a, b, z)$ (a polynomial in z) in D_j, for $1 \leqslant j \leqslant l$.

Let b' be a non-zero point of D', and let α and β be roots of $h_i(a, b', z)$ in D_1 and D_2 respectively. Let $U = \{(x, y) \in \Delta_2 : G(x, y) \neq 0\}$. Then $U = \Delta_2 - T_*$. By Lemma 5.2, there is a path $\Gamma(t) = (\Gamma_x(t), \Gamma_y(t))$ in U from (a, b') to itself such that $\Gamma_{h_i}[\alpha] = \beta$. We can deform Γ in U to a path Γ' in the disc $x = a$, $|y| < s$ (contained in the plane $\{a\} \times \mathbb{C}$) by means of the path homotopy (Munkres, 1975, Chapter 8)

$$H(u, t) = ((1-u)\Gamma_x(t) + ua, \Gamma_y(t))$$

(for which $0 \leqslant u \leqslant 1$, $H(0, t) = \Gamma(t)$ and $H(1, t) = (a, \Gamma_y(t))$). Furthermore, we can deform Γ' in U to a path Γ'' along the circle $x = a$, $|y| = |b'|$ by means of the path homotopy

$$K(u, t) = \left(a, (1-u)\Gamma_y(t) + \frac{u\Gamma_y(t)|b'|}{|\Gamma_y(t)|} \right).$$

Let ϕ'' be the path in $\mathbb{C}^1$ such that $\phi''(0) = \alpha$ and $h_i(\Gamma''(t), \phi''(t)) = 0$ for all $t \in [0, 1]$, given by Lemma 5.1. By Theorem 4.3 of Chapter 8 of Munkres (1975) $\phi''(1) = \beta$ (which implies $\Gamma''_{h_i}[\alpha] = \beta$). Yet, as $\phi''(t)$ is a root of $h_i(\Gamma''(t), z)$ (a polynomial in z) for each fixed $t \in [0, 1]$, we must have that

$$\phi''(t) \in \bigcup_{j=1}^{l} D_j,$$

for each $t \in [0, 1]$. Hence, as ϕ'' is continuous, the D_j are disjoint, and $\phi''(0) \in D_1$, we must have that $\phi''(t) \in D_1$ for every $t \in [0, 1]$. This contradicts $\phi''(1) = \beta$, as β is an element of D_2. Hence, our assumption that $h_i(a, 0, z)$ has more than one distinct root must be false. Thus $h_i(a, 0, z)$ has exactly one distinct root, for each fixed a in the disc $|a| < r$.

We can now show that $h(a, 0, z)$ has exactly one distinct root, for each fixed a in the disc $|x| < r$. There is nothing further to prove if $k = 1$, so assume $k > 1$. Let $1 \leqslant i < j \leqslant k$ and let $R(x, y)$ be the resultant of $h_i(x, y, z)$ and $h_j(x, y, z)$. By Lemma A.3, R is order-invariant in $T_* \cap \Delta_2$ (as R is a factor of F, by Lemma A.1, and F is invariant in $T_* \cap \Delta_2$, by hypothesis). But the zero set of R in Δ_2 is contained in the zero set of F in Δ_2 (as R is a factor of F). Hence, the zero set of R in Δ_2 is equal to $T_* \cap \Delta_2$ (the zero set of R in Δ_2 is non-empty because $R(0, 0) = 0$). Let $|a| < r$ and let α_i and α_j be the unique roots of $h_i(a, 0, z)$ and $h_j(a, 0, z)$ respectively. Then we must have $\alpha_i = \alpha_j$, as $\alpha_i \neq \alpha_j$ would imply $R(a, 0) \neq 0$. Hence, $h(a, 0, z)$ has exactly one distinct root (necessarily of multiplicity m). Theorem 3.3 has been proved. $\square$

6. Examples

As remarked following the definitions of the projection operators $PROJ$ and P, there may be elementary improvements that can be made in practice to reduce the size of the sets $PROJ(A)$ and $P(A)$ (where A is a set of r-variate integral polynomials). In fact, as pointed out by Collins (1975, p. 160), we can always use the following set in place of $psc(red(A))$ in the definition of $PROJ(A)$:

$$\{psc_j(red^k(f), red^l(g)|f, g \in A, f < g, 0 \leqslant k \leqslant \deg(f),$$
$$0 \leqslant l \leqslant \deg(g), 0 \leqslant j < \min(\deg(red^k(f)), \deg(red^l(g)))\},$$

where "$<$" is an arbitrary linear ordering of the elements of A. That is to say, one does not need to compute psc_j's of pairs of different reducta of the same element of A. This is incorporated into Arnon et al.'s (1984a) definition of $PROJ$.

Collins (1975, p. 176) notes that if the first i coefficients of some polynomial f in A can be seen to have only finitely many (or no) common zeros in $\mathbb{R}^{r-1}$, then $red^k(f)$ can be excluded from $red(A)$ for $k \geqslant i$. Moreover, in this case, one need include no more than the first i coefficients of f in $coeff(A)$. Thus, in particular, when the leading coefficient of f is a non-zero integer constant, $red^k(f)$ can be excluded from $red(A)$ for $k \geqslant 1$, and no coefficients of f need be included in $coeff(A)$.

The algorithms CAD from Arnon *et al.* (1984*a*) and $CADR3$ from section 3 of the present paper have been implemented using the SAC-2 computer algebra system. All of the improvements mentioned above have been incorporated into the SAC-2 programs CAD and $CADR3$. We remark that the SAC-2 program CAD, like $CADR3$, computes finest squarefree bases prior to projection.

Slightly modified versions of the standard SAC-2 programs CAD and $CADR3$ have been applied to several examples and observations relating to two of these examples are presented in sections 6.1 and 6.2 respectively. The computing times reported in section 6.1 were measured on a MicroVAX II computer running the MicroVMS operating system, and the times reported in the second subsection were measured on a VAX 11/780 computer running the UNIX operating system.

6.1. CATASTROPHE SURFACE AND SPHERE

Two well-known surfaces are the unit sphere

$$(f(x, y, z) = z^2 + y^2 + x^2 - 1 = 0)$$

and the catastrophe surface

$$(g(x, y, z) = z^3 + xz + y = 0).$$

Each surface by itself would present a quite trivial application of the cad algorithm. However, an interesting example for the algorithm can be made by taking the two surfaces together, that is, taking the input set to be $A = \{f, g\}$.

Regard f and g as polynomials in the main variable z over the ring $\mathbb{Z}[x, y]$. Now A is its own finest squarefree basis. The reduced projection $P(A)$ of A computed by $CADR3$ consists of the discriminant D_f of f, the discriminant D_g of g and the resultant R of f and g:

$$D_f = -4y^2 - 4x^2 + 4$$
$$D_g = -27y^2 - 4x^3$$
$$R = y^6 + 3x^2y^4 - 2xy^4 - 3y^4 + 3x^4y^2 - 4x^3y^2 - 5x^2y^2 + 4xy^2 + 4y^2 +$$
$$x^6 - 2x^5 - 2x^4 + 4x^3 + 2x^2 - 2x - 1.$$

(According to remarks made above it is not necessary to include any coefficients of either f or g in the projection, as f and g are both monic.) The three curves defined by these polynomials are illustrated in Fig. 1. Note that the curve $R = 0$ has an isolated point on the x-axis.

The projection phase of $CADR3$, that is, the computation of the bivariate and univariate projections, took 37·3 seconds. The base phase of the algorithm, that is, the construction of the decomposition of the real line, took 16·4 seconds. The first stage of the extension phase of the algorithm, that is, the construction of the $P(A)$-invariant cad of the plane, took approximately 3 hours and 20 minutes. This cad of the plane is illustrated in Fig. 2. Let the cylinders of this cad be numbered consecutively from left to right, starting at 1. (Then the two-dimensional cylinders or strips have odd numbers and the one-dimensional cylinders or vertical lines have even numbers.) Most of the time for construction of this cad was spent computing sample points for the cells in cylinders 4 and 6.

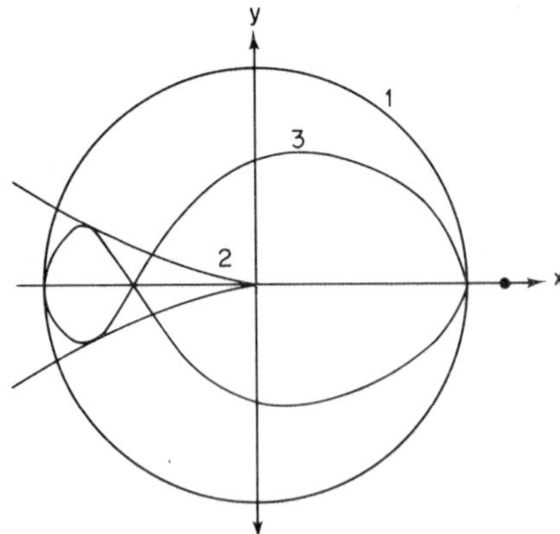

Fig. 1. Where $f(x, y, z) = z^2 + y^2 + x^2 - 1$ and $g(x, y, z) = z^3 + xz + y$, curve 1 is $discr(f) = D_f = 0$, curve 2 is $discr(g) = D_g = 0$, curve 3 is $res(f, g) = R = 0$.

The cad constructed by *CADR3* actually includes several one-dimensional cylinders containing no portions of any curve. These extraneous cylinders are not included in Fig. 2. Two such extraneous cylinders lie between cylinders 14 and 16, and five of them lie to the right of cylinder 16. These cylinders are defined by certain real roots of the polynomials $discr(R)$, $res(D_f, D_g)$, and $res(D_g, R)$, which real roots correspond to non-real intersections of pairs of the complex curves $R = 0$, $\partial R / \partial y = 0$, $D_f = 0$ and $D_g = 0$.

Figures 1 and 2 were drawn with the help of cell adjacency information which was provided by the algorithm from Arnon *et al.* (1984*b*).

We have not yet attempted to run *CADR3* long enough to construct an *A*-invariant cad of $\mathbb{R}^3$.

The projection $PROJ(A)$ of A computed by *CAD* is the set

$$P(A) \cup \{psd_1(g), psc_1(f, g)\},$$

where $psd_1(g) = -6x$ and $psc_1(f, g) = -y^2 - x^2 + x + 1$. (By earlier remarks, no reducta of f or g, except for f and g themselves, need be considered, and no coefficients of f or g need be included in $PROJ(A)$.) The curve $psc_1(f, g) = 0$ is the circle with centre $(1/2, 0)$ and radius $\sqrt{5}/2$. This circle passes through the two real singularities of the curve $R = 0$.

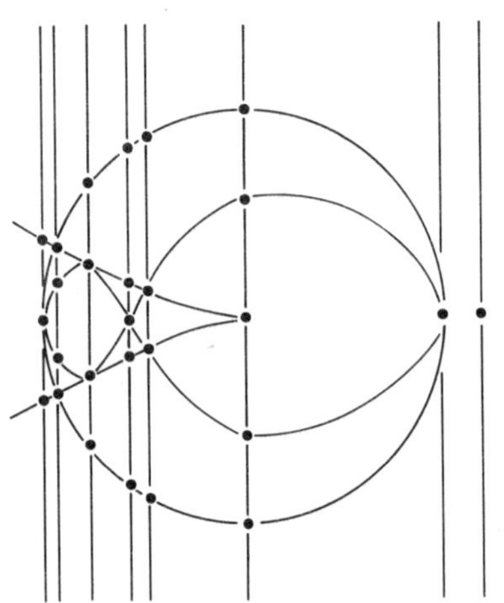

Fig. 2. Cad of the plane for the curves of Fig. 1.

The projection phase of CAD took 45·9 seconds, and the base phase took 12·2 seconds. The construction of the $PROJ(A)$-invariant cad of the plane took approximately 5 hours and 45 minutes. This cad of the plane has one more "meaningful" cylinder, and two more extraneous cylinders than the $P(A)$-invariant cad of the plane.

6.2. RANDOM TRIVARIATE POLYNOMIAL

The following trivariate polynomial of degree 4 in the main variable z, degree 1 in each of the variables x and y, and random integer coefficients from the closed interval $[-1, +1]$ was generated:

$$f(x, y, z) = (y-1)z^4 + xz^3 + x(1-y)z^2 + (y-x-1)z + y.$$

The set $A = \{f\}$ was supplied as input to each of the SAC-2 programs CAD and $CADR3$. The polynomial f is found to be primitive and irreducible, and A is hence its own finest squarefree basis.

The projection $PROJ(A)$ of A, computed by CAD, is the set

$$PROJ(A) = \{ldcf(f), discr(f), psd_1(f), psd_2(f), ldcf(g), discr(g)\},$$

where $ldcf$ denotes "leading coefficient", and g is the reductum of f. (The first two coefficients of f, viz. $y-1$ and x, have only one common zero in the plane. Hence, according to the remarks made at the beginning of section 6, it is not necessary to consider any other reducta of f besides f and g, and the other coefficients of f can be left out of $coeff(A)$. It is also the case that $ldcf(f)$ and $discr(g)$ have only finitely many common zeros in the plane. Hence (Collins, 1975, p. 177), $psd_1(g)$ is also superfluous.)

The characteristics of the polynomials in $PROJ(A)$ are summarized in Table 1. CAD computes the finest squarefree basis B_1 for $prim(PROJ(A))$, constructs the projection of B_1, and then performs a squarefree basis computation. The set U_1 of univariate basis polynomials obtained is described in Table 2. Note that U_1 contains 17 polynomials, including a polynomial of degree 22 (the highest degree present), the maximum length of whose coefficients is 16 decimal digits. CAD isolates the real roots of the polynomials in U_1. A total of 31 real roots is found, yielding a decomposition of the real line into $2 \times 31 + 1 = 63$ cells. The computing times for the various subtasks of CAD discussed so far are given in Table 3.

The reduced projection $P(A)$ of A, computed by $CADR3$, is the set

$$P(A) = \{ldcf(f), discr(f), ldcf(g)\}.$$

$CADR3$ computes the finest squarefree basis B_2 for $prim(P(A))$, constructs the projection of B_2, and then performs a squarefree basis computation. The set U_2 of univariate basis polynomials obtained is described in Table 2. $CADR3$ determines that the polynomials in

Table 1. Composition of $PROJ(A)$ (coefficient length in decimal digits)

Poly	Deg in x	Deg in y	Total deg	Max coeff length
$ldcf(f)$	0	1	1	1
$discr(f)$	6	6	10	4
$psd_1(f)$	4	4	7	3
$psd_2(f)$	2	2	3	2
$ldcf(g)$	1	0	1	1
$discr(g)$	4	4	7	2

Table 2. Composition of U_1 and U_2 (coefficient length in decimal digits)

Degree	U_1		U_2	
	No. polys	Max coeff length	No. polys	Max coeff length
22	1	16	0	
21	1	12	0	
15	1	8	0	
11	1	10	0	
10	1	6	1	6
6	2	7	1	1
4	3	3	2	3
3	3	2	0	
2	1	2	1	2
1	3	1	1	1
	17		6	

Table 3. Computing times in seconds for subtasks

	CAD	CADR3
Computation of basis for input	10·9	10·9
Construction of bivariate proj	303	4·8
Computation of bivariate basis	11·4	2·8
Construction of univariate proj	246	114
Computation of univariate basis	131	24·5
Real root isolation	28·2	3·1
Total	730	160

U_2 have a total of 11 real roots. Thus the decomposition of the real line has $2 \times 11 + 1 = 23$ cells. The computing times for the various subtasks of $CADR3$ are included in Table 3, for comparison with those of CAD.

We have not yet attempted to run either program long enough to complete construction of the cad of the plane. It is to be expected that our current version of the program would take a considerable amount of time and space to do this, due to the high cost of sample point construction.

This paper is based on the author's PhD thesis. I would like to acknowledge the support, guidance and encouragement given to me by my thesis advisor, George Collins. Thanks also to Pierre Milman for helping me to better understand Zariski's theorem. I am grateful to the referees for their helpful comments.

References

Arnon, D. S., Collins, G. E., McCallum, S. (1984a). Cylindrical algebraic decomposition I: the basic algorithm. *SIAM J. Comp.* **13/4**, 865–877.

Arnon, D. S., Collins, G. E., McCallum, S. (1984b). Cylindrical algebraic decomposition II: an adjacency algorithm for the plane. *SIAM J. Comp.* **13/4**, 878–889.

Bochner, S., Martin, W. T. (1948). *Several Complex Variables.* Princeton: Princeton Univ. Press.

Brown, W. S., Traub, J. F. (1971). On Euclid's algorithm and the theory of subresultants. *J. Assoc. Comp. Mach.,* **18/4**, 505–514.

Collins, G. E. (1975). Quantifier elimination for real closed fields by cylindrical algebraic decomposition. In: Second GI Conference on Automata Theory and Formal Languages. *Springer Lec. Notes Comp. Sci.* **33**, 134–183.

Gunning, R. C., Rossi, H. (1965). *Analytic Functions of Several Complex Variables.* Englewood Cliffs: Prentice-Hall.

Kahn, P. J. (1978). Private communication to G. E. Collins.

Kaltofen, E. (1982). Polynomial factorization. In: *Computing,* Supplementum 4: *Computer Algebra—Symbolic and Algebraic Computation.* Vienna and New York: Springer-Verlag.

Kaplan, W. (1966). *Introduction to Analytic Functions.* Reading: Addison-Wesley.

Lang, S. (1984). *Algebra,* 2nd edn. Menlo Park: Addison-Wesley.

Loos, R. G. K. (1982). Computing in algebraic extensions. In: *Computing,* Supplementum 4: *Computer Algebra—Symbolic and Algebraic Computation.* Vienna and New York: Springer-Verlag.

Marden, M. (1966). *Geometry of Polynomials,* 2nd edn. Providence: American Mathematical Society.

McCallum, S. (1984). *An Improved Projection Operation for Cylindrical Algebraic Decomposition.* PhD thesis, Technical Report No. 578, Computer Sciences Department, University of Wisconsin-Madison.

*Munkres, J. (1975). *Topology—A First Course.* Englewood Cliffs: Prentice-Hall.

Tarski, A. (1951). *A Decision Method for Elementary Algebra and Geometry,* 2nd edn. University of California Press.

*Walker, R. T. (1978). *Algebraic Curves.* New York: Springer-Verlag.

Whitney, H. (1972). *Complex Analytic Varieties.* Menlo Park: Addison-Wesley.

Zariski, O. (1935). *Algebraic Surfaces.* New York–Heidelberg–Berlin: Springer-Verlag.

Zariski, O. (1965). Studies in equisingularity II. *Amer. J. Math.* **87/4,** 972–1006.

Zariski, O. (1975). On equimultiple subvarieties of algebroid hypersurfaces. *Proc. Nat. Acad. Sci., USA* **72/4,** 1425–1426.

Appendix

LEMMA A.1. *Let $f(x)$ and $g(x)$ be polynomials of degrees $m > 0$ and $n > 0$ respectively over an integral domain R. Assume that the characteristic of R divides neither m nor n. Then*

$$discr(fg) = discr(f)(res(f, g))^2 discr(g).$$

PROOF.

$$
\begin{aligned}
res(fg, (fg)') &= res(fg, f'g + fg') \\
&= res(f, f'g + fg')\, res(g, f'g + fg') \\
&= res(f, f'g)\, res(g, fg') \\
&= res(f, f')\, res(f, g)\, res(g, f)\, res(g, g')
\end{aligned}
$$

by Theorems 3 and 4 from Loos (1982). The result now follows by Theorem 3.0. $\square$

LEMMA A.2. *Let $f(x, y, z)$ be a primitive polynomial in $\mathbb{Z}[x, y, z]$. Then there exist only finitely many common zeros of the (bivariate) coefficients of $f(x, y, z)$.*

PROOF. By induction on the degree of f in z using Bezout's theorem (Walker, 1978, Theorem III–3.1). $\square$

LEMMA A.3. *Let $K = \mathbb{R}$ or $\mathbb{C}$. Let f and g be analytic functions defined in a connected, open subset U of K^n, and assume that neither f nor g vanishes identically. Let S be a connected subset of U. Then fg is order-invariant in S if and only if both f and g are order-invariant in S.*

PROOF. Now

$$\operatorname{ord}_q fg = \operatorname{ord}_q f + \operatorname{ord}_q g, \tag{A.1}$$

for all $q \in U$. Assume that both f and g are order-invariant in S. Then, by (A.1), fg is order-invariant in S. Conversely, assume that fg is order-invariant in S. Let $p \in S$. Then $\operatorname{ord}_p fg < \infty$, as fg does not vanish identically in U (by the identity theorem). Now there exists a neighbourhood $V \subseteq U$ of p such that $\mathrm{ord}_q f \leqslant \mathrm{ord}_p f$ and $\mathrm{ord}_q g \leqslant \mathrm{ord}_p g$ for every $q \in V$ (by continuity of the partial derivatives of f and g). It now follows from (A.1) that both f and g are order-invariant in $V \cap S$. Hence, by connectedness of S, both f and g are order-invariant in S. $\square$

LEMMA A.4. *Let $E(u, v)$ be real analytic in a neighbourhood of 0 in $\mathbb{R}^2$, let the power series expansion of $E(u, v)$ about 0 be absolutely convergent in the polydisc $\Delta : |u| < r, |v| < s$ about 0 in $\mathbb{C}^2$, and let $E_*(u, v)$ denote the sum of this series (for $(u, v) \in \Delta$), a complex analytic function in Δ. Suppose that E is order-invariant in the real u-axis $T = \{(u, 0)|u \in \mathbb{R}\}$ near the origin. Then E^* is order-invariant in the complex u-axis $T_* = \{(u, 0)|u \in \mathbb{C}\}$ near the origin.*

PROOF. Let $\mathrm{ord}_0 E = m$. If $m = \infty$, then all partial derivatives of E vanish at 0, and hence E_* vanishes identically near 0. So assume $m < \infty$. Let

$$P_*(u, v) = \frac{\partial^{i+j} E_*}{\partial u^i \, \partial v^j}$$

be a partial derivative of $E_*(u, v)$ of order $i+j$. By (the proof of) Theorem 56 of Kaplan (1966) $P_*(u, v)$ is analytic in Δ_1, and its power series expansion about 0, which is absolutely convergent in Δ_1, is obtained by differentiating the power series for E about 0 term by term. Let

$$P(u, v) = \frac{\partial^{i+j} E}{\partial u^i \, \partial v^j}$$

for $(u, v) \in \Delta \cap \mathbb{R}^2$. Clearly, $P(u, v) = P_*(u, v)$ for all $(u, v) \in \Delta \cap \mathbb{R}^2$. Assume $i+j < m$. Then $P(u, 0) = 0$ for all u in some neighbourhood of 0 in $\mathbb{R}^1$. Hence, when one substitutes $v = 0$ into the power series expansion for $P(u, v)$ about 0, one obtains the zero power series in u. It follows that $P_*(u, 0) = 0$ for all u in some neighbourhood of 0 in $\mathbb{C}^1$ (as P and P_* have the same power series about 0). We have shown that every partial derivative of E_* of order less than m vanishes in T_*, near 0.

As $\mathrm{ord}_0 E = \mathrm{ord}_0 E_* = m < \infty$, some partial derivative of E_* of order m does not vanish at 0, and hence does not vanish in a neighbourhood of 0 in $\mathbb{C}^2$. It now follows that, for all points $(u, 0)$ in some neighbourhood of 0 in $\mathbb{C}^2$, $\mathrm{ord}_{(u, 0)} E_* = m$. Thus E_* is order-invariant in T_*, near 0. $\square$

LEMMA A.5. *Let $q(u, v, z)$ be a nowhere-vanishing function in the polydisc $\Delta' : |u| < r_2$, $|v| < s_2$, $|z| < \varepsilon$, let $g(u, v, z)$ be a pseudopolynomial of degree $d \geqslant 1$ in the polydisc $\Delta_2 : |u| < r_2, |v| < s_2$, and let $h(u, v, z)$ be a Weierstrass polynomial of degree $m \geqslant 1$ in Δ_2. Assume that the relation*

$$g(u, v, z) = q(u, v, z)h(u, v, z) \tag{A.2}$$

holds for all $(u, v, z) \in \Delta'$, and that for each fixed point (u, v) of Δ_2, every root of $h(u, v, z)$ (a polynomial in z alone) is contained in the disc $|z| < \varepsilon$. Then $q(u, v, z)$ is a pseudopolynomial in Δ_2.

PROOF. Let (u, v) be a fixed point of Δ_2. Let ξ be a root of $h(u, v, z)$ (a polynomial in z alone). Then $|\xi| < \varepsilon$, by hypothesis. Hence, as (A.2) holds near the point ξ in the z-plane, and as $q(u, v, z) \neq 0$ near ξ, $g(u, v, \xi) = 0$ and the multiplicity of the root ξ of $h(u, v, z)$ (a polynomial in z alone) is equal to the multiplicity of the root ξ of $g(u, v, z)$ (a polynomial in z alone). It follows that $h(u, v, z)|g(u, v, z)$ in $\mathbb{C}[z]$ and hence that the quotient $q(u, v, z)$

is a polynomial in z. Let l be the degree of $g(u, v, z)$. Then $m \leqslant l \leqslant d$, and the degree of $q(u, v, z)$ is $l - m$. Thus $q(u, v, z)$ can be expressed as follows:

$$q(u, v, z) = q_0(u, v)z^{d-m} + q_1(u, v)z^{d-m-1} + \ldots + q_{d-m}(u, v),$$

where $q_0(u, v) = \ldots = q_{d-l-1}(u, v) = 0$, and $q_{d-l}(u, v) \neq 0$. Letting (u, v) now be a variable point of Δ_2, the coefficients $q_j(u, v)$ of $q(u, v, z)$ can be regarded as functions defined in Δ_2. That these functions are in fact analytic in Δ_2 can be seen by equating coefficients in (A.2) $\square$

LEMMA A.6. *Let $F(x, y)$ be analytic and not identically vanishing in the polydisc $\Delta: |x| < r, |y| < s$ about 0 in $\mathbb{C}^2$. Assume that F is order-invariant in $T* \cap \Delta$, where $T*$ is the complex x-axis. Then there exists a polydisc $\Delta' \subseteq \Delta$ about 0 such that the zero set of F in Δ' is either empty or equal to $T* \cap \Delta'$.*

PROOF. Let $\mathrm{ord}_0 F = m$. Let

$$F(x, y) = F_0(x) + F_1(x)y + F_2(x)y^2 + \ldots$$

be the power series expansion for F about 0 arranged as an iterated series. Let $|x_0| < r$. The power series expansion for F about $(x_0, 0)$ is

$$F_0'(x - x_0) + F_1'(x - x_0)y + F_2'(x - x_0)y^2 + \ldots$$

where $F_i'(x - x_0)$ is the power series expansion of $F_i(x)$ about x_0. Now $\mathrm{ord}_{(x_0, 0)} F = m$ as F is order-invariant in $T* \cap \Delta$. Therefore, $F_i'(0) = 0$ for $0 \leqslant i \leqslant m-1$. Hence, $F_i(x_0) = 0$ for $0 \leqslant i \leqslant m-1$. We have shown that the functions F_i are identically zero in the disc $|x| < r$, for $0 \leqslant i \leqslant m-1$. Hence,

$$F(x, y) = F_m(x)y^m + F_{m+1}(x)y^{m+1} + \ldots,$$

where $F_m(0) \neq 0$ (as $\mathrm{ord}_0 F = m < \infty$). Thus, if we set

$$N(x, y) = F_m(x) + F_{m+1}(x)y + F_{m+2}(x)y^2 + \ldots,$$

then $F(x, y) = y^m N(x, y)$ for all $(x, y) \in \Delta$. As $N(0, 0) = F_m(0) \neq 0$, there exists a polydisc $\Delta' \subseteq \Delta$ about 0 such that $N(x, y) \neq 0$ for all $(x, y) \in \Delta'$. The zero set of F in Δ' is either empty (when $m = 0$) or equal to $T* \cap \Delta'$ (when $m > 0$). $\square$

LEMMA A.7. *Let x denote the $(r-1)$-tuple $(x_1, \ldots, x_{r-1})$, let f and g be polynomials in $\mathbb{R}[x, x_r]$, and let S be a connected subset of $\mathbb{R}^{r-1}$. Suppose that the product $h = fg$ is delineable on S. Then both f and g are delineable on S.*

PROOF. Let s be a section of h on S. Then s is the graph of some continuous function ϕ from s to $\mathbb{R}$. Suppose that $f(a, \alpha) = 0$, for some $(a, \alpha) \in s$. We will show that $\phi(b)$ is a root of $f(b, x_r)$ of the same multiplicity as that of the root α of $f(a, x_r)$, for all $b \in S$. Let m (respectively m_1, m_2) be the multiplicity of the root α of $h(a, x_r)(f(a, x_r), g(a, x_r)$ respectively). (Note: if α is not a root of $g(a, x_r)$, then α is said to have multiplicity 0.) Then $m = m_1 + m_2$. We claim that there exists a neighbourhood N of a such that if $b \in S \cap N$, then the multiplicity of the root $\phi(b)$ of $f(b, x_r)$ (respectively $g(b, x_r)$) is less than or equal to m_1 (respectively m_2).

It follows by the delineability of h on S that for all $b \in S \cap N$ the multiplicity of the root $\phi(b)$ of $f(b, x_r)$ (respectively $g(b, x_r)$) equals m_1 (respectively m_2). Hence, by connectedness of S, the multiplicity of the root $\phi(b)$ of $f(b, x_r)$ equals m_1 for all $b \in S$. $\square$

J. Symbolic Computation (1988) **5**, 163–187

An Adjacency Algorithm for Cylindrical Algebraic Decompositions of Three-dimensional Space[*]

DENNIS S. ARNON, GEORGE E. COLLINS[†]

AND SCOTT McCALLUM[‡]

Xerox PARC, 3333 Coyote Hill Road, Palo Alto, California 94304, U.S.A.
† *Department of Computer and Information Science, The Ohio State University, Columbus, Ohio 43210 U.S.A.*
‡ *Research School of Physical Science, Australian National University, Canberra ACT 2601, Australia*

(Received 20 April 1985, and in revised form 15 November 1987)

Let $A \subset \mathbf{Z}\,[x_1, \ldots, x_r]$ be a finite set. An *A-invariant cylindrical algebraic decomposition (cad)* is a certain partition of r-dimensional euclidean space E^r into semi-algebraic cells such that the value of each $A_i \in A$ has constant sign (positive, negative, or zero) throughout each cell. Two cells are *adjacent* if their union is connected. We give an algorithm that determines the adjacent pairs of cells as it constructs a cad of E^3. The general technique employed for E^3 adjacency determination is "projection" into E^2, followed by application of an existing E^2 adjacency algorithm (Arnon, Collins, McCallum, 1984). Our algorithm has the following properties: (1) it requires no coordinate changes, and (2) in any cad of E^1, E^2, or E^3 that it builds, the boundary of each cell is a (disjoint) union of lower-dimensional cells.

1 Introduction

In this paper we give an algorithm which determines the pairs of adjacent cells as it constructs a cylindrical algebraic decomposition (cad) of three-dimensional euclidean space E^3. Our algorithm is an extension to E^3 of the work of Arnon *et al.* (1984b), where we gave an algorithm that determines the pairs of adjacent cells as it constructs a cad of the plane E^2.

We begin by formalizing some notions relating to cell boundaries in cad's (the reader may wish to refer to Arnon *et al.*, 1984a, 1984b, for terminology that we do not redefine here). Recall that we call a connected subset of E^r, $r \geq 1$ a *region*, and we say that the *boundary* of a region $X \subset E^r$, written ∂X, is the set of all limit points of X not in X. It is not hard to see that two disjoint regions are adjacent if and only if one contains

[*]This work was supported by the National Science Foundation (Grant MCS-8009357 to the University of Wisconsin-Madison), the Purdue Research Foundation, and the Xerox Corporation. This paper was typeset at Xerox PARC using TeX in the Cedar environment.

0747–1717/88/010163 + 25 $03.00/0　　　　　　　

a boundary point of the other. We say that a cad D of E^r, has the *boundary property* if (1) for each cell c of D, ∂c is the union of (zero or more) lower dimensional cells of D, and (2) if $r \geq 2$, then the induced cad D' of E^{r-1} has the boundary property. It is not hard to see that the following is an equivalent statement of (1): if cells c and d of D are adjacent, then they have different dimensions, and the smaller-dimensional cell is contained in the boundary of the larger-dimensional cell. It is easy to see that any cad of E^1 has the boundary property. The cad's of E^2 that the algorithms we give in this paper will build are *proper* in the sense defined by Arnon *et al.* (1984b). From the fact that any cad of E^1 has the boundary property, and Corollary 2.5 of Arnon *et al.* (1984b), if follows that any proper cad of E^2 has the boundary property. It is a prerequisite for the E^3 adjacency algorithms that we give in this paper that the induced cad of E^2 have the boundary property.

Let D be a cad of E^r, $r \geq 2$. Recall that D consists of stacks over the cells of D', the induced cad of E^{r-1}, and that for any cell $c \in D'$, $S(c)$ denotes the unique stack over c which is part of D. We say that stacks $S(c)$ and $S(d)$ of D are *adjacent* if $\{c, d\}$ is an adjacency of D'. If D' has the boundary property, then clearly any pair of adjacent stacks of D have different dimensions (the dimension of a stack is defined to be the dimension of its sectors), cells of the larger-dimensional stack have boundary points in (cells of) the smaller-dimensional stack, and no cell of the smaller-dimensional stack has any boundary points in (cells of) the larger-dimensional stack. Indeed, if D' has the boundary property, then it is not hard to see that given any stack $S(c)$ of D, and any element s of $S(c)$, then all boundary points of s lie in (certain cells of) $S(c)$ and in (certain cells of) the stacks $S(d)$, where d ranges over all cells in ∂c. In other words, each boundary point of s is either in the same stack as s, or in some adjacent, lower-dimensional stack. Hence all cells of D that are adjacent to s either belong to the same stack as s, or are contained in some adjacent, lower-dimensional stack.

The considerations of the above paragraph give us our strategy for cad and adjacency construction in E^3: we build a certain cad D of E^3 such that the induced cad D' of E^2 has the boundary property, then for each adjacency $\{c, d\}$ of D', we find all interstack adjacencies among the cells of $S(c)$ and $S(d)$ (interstack adjacencies are those in which the two cells involved belong to different stacks). As always, it is trivial to determine the intrastack adjacencies of D (intrastack adjacencies are those in which the two cells involved belong to the same stack). Clearly if a cad of E^2 has the boundary property, then the possible multisets of dimensions of a pair of adjacent cells in it are $\{1, 2\}$, $\{0, 1\}$, and $\{0, 2\}$. Let us be sure that this notation is clear: the multiset $\{0, 1\}$, for example, refers to an adjacent pair of cells where one is a 0-cell and the other a 1-cell.

Given a set of r-variate polynomials $A \subset I_r = \mathbf{Z}\,[x_1, \ldots, x_r]$, a set X in E^{r-1} is *nullifying* (for A) if some $A_i \in A$ is *nullified* on X, i.e. $A_i(\alpha, x_r) = 0$ for all α in X. Previous papers have used the terms "A_i is identically zero on X" (Arnon *et al*, 1984a) and "A_i vanishes identically on X" (McCallum, 1988) as synonyms for "A_i is nullified on X". The essential point of the notion of a proper cad of E^2 is that no cell of the induced cad of the line nullifies the defining polynomial of the cad of E^2. For cad's of E^3, as we will show in Section 2, we must in general change coordinates in order to have a trivariate defining polynomial for the cad that is not nullified by any cell of the induced cad of the plane. The algorithm we present in this paper for constructing cad's of E^3 does not change coordinates. One reason for this is that if we change coordinates, construct a cad, and then return to the original coordinate system, the cells of the decomposition are no longer arranged into cylinders with respect to the original coordinate system.

Such cylindrical arrangement of cells, with respect to the original coordinate system, is crucial to the use of cad's for quantifier elimination (see Collins, 1975). It is possible to put extra cells into the decomposition so that it is cylindrical with respect to two coordinate systems, but this seems likely to be expensive.

It is the case (see Section 2) that for the cad's of three space that our algorithm constructs, any nullifying cell in the induced cad of the plane is 0-dimensional. Such nullifying 0-cells have the interesting property of permitting choice in the exact makeup of the stack in E^3 that is constructed over them. Consider, for example $F(x, y, z) = xz - y$. If we are constructing an F-invariant cad of E^3, the point $< 0, 0 >$ will be a nullifying 0-cell in the induced cad of the plane. Any stack over it is F-invariant, so the question arises - does it matter what stack we choose to use as part of our F-invariant cad of E^3? The answer is yes. We will specify precisely how the stacks over nullifying 0-cells are to be constructed. The requirements we impose yield cad's of E^3 which have the boundary property.

That our cad's of E^3 have the boundary property is useful for two reasons. First, as was the case in Arnon *et al.* (1984b), it implies that there is a certain proper subset of interstack adjacencies (essentially, the so-called "section-section" adjacencies), which, once found, allow us to infer all other interstack adjacencies. Second, it helps us to actually find the adjacencies of this sufficient subset, by guaranteeing us that when we perform certain "projections" of an E^3 adjacency situation into E^2 (e.g. slicing by a certain plane, or computing a certain resultant), the adjacencies we then see in E^2 have a definite correspondence with the actual adjacencies present in E^3, and we can "read off" the E^3 adjacencies from the E^2 data. Sections 3-6 amplify and substantiate these preliminary remarks.

Altogether, we will distinguish four categories of adjacencies in the (induced) cad's of E^2 that we build: $\{1, 2\}$, $\{0, 1\}$, nonnullifying $\{0, 2\}$ (*i.e.* a $\{0, 2\}$ adjacency in which the 0-cell is nonnullifying), and nullifying $\{0, 2\}$ (*i.e.* a $\{0, 2\}$ adjacency in which the 0-cell is nullifying). In Sections 3-6 we consider each of these cases individually. Algorithm SSADJ3 of Section 4 is the key subalgorithm for E^3 adjacency determination: a method for reducing an E^3 adjacency determination to an E^2 determination that we can do with algorithm SSADJ2 of Arnon *et al.* (1984b). In Section 7 we summarize previous sections with a main algorithm (called CADA3) that constructs a cad of E^3 and its adjacencies, and we complete the proof that a cad of E^3 contructed by CADA3 has the boundary property. Section 8 traces CADA3 on an example. We have not yet done extensive study of our implementation of CADA3, but typically the cost of adjacency determination seems not to exceed the cost of cad construction. The examples considered in Arnon (1988) provide further data on CADA3's behavior in practice.

In Arnon *et al.* (1984b) we constructed proper cad's of E^2. In the present paper we construct *basis-determined* cads of E^3 (definition given in Section 2). A cad of E^2 is proper if and only if it is basis-determined, however for euclidean space of dimension three or more, the notion of basis-determined cad is slightly more general. In Section 2 we prove the sameness of the two notions for E^2 (which allows us to use the results and algorithms of Arnon *et al.*, 1984b), and explain the need for basis-determined cad's of E^3.

At the time the work reported in this paper was carried out (1979-1981), the prior work on incidence or adjacency algorithms that we knew of was that of McCallum (1979) for triangulation of real algebraic curves and surfaces, and our own adjacency algorithm for cad's of the plane (Arnon *et. al.*, 1984b). It was also known that adjacency of cells in a cad of E^r, for any r, is decidable (Arnon, 1979). Subsequent to our work, other cad adjacency algorithms have been given (Prill, 1986; Kozen & Yap, 1985; Schwartz & Sharir, 1983). These algorithms can be used for cad's of E^r for any r, and avoid nullifying cells in the induced cad by changing coordinates. We are not aware that any of them has been implemented.

2   Basis-determined cylindrical algebraic decompositions

We say that the sign of an element of I_0 is its sign as an integer, and that for $r \geq 1$, the sign of an element of $I_r = I_{r-1}(x_r)$, is the sign of its leading coefficient, an element of I_{r-1}. An element F of I_r is *positive* if $sign(F)$ is positive.

Recall that for any $F \in I_r = I_{r-1}(x_r)$, the *content* of F, written $content(F)$, is the greatest common divisor of F's coefficients. We adopt the convention that $sign(content(F))$ is chosen to be $sign(F)$. Any F for which $content(F) = \pm 1$ is said to be *primitive*. The *primitive part* of F, written $pp(F)$, is $F/content(F)$. Hence if $F \neq 0$, then $pp(F)$ is primitive and positive. We define $content(0) = 1$ and $pp(0) = 0$. Also, again viewing I_r as $I_{r-1}(x_r)$, we define the *degree* of $F \in I_r$, written $deg(F)$, to be its degree in x_r. If we are interested in the degree of F with respect to some other variable x_i, we shall say "degree of F in x_i".

$B \subset I_r$ is a *basis* if each element of B is primitive, positive, and of positive degree, and if the elements of B are pairwise relatively prime. Let $A \subset I_r$. A basis $B \subset I_r$ is a *basis for A* if (1) For each $A_i \in A$ that is of positive degree, there exist (not necessarily distinct) $B_1, \ldots, B_k \in B$, $k \geq 1$, such that $A_i = content(A_i) \prod_{j=1}^{k} B_j$, and (2) Each $B_j \in B$ divides some $A_i \in A$. Recall that for any $F \in I_r$, $V(F)$ denotes the real variety of F, i.e. the set of all real zeros of F, a subset of E^r. A cad D of E^r, $r \geq 1$, is *basis-determined* if there exists a basis $B \subset I_r$ such that (1) D is B-invariant, (2) every section of D is contained in $V(F)$ for some $F \in B$, and (3) if $r > 1$, then D' is basis-determined. We say that B is a *basis for D*.

THEOREM **2.1** *A cad of E^2 is basis-determined if and only if it is proper (in the sense of Arnon* et al., *1984b).*

PROOF. Suppose D is a proper cad of E^2, with defining polynomial F. Since $V(F)$ is equal to the union of the sections of D, F is not the zero polynomial, and $V(F) = V(pp(F))$. Without loss of generality we may assume that F is also of positive degree (if not, then we may multiply it by $y^2 + 1$). Thus $\{pp(F)\}$ is a basis. Clearly D is $pp(F)$-invariant, and every section of D is contained in $V(pp(F))$. If $G \in I_1$ is a defining polynomial for D', then G is not the zero polynomial, $V(G) = V(pp(G))$, $\{pp(G)\}$ is a basis, it is easy to see that $\{pp(G)\}$ is a basis for D', hence D' is basis-determined, and so by Theorem 2.1 of Arnon *et al.* (1984b), D is basis-determined.

Suppose D is a basis-determined cad of E^2, with basis B. Let B' be a basis for D'. Set $F = \prod B$, and $G = \prod B'$; both F and G are primitive. It is clear that $V(G)$ is the union of all the sections of D', and from the fact that F is primitive it follows that $V(F)$ is the union of all the sections of D. Hence D is proper $\square$

It follows immediately from the next theorem that in a basis-determined cad of E^3, any nullifying cell in the induced cad of E^2 is a 0-cell.

THEOREM **2.2** *Let $F \in I_3$ be primitive and of positive degree. Let R be a region in E^2 such that F is nullified on R. Then R consists of a single point.*

PROOF. As McCallum (1988) notes, this can be proved by induction on the degree of F using Bezout's Theorem (see e.g. Walker, 1978). $\square$

It is the existence of nullifying cells that forces us to have a set of polynomials (the basis) associated with a cad of E^3, and not just the single defining polynomial we had for a proper cad of E^2. Let us see why. When we are given $A \subset I_2$, with each element of A primitive, then it doesn't matter whether we work with the elements of A individually, or $\prod A$, to determine a cad of E^2 (given an appropriate induced cad of E^1); either way we will get the same A-invariant cad. However if $A \subset I_3$, and each element of A is primitive, but some element of A is nullified at a point $p \in E^2$, then we may not be able to construct our cad using just $\prod A$. For $\prod A$ is nullified on p, but some $F \in A$ may be delineable over p, and its sections must be included in the stack we build over p in order to obtain an A-invariant cad (see either Arnon *et al.*, 1984a, or McCallum, 1988, for definitions of delineability; although slightly different, these two definitions are equivalent for the purposes of this paper). Thus we have to work with *sets* of "defining polynomials", rather than single defining polynomials, to build cad's of E^3.

McCallum (1988) and Arnon *et al.* (1984a) have discussed the projection phase of the cad algorithm, and defined a particular projection operator $PROJ$ that maps the input polynomials $A \subset I_r$ to $PROJ(A) \subset I_{r-1}$. Given $A \subset I_r$, we use in this paper a projection operator that is slightly different from $PROJ(A)$, to construct an A-invariant basis-determined cad of E^r. As in McCallum (1988), for any $A \subset I_r$, $r \geq 1$, let $cont(A)$ denote the set of non-zero non-unit contents of the elements of A, and $prim(A)$ the set of primitive parts of those elements of A which have positive degree. Theorem 2.3 introduces and establishes the validity of our new projection operator.

THEOREM **2.3** *Let A be a subset of I_r, $r \geq 2$, and let B be a basis for $prim(A)$. Let R be a region in E^{r-1} which is both $PROJ(B)$-invariant and $cont(A)$-invariant. Then*

(1) Every element of B is either delineable or nullified on R.

(2) Where $B_R \in I_r$ denotes the product of all elements of B which are delineable on R, B_R is delineable on R.

(3) Where $S(B_R, R)$ denotes the (B-invariant) stack over R induced by B_R (see Arnon et al., 1984a, for further discussion of this notation), $S(B_R, R)$ is an A-invariant stack over R.

PROOF. (1) is just a restatement of part of Theorem 3.4 of Arnon *et al.* (1984a). Recall that for any subset $X \subset E^{r-1}$, $Z(X)$ denotes $X \times E$, the cylinder over X. By the same Theorem 3.4, for any $F, G \in B$, each F-section and each G-section of $Z(R)$ are either disjoint or identical, from which it follows that B_R is delineable on R. Clearly $S(B_R, R)$ is B-invariant. Since each element of A is equal to its content times a product of elements of B, and since R is $cont(A)$-invariant, it follows that $S(B_R, R)$ is A-invariant. $\square$

The proof of the following Corollary is straightforward.

COROLLARY **2.4** *Let $A \subset I_r$, let B be a basis for $prim(A)$, and let D' be a $PROJ(B)$-invariant and $cont(A)$-invariant basis-determined cad of E^{r-1}. Then $\cup_{c \in D'} S(B_c, c)$ is an A-invariant, basis-determined cad of E^r (with basis B).*

The algorithm CADA3 we will develop in subsequent sections of this paper essentially constructs the cad of E^3 indicated by Corollary 2.4. However, as we indicated in Section 1, the adjacency algorithms we develop in Sections 3-6 require that CADA3 put additional sections in the stacks over nullifying 0-cells in the plane (it is easy to see that after doing so, we still have an A-invariant, basis-determined cad of E^3). Fig. 2 gives the actual algorithm *ExtendCellToStack* that we use for cad extension; it calls algorithm *CellExtensionPolynomial* given in Fig. 1. At this point we have not fully motivated all the particulars of these two algorithms, but such motivation will be given by the end of Section 6.

Let us now define certain notations and terms used in these and subsequent algorithms. Throughout this and subsequent sections we freely use the variables x, y, z interchangably with x_1, x_2, x_3. For $F \in I_3$, let F_x, F_y, and F_z denote the partial derivatives of F with respect to x, y, and z. Let F and G be elements of I_3. If both F and G have positive degree in y, then $Res_y(F, G)$ denotes the resultant of F and G with respect to y (see Walker (1978) for the definition of resultant). If only one of F and G has positive degree in y, say F, then $Res_y(F, G)$ denotes $G(x, 0, z)$. If neither F nor G has positive degree in y, then $Res_y(F, G)$ is undefined. $Res_x(F, G)$ and $Res_z(F, G)$ are defined similarly. We say that a polynomial $f(w)$, with coefficients in any unique factorization domain, is *squarefree* if it has no multiple factors. The *greatest squarefree divisor* of f, written $gsfd(f)$, is defined to be f divided by $gcd(f, f')$. It is easy to see that $gsfd(f)$ is a squarefree polynomial whose roots are the same as those of f. Kaltofen (1982) and Collins & Loos (1982) discuss squarefree factorization algorithms. We use the phrase *i-section* (*i-sector*) as a shorthand for "i-dimensional cell which is a section (sector) in some cad". When an algorithm has a "cell" as an input (e.g. input c of algorithm *CellExtensionPolynomial*), it is assumed that the index, and a sample point, for that cell are provided in any actual call to the algorithm.

3   Adjacencies over a $\{1, 2\}$ adjacency in the plane

Let S be a stack. Recall from Arnon *et al.* (1984b) that S^*, the *extension* of S, is S together with two infinite sections (at positive and negative infinity). Also, we may speak of the *underlying cylinder* of S, written Z_S, which is just the union of (the elements of) S. Z_S^*, the extension of Z_S, is Z_S together with two infinite sections. Given two adjacent stacks S and T in E^r, $r \geq 2$, we say that S^* has the *unique section boundary property (USBP)* in T^* if for any section s of S^*, the boundary points of s in Z_T^* constitute a section t of T^* such that $dimension(t) < dimension(s)$. Theorem 2.2 of Arnon *et al.* (1984b) establishes that in a proper cad of the plane, any 2-stack has the USBP in each of the two 1-stacks adjacent to it, and so shows us how to determine all section-section adjacencies in a (proper) cad of the plane. "Section-section" adjacencies are those in which both of the cells involved are sections of their respective stacks.

Theorem 2.3 of Arnon *et al.* (1984b) showed us how to infer all interstack adjacencies between two adjacent stacks of a proper cad of the plane, from knowledge of their section-section interstack adjacencies. The proof of Theorem 2.3 is easily generalized to show that in a cad of r-space, $r \geq 2$, if S^* has the USBP in a stack T^*, then the boundary of any sector of S in Z_T^* is an "interval" $[u, v]$ of Z_T^*, such that u and v are sections of T^* (this interval notation is from Arnon *et al.*, 1984b), . Thus in a cad of r-space, $r \geq 2$, when one stack has the USBP in another, determination of the section-section interstack adjacencies between the two stacks suffices for determination of all interstack adjacencies between them.

$$g \leftarrow \textbf{CellExtensionPolynomial}\ (c, B', B)$$

Inputs: c is a cell in a basis-determined cad D' of E^{r-1}, $r \geq 2$. $B' \subset I_{r-1}$ is a basis for D'. $B \subset I_r$ is a basis, such that each element of B is either delineable or nullified on c.

Outputs: Let $p = < p_1, ..., p_{r-1} >$ be the sample point for c, and suppose the real algebraic number γ is a primitive element for p, i.e. each $p_i \in Q(\gamma)$. g is a nonzero squarefree univariate polynomial $g(x_r)$ with coefficients in the field $Q(\gamma)$, and whose real roots are in one-one correspondence with the sections of a B-invariant stack S over c.

(1) [Get initial g, check for nullifying 0-cell in E^2.] Initialize the set Γ to be $\{B_c(p_1, ..., p_{r-1}, x_r)\}$. If c is a 0-cell in E^2 that nullifies some element of B (i.e. $r = 3$ and $B_c \neq \prod B$), then go to step (2), else go to step (3).

(2) [Add more factors of g if nullifying 0-cell in E^2.] Let B_N be the set of elements of B that are nullified on c, i.e. the elements of B that c nullifies. For each $F \in B_N$, do the following steps (2.1) - (2.3):

(2.1) [Possible boundary 0-sections in $S^*(c)$ of $V(F)$ over sections of D' adjacent to c.] Set $\hat{B}(x, y) \in I_2$ to be $\prod B'$. Set $G(x, z) \in I_2$ to be $pp(Res_y(\hat{B}, F))$. Add $G(p_1, z)$ to Γ.

(2.2) [Possible boundary 0-sections in $S^*(c)$ of $V(F)$ over sectors of D' adjacent to c.] Let $M(x) \in I_1$ be the minimal polynomial of p_1. Set $G(y, z) \in I_2$ to be $pp(Res_x(M, F))$. Add $G(p_2, z)$ to Γ.

(2.3) [Possible boundary 0-sections in $S^*(c)$ of $V(F)$ over interior sections of partial derivative projections (cf. Section 6).] For each element $T(x, y)$ of the set $\{\ pp(Res_z(F, F_y)), pp(Res_z(F, F_z))\ \}$, do the following loop: if T of positive degree in y, and if $T(p_1, p_2) = 0$, then set $G(x, z) \leftarrow pp(Res_y(T, F))$, and add $G(p_1, z)$ to Γ.

(3) [Make g.] $g(x_r) \leftarrow gsfd(\prod \Gamma)$. $\square$

Figure 1: Algorithm CellExtensionPolynomial.

$$\textbf{ExtendCellToStack}\ (c, B', B; g, J, I, L)$$

Inputs: c is a cell in a basis-determined cad D' of E^{r-1}, $r \geq 2$. $B' \subset I_{r-1}$ is a basis for D'. $B \subset I_r$ is a basis, such that each element of B is either delineable or nullified on c.

Outputs: Let p be the sample point for c, and suppose the real algebraic number γ is a primitive element for p (see Arnon *et al.* (1984a) for this terminology). g is a nonzero univariate polynomial $g(x_r)$ with coefficients in the field $Q(\gamma)$, and whose real roots are in one-one correspondence with the sections of a B-invariant stack S over c. J is a list of isolating intervals for the real roots of g. I is a list of cell indices for the elements of S (cell indices are defined in Section 4 of Arnon *et al.*, 1984a). L is a list of the intrastack adjacencies of S.

(1) [Get g, isolate roots, make up outputs.] $g(x_r) \leftarrow CellExtensionPolynomial(c, B', B)$. Isolate the real roots of g to get J. Assuming we know the cell index of c, then once we know how many real roots g has, we know the indices for the cells that comprise S, and so it is easy to make up I and L. $\square$

Figure 2: Algorithm ExtendCellToStack.

We now establish that in a basis-determined cad of E^3, we have the USBP over a $\{1,2\}$ adjacency in the induced cad of the plane.

THEOREM **3.1** *Let D be a basis-determined cad of E^3. Let $\{c^1, c^2\}$ be a $\{1,2\}$ adjacency of D'. Then $S^*(c^2)$ has the unique section boundary property in $S^*(c^1)$.*

PROOF. If s is an infinite section of $S^*(c^2)$, then $\partial s \cap Z^*(c^1)$ is the corresponding infinite section of $S^*(c^1)$. Suppose that s is a finite section of $S^*(c^2)$, and that it has a finite boundary point p in $Z(c^1)$. Let $B \subset I_r$ be a basis for D. There is some $F \in B$ such that $s \subset V(F)$, and since $V(F)$ is closed, $p \in V(F)$. By Theorem 2.2, F doesn't vanish on $Z(c^1)$, hence by the F-invariance of $S(c^1)$, p lies on a section t of $S(c^1)$, and $t \subset V(F)$. We claim there is a relative open ball U of p in t, such that $U \subset \partial s$. Suppose to the contrary that for every open ball U of p in t, there exists $q \in U$ such that $q \notin \partial s$. For any such q, $q = <\alpha, \beta>$, $\alpha \in c^1$, $\beta \in E$, s has a limit point $r \in Z^*(\alpha)$. Since $r \in V(F)$, r lies on a section u of $S^*(c^1)$. Suppose that s is an f-section, for a continuous function $f : c^1 \to E$. As the radius of U goes to 0, by the continuity of f, r approaches p. But then u and t intersect at p, a contradiction. Hence for any finite boundary point p of s in $Z(c^1)$, lying on section t of $S(c^1)$, there is a relative open ball U of P in t such that $U \subset \partial s$. If there exists a point $q \in t$ which is not in ∂s, then clearly there is a relative open ball of q in t which is disjoint from ∂s. Hence t can be separated into disjoint nonempty open (in the relative topology on t) subsets, contradicting its connectedness. Hence if s has a finite boundary point in $Z(c^1)$, then there is a finite section of $S(c^1)$ contained in ∂s, and clearly that is all of $\partial s \cap Z^*(c^1)$. If s has no finite boundary points in $Z(c^1)$, then it is not difficult to see that $\partial s \cap Z^*(c^1)$ is an infinite section of $Z^*(c^1)$. Obviously t has dimension one and s has dimension two, i.e. t is of lower dimension. $\square$

We now sketch an algorithm for determining the section-section adjacencies over a $\{1,2\}$ adjacency $\{c^1, c^2\}$ of a basis-determined cad D' of E^2. Suppose first that c^1 is a section of D', with $c^1 \subset V(G)$ for some $G(x,y) \in B'$, where B' is a basis for D'. Then there exists a 1-cell d in the induced cad of E^1, such that $c^1 \subset Z(d)$, and c^2 is the sector of $Z(d)$ either immediately above or immediately below c^1. d is an open interval, say (u_1, u_2). Fig. 3 illustrates the way things might look in the xy-plane. Our strategy is to pick some rational $a \in (u_1, u_2)$ and take a two-dimensional "slice" of the cad of E^3 by the plane $x = a$. We then apply algorithm SSADJ2 of Arnon *et al.* (1984b) to determine the adjacencies present in the slicing plane, between the "slices" of the two stacks of interest. The USBP we established in Theorem 3.1 guarantees us that regardless of what particular $a \in (u_1, u_2)$ we pick, we will see the same adjacencies in the slicing plane, and that the adjacencies we see there are in one-one correspondence with the adjacencies in E^3. If c^1 is a sector, then it has the form $\{\alpha\} \times (v_1, v_2)$, for real algebraic numbers α, v_1, and v_2, such that $\alpha = c^0$ is a 0-cell in the induced cad of E_1. We proceed in a fashion entirely analogous to the section case, picking a rational $b \in (v_1, v_2)$, and slicing by the plane $y = b$. The complete algorithm is given in Fig. 4.

4   Adjacencies over a $\{0,1\}$ adjacency in the plane

In this section, we introduce the main idea of subalgorithm SSADJ3 with an example, prove a theorem to establish the general validity of this idea, and conclude with our algorithm for determination of interstack adjacencies over a $\{0,1\}$ adjacency in the plane. At the outset, it appears that the cases of a nonnullifying $\{0,1\}$ adjacency and

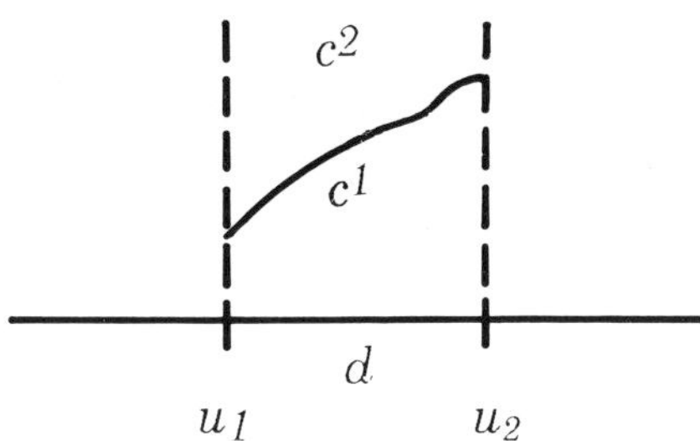

Figure 3: Typical $\{1, 2\}$ adjacency situation in the plane.

$$L \leftarrow \textbf{AdjacenciesOver12}\ (c^1, c^2, B)$$

Inputs: c^1 is a 1-cell, and c^2 a 2-cell, in a cad D' of E^2 induced by a basis-determined cad D of E^3, such that $\{c^1, c^2\}$ is an adjacency of D'. B is a basis for D.

Output: L is a list of all interstack adjacencies between $S^*(c^1)$ and $S(c^2)$.

(1) [Section-section interstack adjacencies: c^1 a section.] Suppose that c^1 is a section. Then c^1 has index (i, j), i odd and j even, and sample point $< a, \beta >$, where a is rational, and β is an algebraic number which is also a primitive element for the point. Furthermore, c^2 is either cell $(i, j - 1)$ or cell $(i, j + 1)$. Let $< a, b_1 >$ and $< a, b_2 >$ be the respective sample points for cells $(i, j - 1)$ and $(i, j + 1)$; assume that b_1 and b_2 are rational numbers. Set $B^* \leftarrow \prod B$, and let $H(y, z)$ be an element of I_2 obtained by multiplying $B^*(a, y, z)$ by a suitable integer; it follows that $[b_1, \beta)$ and $(\beta, b_2]$ are nonempty intervals on each of which $H(y, z)$ is delineable. Call algorithm SSADJ2 of Arnon *et al.* (1984a) with inputs $H(y, z)$, β, b_1, and b_2, and according to whether c^2 is cell $(i, j - 1)$ or cell $(i, j + 1)$, add the adjacencies of output L_1 or output L_2 of SSADJ2 to L (after modifying their indices in much the same fashion as was done in step (3) of algorithm CADA2). Exit.

(2) [Section-section interstack adjacencies: c^1 a sector.] Suppose that c^1 is a sector. Then c^1 has index (i, j), with i even and j odd, and sample point $< \alpha, b >$, where b is rational, and α is an algebraic number which is also a primitive element for this sample point. Construct an element of I_2 by multiplying $B^*(x, b, z)$ by a suitable integer, and let $H(x, z)$ be its primitive part. Compute $PROJ(\{H\})$, a collection of univariate polynomials in x, and by isolating the real roots of its elements, and by separating (i.e. refining if necessary) these isolating intervals from the isolating interval for α that is part of the representation of c^1's sample point, determine rational numbers a_1 and a_2 such that $[a_1, \alpha)$ and $(\alpha, a_2]$ are nonempty intervals on each of which $H(x, z)$ is delineable. The index of c^2 is either of the form $(i - 1, k)$, for some odd k, or $(i + 1, l)$, for some odd l. Call SSADJ2 with inputs $H(x, z)$, α, a_1, and a_2, and according to whether c^2 is cell $(i - 1, k)$ or cell $(i + 1, l)$, add the adjacencies of either output L_1 or output L_2 of SSADJ2 to L (the section numbers that occur in the adjacencies of L_1 or L_2 must be converted into indices of the corresponding cells of D, as in algorithm CADA2 of Arnon *et al.*, 1984b).

(3) [Infer remaining interstack adjacencies.] Use the current contents of L to infer the remaining interstack adjacencies between $S^*(c^1)$ and $S(c^2)$, and add them to L. $\square$

Figure 4: Algorithm AdjacenciesOver12.

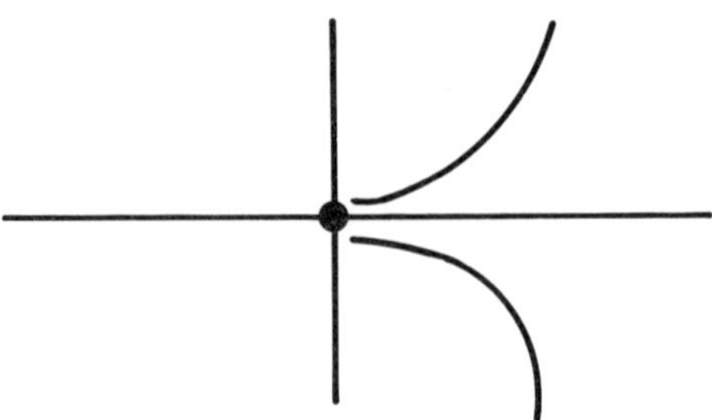

Figure 5: Induced decomposition of the plane for catastrophe surface.

a nullifying $\{0,1\}$ adjacency must be treated separately. Let us restrict attention at first to the case of a nonnullifying $\{0,1\}$ adjacency. The argument used in the proof of Theorem 2.2 in Arnon *et al.* (1984b), with obvious small changes, yields the following:

THEOREM **4.1** *Let D be a basis-determined cad of E^3. Let $\{c^0, c^1\}$ be a nonnullifying $\{0,1\}$ adjacency of D'. Then $S^*(c^1)$ has the unique section boundary property in $S^*(c^0)$.*

Let us now consider an example. Suppose that we wish to construct an F-invariant cad D of E^3 for $F(x,y,z) = z^3 - xz + y$ (F defines a well-known catastrophe surface). The set $\{F\}$ is a basis for itself, $cont(F)$ is trivial, and we have $PROJ(\{F\}) \cup cont(\{F\}) = \{x, y, -27y^2 + 4x^3\}$, so the induced cad D' of E^2 is as shown in Fig. 5. Suppose now that we have extended D' to D, and we want to determine the adjacencies of D. Consider, for example, determination of the section-section adjacencies between the stacks in E^3 over the 0-section in the plane whose cell index is $(2,2)$ (this is the point $< 0, 0 >$ in the plane), and over the 1-section that is adjacent to cell $(2,2)$ and has index $(3,2)$. Note that this $\{0,1\}$ adjacency of D' is nonnullifying. Let $G(x,y) = -27y^2 + 4x^3$; both cell $(3,2)$ and cell $(2,2)$ lie in $V(G)$ (in general, for any adjacent 0-cell and 1-section of D', there will be some such $G(x,y)$ in the basis for the E^2 cad whose variety contains both). Our strategy is to project these two stacks in E^3 into the xz-plane, apply algorithm SSADJ2 of Arnon *et al.* (1984b) to determine the adjacencies between their projections, and extract from the information produced by SSADJ2 the adjacencies in D between the original stacks. We project by computing the resultant $R(x,z) = Res_y(G,F)$, and obtain:

$$R(x,z) \;=\; -27z^6 + 54xz^4 - 27x^2z^2 + 4x^3 \;=\; -\,(3z^2 - 4x)(3z^2 - x)^2\,,$$

which gives us the cad of the (x,z)-plane shown in Fig. 6. We find suitable rational numbers a_1 and a_2 just to the left and right of $x = 0$ respectively (see step (2) of algorithm SSADJ3 in Fig. 7), and call SSADJ2 with inputs R, a_1, 0, and a_2 to obtain the four section-section adjacencies apparent in Fig. 6. We extract from this data (see step (3) of SSADJ3) the section-section adjacencies in E^3 between the stacks over cells $(3,2)$ and $(2,2)$ of D', which turn out to be: $\{(2,2,2),(3,2,2)\}$, and $\{(2,2,2),(3,2,4)\}$.

Another description of this method is: $G(x,y)$ defines a surface in E^3 (the cylinder over the plane curve $G(x,y) = 0$); this surface and $V(F)$ have an intersection curve in E^3; the zeros of the resultant we compute are the projection into the xz-plane of this

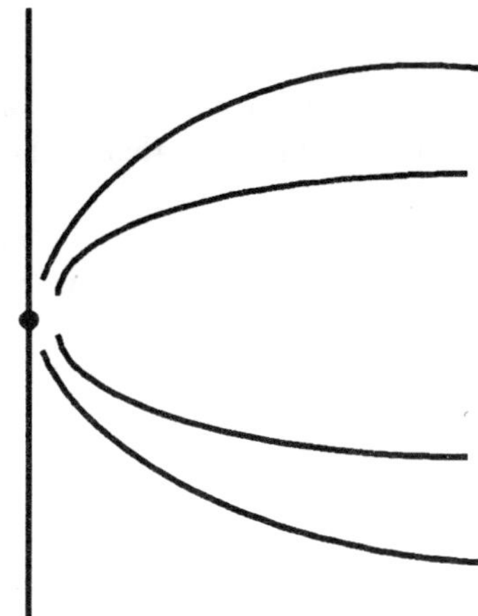

Figure 6: Projection into xz-plane of decomposition of E^3.

intersection curve. In other words, $V(R)$ contains the projection (in the xz-plane) of the points of $V(F)$ which lie over the curve (in the xy-plane) $G(x, y) = 0$. Thus the projections of F-sections over cell (3,6) of D', as well as over cells (3,2) and (2,2) of D', are part of $V(R)$. This explains why, although there are only two section-section adjacencies between the stacks over $(3, 2)$ and $(2, 2)$, there are four 1-sections of $V(R)$ in Fig. 6. Thus when we call SSADJ2, it reports (the projections of) the adjacencies between the stacks over cells (3,6) and (2,2), as well as between the stacks over cells (3,2) and (2,2). Step (3) of SSADJ3 gives a method of extracting just those section-section adjacencies that involve elements of the stacks over (3,2) and (2,2).

If c^1 is a sector, we proceed similarly, except the resultant we compute will eliminate x instead of eliminating y (for any adjacent 0-cell and 1-sector of D', there is some $G(x) \in I_1$ in the basis for the induced cad of E^1 whose variety in the xy-plane, which is a set of vertical lines, contains both cells).

Let us now state and prove a theorem justifying the above method. We write π_y to denote a projection map (either $E^3 \to E^2$, or $E^2 \to E^1$) which eliminates y.

THEOREM **4.2** *Suppose that c^0 and c^1 are an adjacent 0-cell and 1-section in a basis-determined cad D of E^2, that $B \subset I_2$ is a basis for D, that both c^0 and c^1 are contained in $V(G(x, y))$, $G \in B$, that $F(x, y, z) \in I_3$ is delineable on c^1, that s^1 is an F-section of $Z(c^1)$, and that s^1 has a unique limit point s^0 in $Z^*(c^0)$. Let $R(x, z) = Res_y(G, F)$, $P(x, z) = pp(R(x, z))$, and $C(x) = content(P(x, z))$. Then*

(1) $t^1 = \pi_y(s^1)$ is contained in $V(P)$.

(2) Supposing that $c^0 = < \alpha, \beta >$, t^1 has a unique limit point t^0 in $Z^*(\alpha)$, and if t^0 is finite, then it is contained in $V(P)$.

(3) $t^0 = \pi_y(s^0)$.

PROOF. Since c^1 is a section of D', $\pi_y(c^1)$ is a 1-cell in E^1, and hence t^1 is a section of the cylinder in the xz-plane over that 1-cell. By a basic property of resultants (see e.g. Theorem 5 of Collins (1971)), $t^1 \subset V(R)$. In the xz-plane, $V(C(x))$ is a finite set of vertical lines, hence t^1 meets $V(C)$ at finitely many points, and so $t^1 \subset V(P)$.

Then since $P(x, z)$ is primitive, our present context essentially satisfies the hypotheses of Theorem 2.2 of Arnon *et al.* (1984b) (with P in place of the polynomial $F(x, y)$ that occurs there), hence by the argument used in the proof of Theorem 2.2, t^1 has a unique limit point t^0 in $Z^*(\alpha)$. If t^0 is finite, then it is in $V(P)$ since $V(P)$ is closed. $\pi_y(s^0)$ is a limit point of t^1, and if it is not equal to t^0, then t^1 has two distinct limit points in $Z^*(\alpha)$, which is impossible. $\square$

Clearly there is a similar theorem for when c^1 is a sector. In that case, $G(x, y)$ has positive degree in x (in fact is univariate in x), and we compute $R(y, z) = Res_x(G, F)$ instead of $R(x, z) = Res_y(G, F)$.

The hypotheses of Theorem 4.2 are clearly satisfied over any nonnullifying $\{0, 1\}$ adjacency of D', hence given any such adjacency $\{c^0, c^1\}$, we can apply SSADJ3 to determine the adjacencies between $S^*(c^0)$ and $S^*(c^1)$. The precise specifications of SSADJ3's inputs are a consequence of the diverse contexts in which it is invoked.

Let us now note that the hypotheses of Theorem 4.2 do not exclude the possibility that c^0 nullifies F. However, before we can be sure of the validity of the theorem for the case of a nullifying $\{0, 1\}$ adjacency, we must verify that the hypothesis "s^1 has a unique limit point s^0 in $Z^*(c^0)$" holds in such a case. It is not hard to see that it must, for clearly s^1 still has at least one limit point in $Z^*(c^0)$, and if it had more than one, then so would $\pi_y(s^1)$ have more than one limit point in $\pi_y(Z^*(c^0))$, which is clearly impossible. We may further observe that Theorem 4.2 tells us precisely what (the z-coordinate of) s^1's unique limit point in $Z^*(c^0)$ is.

It is enlightening to observe exactly how it happens that the conclusion of Theorem 4.2 still holds in the case that c^0 nullifies F. The key is the replacement of $R(x, z)$ by its primitive part $P(x, z)$. It is not hard to convince oneself that for any primitive bivariate polynomial $K(x, z)$, there is no value of x for which it is nullified. Thus, if c^0 nullifies F, then $V(F)$ contains a "vertical line" over c^0 which we may think of "noise" obscuring the limit point of s^1 in $Z^*(c^0)$, $V(R)$ contains the projection of this "noise" (and so R is necessarily an imprimitive polynomial), but replacing R by its primitive part P serves to filter out the "noise", and then since $\pi_y(s^1) \subset V(P)$ we can find s^1's limit point in $Z^*(c^0)$ using SSADJ2.

Thus Theorem 4.2 yields the following corollary:

THEOREM **4.3** *Suppose that D is a basis-determined cad of E^3 with basis B, and that $\{c^0, c^1\}$ is a nullifying $\{0, 1\}$ adjacency of D' such that $c^0 = <\alpha, \beta>$, and c^1 is a section contained in $V(G(x, y))$ for $G(x, y)$ in the basis for D'. Suppose further, for every $F \in B$ which is nullified on c^0, and every real root γ of $pp(Res_y(G, F))(\alpha, z)$, that $<\alpha, \beta, \gamma>$ is a section of the stack $S(c^0)$ in D. Then $S^*(c^1)$ has the unique section boundary property in $S^*(c^0)$.*

There is a similar theorem for the case of c^1 a sector. Thus if D is a cad of E^3 whose stacks were determined by algorithm *ExtendCellToStack*, i.e. by algorithm *CellExtensionPolynomial*, of Section 2, and if $\{c^0, c^1\}$ is a nullifying $\{0, 1\}$ adjacency of D', then $S^*(c^1)$ has the USBP in $S^*(c^0)$. Thus we can apply SSADJ3 in the nullifying $\{0, 1\}$ case to determine the adjacencies between $S^*(c^0)$ and $S^*(c^1)$, just as in the nonnullifying case. Altogether we have the algorithm *AdjacenciesOver01* given in Fig. 8.

$$L \leftarrow \textbf{SSADJ3}\ (c^0, h(z), G(x,y), i, leftOrBelow, F(x,y,z))$$

Inputs: $c^0 = <\alpha, \beta>$, $\alpha, \beta \in Q(\gamma)$, is a 0-cell in E^2. $h(z)$ is a squarefree element of $Q(\gamma)[z]$ whose roots define a stack T over c^0. $G(x,y)$ is a primitive element of I_2 of positive degree in x or y or both, such that $c^0 \subset V(G)$, and such that there is some 1-cell $c^1 \subset V(G)$ which is adjacent to c^0. i is a positive integer, and $leftOrBelow$ is a boolean. Taken together, i and $leftOrBelow$ will tell us which "branch of $V(G)$ in the neighborhood of c^0" c^1 is. c^1 is a sector of the cylinder $Z(\alpha) \subset E^2$ if and only if G has degree 0 in y (and hence positive degree in x alone). When c^1 is such a sector, then α is the ith real root of $G(x,\beta)$, and c^1 is the sector immediately below c^0 if $leftOrBelow$ is true, and the sector immediately above c^0 if $leftOrBelow$ is false. Suppose that G has positive degree in y, and let $\epsilon > 0$ be sufficiently small so that G is delineable on both $X = [\alpha - \epsilon, \alpha)$ and $Y = (\alpha, \alpha + \epsilon]$. If $leftOrBelow$ is true, then c^1 is the ith section of $S(G, X)$, else c^1 is the ith section of $S(G, Y)$. $F(x,y,z)$ is an element of I_3 which is delineable on c^1, such that $S^*(F, c^1)$ has the unique section boundary property in T^*.

Output: L is the set of all section-section interstack adjacencies between cells of T^* and cells of S^*. We follow the convention that the sections occurring in these adjacencies are numbered consecutively from bottom to top, starting with section 1 (the lowest finite section), 2, ..., n (the highest finite section). The sections of extended stacks are numbered starting with section 0 (the $-\infty$-section), 1 (the lowest finite section), 2, ..., n (the highest finite section), $n + 1$ (the $+\infty$-section).

(1) [Project E^3 into a plane by resultant computation.] If $deg_y(G) > 0$, then set $P(x,z) \leftarrow pp(Res_y(G, F))$, else if $deg_x(G) > 0$, then set $P(y,z) \leftarrow pp(Res_x(G, F))$. Assume from now on, without loss of generality, that G has positive degree in y, i.e. c^1 is a section, and thus that we have computed $P(x,z)$ in this step. Small changes to steps (2) and (3) are needed for the case c^1 a sector.

(2) [Determine adjacencies in the plane.] Set $J = PROJ(\{P(x,z)\})$. Isolate the real roots of the elements of J (i.e. isolate the roots of $\prod J$) to determine rational numbers a_1 and a_2 such that P is delineable on $[a_1, \alpha)$, and on $(\alpha, a_2]$. Call algorithm SSADJ2 (of Arnon *et al.*, 1984b) with inputs P, a_1, α, and a_2 to obtain lists L_1 and L_2 of adjacencies.

(3) [Extract E^3 adjacencies from adjacencies in the plane.] Assume, without loss of generality, that $leftOrBelow$ is false, i.e. that c^1 is "to the right of c^0". Choose a rational $\hat{a}$, $\alpha < \hat{a} \le a_2$, such that G (and *a fortiori* P) are delineable on $(\alpha, \hat{a}]$ (this can be done, for example, by computing $PROJ(\{G\})$, isolating its real roots, and isolating those roots from the real roots of $M(x)$, where M is the minimal polynomial of α). Isolate the real roots of $P(\alpha, z)$ and $P(\hat{a}, z)$. Let δ be the unique real algebraic number such that $< \hat{a}, \delta > \in c^1$. ($\delta$ is the ith root of $G(\hat{a}, y)$). For each real root of $F(\hat{a}, \delta, z)$ (corresponding to a section s^1 of S), do the following two steps: first, shrink the isolating interval for this root until it overlaps a unique isolating interval for a root of $P(\hat{a}, z)$, to determine the unique projection t^1 of s^1; second, set t^0 to the (unique) boundary section of t^1 in $Z^*(\alpha)$, as determined from the output L_2 of SSADJ2. If t^0 is infinite, then record that the boundary section of s^1 in T^* is the appropriate infinite section of T^*, else if t^0 is finite, then refine isolating intervals for the roots of $P(\alpha, z)$ and $h(z)$ until the interval about the root of $P(\alpha, z)$ corresponding to t^0 overlaps a unique interval about a root of $h(z)$, whose corresponding section in T^* is the boundary section of s^1 (in T^*). $\square$

Figure 7: Algorithm SSADJ3.

$$L \leftarrow \textbf{AdjacenciesOver01} \ (c^0, c^1, B', B)$$

Inputs: $c^0 = \ <\alpha, \beta>, \ \alpha, \beta \in Q(\gamma)$, is a 0-cell, and c^1 a 1-cell, in a cad D' of E^2 induced by a basis-determined cad D of E^3, such that $\{c^0, c^1\}$ is an adjacency of D'. B' is a basis for D', and B is a basis for D.

Output: L is a list of all interstack adjacencies between $S^*(c^0)$ and $S(c^1)$.

(1) [Section-section interstack adjacencies: c^1 a section.] Set $h(z)$ to the result of calling $CellExtensionPolynomial(c^0, B', B)$. If c^1 is a sector, then go to Step (2). Set $\hat{B} \leftarrow \prod B'$. Call SSADJ3 with inputs c^0, $h(z)$, $\hat{B}(x, y)$, the appropriate value of i, the appropriate value of $leftOrBelow$, and $\prod B = B_{c^1}$, to obtain output L^*. Convert the section numbers that occur in the adjacencies of L^* to indices of the corresponding cells of D, and add the resulting adjacencies of D to L. Go to Step (3).

(2) [Section-section interstack adjacencies: c^1 a sector.] Let $M(x)$ be the minimal polynomial of α. Set $h(z) \leftarrow CellExtensionPolynomial(c^0, B', B)$. Call SSADJ3 with inputs c^0, $h(z)$, $M(x)$, the appropriate value of i, the appropriate value of $leftOrBelow$, and $\prod B = B_{c^1}$, to obtain output L^*. Convert the section numbers that occur in the adjacencies of L^* to indices of the corresponding cells of D, and add the resulting adjacencies of D to L.

(3) [Infer remaining interstack adjacencies.] Use the current contents of L to infer the remaining interstack adjacencies between $S^*(c^0)$ and $S(c^1)$, and add them to L. $\square$

Figure 8: Algorithm AdjacenciesOver01.

5   Adjacencies over a nonnullifying $\{0, 2\}$ adjacency in the plane

As for Theorem 4.1, the argument used to prove Theorem 2.2 of Arnon *et al.* (1984b) can be adapted to prove the following:

THEOREM **5.1** *Let D be a basis-determined cad of E^3. Let $\{c^0, c^2\}$ be a nonnullifying $\{0, 2\}$ adjacency of D'. Then $S^*(c^2)$ has the unique section boundary property in $S^*(c^0)$.*

The following lemma will be useful in this and subsequent sections.

LEMMA **5.2** *Let c^0 and c^2 be respectively a 0-cell and a 2-cell in some cad of E^2, and suppose that c^0 is adjacent to c^2. Then there exist exactly two 1-cells in the cad which are adjacent to both c^0 and c^2.*

PROOF. Use a variation of the proof of Lemma 5.10 in Massey (1978), p. 137-138. $\square$

We call the two 1-cells of Lemma 5.2 the "boundary 1-cells" of c^2 (with respect to c^0). Fig. 9 gives an algorithm to find them.

Suppose now that D is a basis-determined cad of E^3, and that $\{c^0, c^2\}$ is a nonnullifying $\{0, 2\}$ adjacency of D'. Let c^1 be a boundary 1-cell of c^2 with respect to c^0. We have shown in Section 3 that $S^*(c^2)$ has the USBP in $S^*(c^1)$, and in Section 4 that $S^*(c^1)$ has the USBP in $S^*(c^0)$. Thus if s^2 is a section of $S(c^2)$, then s^2 has a unique boundary section s^1 in $S^*(c^1)$, and s^1 has a unique boundary section s^0 in $S^*(c^0)$. It follows that s^0 is adjacent to s^2, hence it must be the unique boundary section of s^2 in $S^*(c^0)$. Thus to determine the section-section adjacencies between $S(c^0)$ and $S(c^2)$, it

$$(c_1^1, c_2^1) \leftarrow \textbf{BoundaryOneCells } (c^0, c^2)$$

Inputs: c^0 is a 0-cell, and c^2 a 2-cell, in a cad of E^2.

Output: c_1^1 and c_2^1 are the boundary 1-cells of c^2 with respect to c^0.

(1) [Find "upper" boundary 1-cell] If the unique 1-section in the same stack as, and directly above, c^2 (this 1-section may be either finite or infinite) has c^0 as its boundary 0-section in the (extension of the) stack containing c^0, then set c_1^1 to be this 1-section. Otherwise, set c_1^1 to be the unique 1-sector that is in the same stack as, and directly above, c^0, and that is adjacent to both c^0 and c^2.

(2) [Find "lower" boundary 1-cell] If the unique 1-section in the same stack as, and directly below, c^2 (this 1-section may be either finite or infinite) has c^0 as its boundary 0-section in the (extension of the) stack containing c^0, then set c_1^1 to be this 1-section. Otherwise, set c_1^1 to be the unique 1-sector that is in the same stack as, and directly below, c^0, and that is adjacent to both c^0 and c^2. $\square$

Figure 9: Algorithm BoundaryOneCells.

suffices to pick a boundary 1-cell c^1 of c^2 with respect to c^0, and then find (in already constructed adjacency information) all sections of $S(c^1)$ which are adjacent both to a section of $S(c^0)$ and to a section of $S(c^2)$. Fig. 10 gives the complete algorithm.

6 Adjacencies over a nullifying $\{0, 2\}$ adjacency in the plane

We would like for a stack in E^3 (over a cell $c \subset E^2$) to have the USBP in each adjacent lower-dimensional stack (more specifically, in each stack over a cell d which meets ∂c). As we have seen, we can achieve this goal over $\{1, 2\}$, nonnullifying $\{0, 1\}$, nullifying $\{0, 1\}$, and nonnullifying $\{0, 2\}$ adjacencies in E^2. It turns out that we cannot achieve this goal over nullifying $\{0, 2\}$ adjacencies. For example, consider again $F(x, y, z) = xz - y$. In a typical F-invariant cad of E^3, there will be a 2-section (whose base is the 2-cell "$x > 0 \; AND \; y > 0$" in the induced cad of the plane) which has infinitely many boundary points in the z-axis. The unique section boundary property cannot possibly be made to hold in such a case. However, there is a weaker, but still useful, property that we can make hold. Given adjacent stacks S and T in E^r, we say that S^* has the *section boundary property (SBP)* in T^*, if for any section s of S^*, the set of boundary points of s in the underlying cylinder of T^* is equal to the union of one or more lower-dimensional elements of T^*. For a nullifying $\{0, 2\}$ adjacency in E^2, if the stacks S and T over the 2-cell and 0-cell respectively are constructed according to the specifications we will give, then we can show that S^* has the SBP in T^* (Theorem 6.4). By extending the argument used to prove Theorem 2.3 of Arnon *et al.* (1984b), the following can then be shown. Suppose a stack S^* has the SBP in a stack T^*, let u be a sector of S^*, and suppose that s_1 and s_2 are the respective sections of S^* immediately below and above u. Let t_1 be the "lowest" element of T^* which is in the boundary of s^1, and let t_2 be the "highest" element of T^* which is in the boundary of s_2. Then the set of boundary points of u in the underlying cylinder of T^* is equal to the union of all elements of T^*

$L \leftarrow$ **AdjacenciesOverNonNullifying02** (c^0, c^2, B', B, L')

Inputs: c^0 is a 0-cell, and c^2 a 2-cell, in a cad D' of E^2 induced by a basis-determined cad D of E^3, such that $\{c^0, c^2\}$ is an adjacency of D'. B' is a basis for D', and B is a basis for D. Let c_1^1 and c_2^1 be the boundary 1-cells of c^2 with respect to c^0. L' is a collection of adjacencies of D that includes the interstack adjacencies over the adjacencies $\{c^0, c_1^1\}$ and $\{c^0, c_2^1\}$ of D', and the interstack adjacencies over the adjacencies $\{c_1^1, c^2\}$ and $\{c_2^1, c^2\}$ of D' (if L' does not contain these adjacencies, then algorithms *BoundaryOneCells*, *AdjacenciesOver01* and *AdjacenciesOver12* may be used to add them to it).

Output: L is a list of all interstack adjacencies between $S^*(c^0)$ and $S(c^2)$.

(1) [Step through sections of $S(c^2)$ from bottom to top.] Set L to L'. $(c_1^1, c_2^1) \leftarrow$ *BoundaryOneCells*(c^0, c^2). For each section s^2 of $S(c^2)$, do: find the boundary section t^1 of s^2 in $S^*(c_1^1)$ by querying L'; find the boundary section t^0 of t^1 in $S^*(c^0)$ by querying L'; then $\{t^0, s^2\}$ is the unique (section-section) adjacency between s^2 and a cell of $S^*(c^0)$; add it to L.

(2) [Infer remaining interstack adjacencies.] Use the current contents of L to infer the remaining interstack adjacencies between $S^*(c^0)$ and $S(c^2)$, and add them to L. $\square$

Figure 10: Algorithm AdjacenciesOverNonNullifying02.

between t_1 and t_2 inclusive. It now is not hard to see that when S^* has the SBP in T^*, we can infer all interstack adjacencies between S^* and T^* from knowledge of those which involve sections of S.

We begin the development for Theorem 6.4 with the following lemma.

THEOREM **6.1** *Let c^0 and c^2 be respectively a 0-cell and a 2-cell in some cad of E^2, and suppose that c^0 is adjacent to c^2. Let s be a section of $Z(c^2)$. If p and q are limit points of s in $Z^*(c^0)$, then every point of $Z^*(c^0)$ between p and q is also a limit point of s.*

PROOF. There is a sequence of points $\{p_i\}$ in s converging to p, and a sequence $\{q_i\}$ in s converging to q. Let π_z denote the projection map $E^3 \to E^2$ which eliminates z. c^2 is connected, hence path-connected, so for each i, $\pi_z(p_i)$ and $\pi_z(q_i)$ can be joined by a path lying in c^2. Suppose without loss of generality that p and q are both finite, let r be a point of $Z(c^0)$ between p and q, and without loss of generality, suppose r to be halfway between p and q. Suppose also that s is an f-section, for a continous map $f : c^2 \to E$. By the Intermediate Value Theorem, for each i, there is some point e_i on the path joining $\pi_z(p_i)$ and $\pi_z(q_i)$ at which the value of f is halfway between its values at $\pi_z(p_i)$ and $\pi_z(q_i)$. Then the sequence of points $\{< e_i, f(e_i) >\}$ in s converges to r. $\square$

Thus if $\{c^0, c^2\}$ is a nullifying $\{0, 2\}$ adjacency in the plane, and s is a section of $S(c^2)$, then $\partial s \cap Z^*(c^0)$ is a "closed interval" of $Z^*(c^0)$, and our task is to compute its endpoints.

For $X \subset E^2$ and f a differentiable function $X \to E$, let f_x denote the partial derivative of f with respect to x, and f_y the partial derivative of f with respect to y. The next theorem introduces the essential idea for Theorem 6.4, and for the algorithms based on it.

THEOREM **6.2** *Let c^2 be a 2-cell in some cad of E^2, let s be an f-section of $Z(c^2)$, for some continuous $f : c^2 \to E$, and let $F \in I_3$ be such that $F(x, y, f(x, y)) = 0$ for* *all $< x, y > \in c^2$. If neither F_x nor F_y nor F_z vanishes at any point of s, then f is differentiable, and the values of the functions f_x and f_y each are of constant nonzero sign throughout c^2.*

PROOF. Since F_z does not vanish at any point of s, by the Implicit Function Theorem f is differentiable on c^2, and

$$f_x = \frac{-F_x}{F_z}$$

on c^2. Since neither F_x nor F_z vanishes at any point of s, s is F_x-invariant, and F_z-invariant. Hence f_x is sign-invariant and nonzero on c^2. The same argument applies for f_y. $\square$

Assume for a moment the hypotheses of Theorem 6.2. If F_x vanishes at a point $< \alpha, \beta, \gamma >$ of s, then by a basic property of resultants (see e.g. Theorem 5 of Collins, 1971), $Res_z(F, F_x)$ vanishes at $< \alpha, \beta >$. Hence if we knew that $Res_z(F, F_y)$ does not vanish at any point of c^2, then we would know that F_y does not vanish at any point of s. The same holds for $Res_z(F, F_z)$ and F_z. We write $PDP(F)$ to denote the set $\{Res_z(F, F_y), Res_z(F, F_z)\}$; PDP stands for "partial derivative projection".

THEOREM **6.3** *Assume the hypotheses of Theorems 6.1 and 6.2. Let c_1^1 and c_2^1 be the boundary 1-cells of c^2 with respect to c^0. Suppose that s has unique boundary sections t_1^1 and t_2^1 in $Z^*(c_1^1)$ and $Z^*(c_2^1)$ respectively, and let p_1^0 and p_2^0 denote the (respective) limit points of t_1^1 and t_2^1 in $Z^*(c^0)$. If neither element of $PDP(F)$ vanishes at any point of c^2, then $\partial s \cap Z^*(c^0)$ is all points of $Z^*(c^0)$ between p_1^0 and p_2^0 inclusive.*

PROOF. Suppose first that both c_1^1 and c_2^1 are sections. Obviously p_1^0 and p_2^0 are boundary points of s, and hence by Theorem 6.1, so are all points of $Z^*(c^0)$ between them. If point p of $Z^*(c^0)$ is a boundary point of s, then there is a sequence of points $\{p_i\} = \{x_i, y_i, f(x_i, y_i)\}$ in s converging to p. $\{\pi_z(p_i)\} = \{x_i, y_i\}$ is a sequence in c^2. Suppose that c_1^1 is a g_1-section, c_2^1 is a g_2-section, and without loss of generality, that $g_1 < g_2$. Our hypothesis on $PDP(F)$ implies, as observed a moment ago, that neither F_y nor F_z vanishes at any point of s. Hence by Theorem 6.2, f_y is either positive or negative on c^2; assume without loss of generality positive. Suppose that t_1^1 is an f_1-section, and that t_2^1 is an f_2-section. The sequence $< x_i, g_1(x_i), f_1(x_i, g_1(x_i)) >$ converges to p_1^0, and the sequence $< x_i, g_2(x_i), f_2(x_i, g_2(x_i)) >$ converges to p_2^0. Clearly $f_1(x_i, g_1(x_i)) < f(x_i, y_i) < f_2(x_i, g_2(x_i))$ for each i, and so p is between p_1^0 and p_2^0.

Suppose that c_1^1 is a sector, and c_2^1 a section. Imagine inserting a new 1-dimensional section which is in c^2, adjacent to c^0, and very close to c_1^1. Then we can apply the above argument. As this new section approaches c_1^1, the boundary point of s over this section in $Z^*(c^0)$ must approach the boundary point of t_1^1 in $Z^*(c^0)$, by the continuity of F. Hence the conclusion of the theorem remains valid. The same argument can be used if c_2^1 is a sector, or if both c_1^1 and c_2^1 are sectors. $\square$

Let us see how to make use of Theorem 6.3. Assume its hypotheses, and suppose further that $\{c^0, c^2\}$ is a nullifying $\{0, 2\}$ adjacency in the plane. We are interested in the behavior of s close to $Z^*(c^0)$, so actually it is not necessary that the elements of $PDP(F)$ be nonvanishing at every point of c^2. It suffices that there be some ball in the plane, centered at c^0, such that the elements of $PDP(F)$ are nonvanishing at all points of the portion of c^2 inside the ball. If such a ball exists, then p_1^0 and p_2^0 are the endpoints of the interval of boundary points of s in $Z^*(c^0)$, and if p_1^0 and p_2^0 have been made sections of $S^*(c^0)$ (by CADA3), then we can find them with two applications of

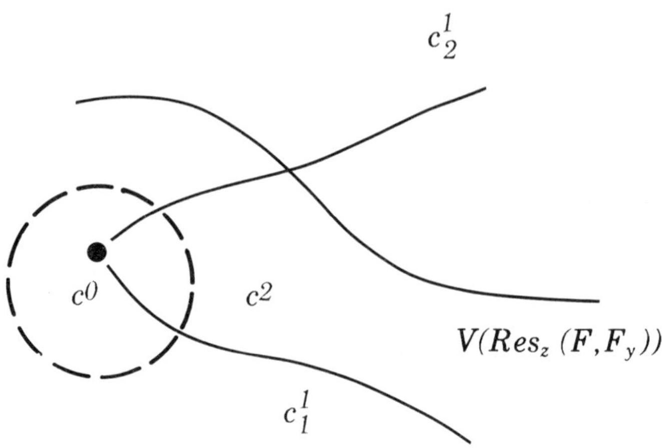

Figure 11: Ball of the desired kind exists.

SSADJ3. Fig. 11 illustrates the situation we might have in the plane when a ball of the desired kind exists. A ball of the desired kind fails to exist when either $V(Res_z(F, F_y))$ or $V(Res_z(F, F_z))$ or both have 1-sections that are between c_1^1 and c_2^1, and adjacent to c^0, as depicted in Fig. 12. We handle this case by thinking of c^2 as being partitioned, in the neighborhood of c^0, into 2-dimensional "subsectors" separated by these sections of $V(Res_z(F, F_y))$ and $V(Res_z(F, F_z))$. This partition of c^2 induces a partition of s into 2-dimensional "subsections", separated by the 1-dimensional "slices" of s which lie over the sections of $V(Res_z(F, F_y))$ and $V(Res_z(F, F_z))$. If we make the limit point in $Z^*(c^0)$ of each "slice" of s a section of $S^*(c^0)$, then by two applications of SSADJ3 for each "subsection" of s, we can find the boundary interval of that "slice" of s in $Z^*(c^0)$. This is exactly what our algorithms (specifically, algorithm *CellExtensionPolynomial* of Section 2 and algorithm *AdjacenciesOverNullifying02* given below) do. Clearly the boundary interval of s in $Z^*(c^0)$ is the union of the boundary intervals of its "subsections". *AdjacenciesOverNullifying02* uses algorithm *InteriorSections*, given in Fig. 13, to determine whether $V(Res_z(F, F_y))$ or $V(Res_z(F, F_z))$ has 1-sections that lie between c_1^1 and c_2^1 and are adjacent to c^0.

Theorem 6.4 summarizes our development.

THEOREM **6.4** *Suppose that D is a basis-determined cad of E^3 with basis B, that $\{c^0, c^2\}$ is a nullifying $\{0, 2\}$ adjacency of D' with $c^0 =< \alpha, \beta >$, that the boundary 1-cells of c^2 with respect to c^0 are sections c_1^1 and c_2^1 of D', contained respectively in $V(G_1(x, y))$ and $V(G_2(x, y))$, for elements G_1 and G_2 of the basis for D', and that for every $F \in B$ which is nullified on c^0, and for all real roots γ of*

(1) $pp(Res_y(F, G_1))(\alpha, z)$,

(2) $pp(Res_y(F, G_2))(\alpha, z)$,

(3) $pp(Res_y(F, Res_z(F, F_y)))(\alpha, z)$, and

(4) $pp(Res_y(F, Res_z(F, F_z)))(\alpha, z)$,

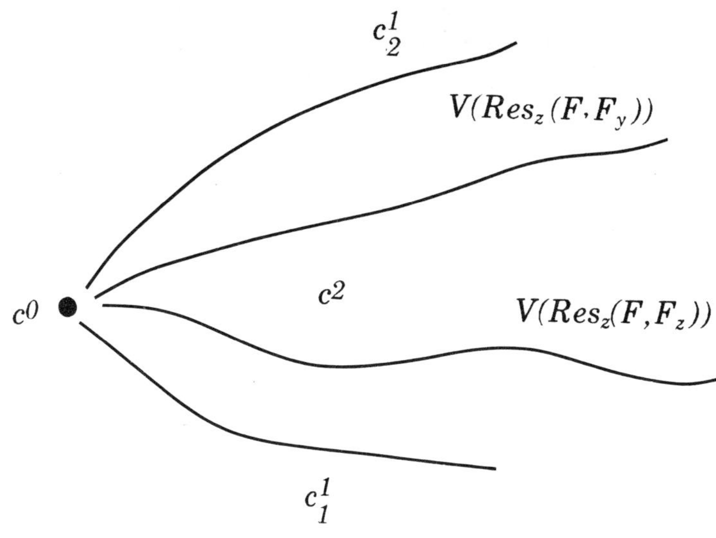

Figure 12: Ball of the desired kind fails to exist.

$$(i,j) \leftarrow \textbf{InteriorSections } (c^0, c^2, F(x,y))$$

Inputs: $c^0 = <\alpha, \beta>$ is a 0-cell, and c^2 a 2-cell, in a cad of E^2. $F(x,y)$ is a primitive element of I_2.

Output: (i,j) is a pair of two non-negative integers. Assume, without loss of generality, that c^2 is "to the right of" c^0. If $i > 0$, then for sufficiently small ϵ, real roots $i, i+1, ..., i+j$ of $F(\alpha + \epsilon, y)$ correspond to 1-cells contained in $V(F)$, lying within c^2, that are between the boundary 1-cells of c^2 with respect to c^0, and that are adjacent to c^0. If c^2 is "to the left of" c^0, then $L = (i,j)$ refers to real roots of $F(\alpha - \epsilon, y)$, and small changes are needed to the steps below.

(1) [Exit if F not of positive degree in y, or c^0 not contained in $V(F)$.] If F not of positive degree in y, or if $F(\alpha, \beta) \neq 0$, then RETURN[$(0, 0)$].

(2) [Find F-sections with required properties.] Do $(c_1^1, c_2^1) \leftarrow BoundaryOneCells(c^0, c^2)$ to get the boundary 1-cells of c^2 with respect to c^0. Assume that both c_1^1 and c_2^1 are sections; small adjustments are needed if one or both are sectors. Suppose that c_1^1 is real root m_1 of $G_1(\alpha + \epsilon, y)$, and c_2^1 is real root m_2 of $G_2(\alpha + \epsilon, y)$, for primitive polynomials $G_1(x,y), G_2(x,y)$ of positive degree in y. Compute $P = PROJ(\{F, G_1, G_2\})$, and choose ϵ so that there are no real roots of $\prod P$ in the interval $J = (\alpha, \alpha + \epsilon]$. It follows (Arnon *et al.*, 1984a) that F, G_1, G_2, and FG_1G_2, are all delineable on J. If real roots $i, i+1, ..., i+j$ of $F(\alpha + \epsilon, y)$ lie between root m_1 of $G_1(\alpha + \epsilon, y)$, and root m_2 of $G_2(\alpha + \epsilon, y)$, then RETURN[(i, j)], else RETURN[$(0, 0)$]. $\square$

Figure 13: Algorithm InteriorSections.

$< \alpha, \beta, \gamma >$ *is a section of* $S(c^0)$. *Then* $S^*(c^2)$ *has the section boundary property in* $S^*(c^0)$.

When one (or both) of the boundary 1-cells, say c_1^1, of c^2 is a sector, a variant of Theorem 6.4 is needed. Given that $c^0 = < \alpha, \beta >$, let $M(x) \in I_1$ be the minimal polynomial of α. The variant is obtained by replacing $pp(Res_y(F, G_1))(\alpha, z)$ with $pp(Res_x(F, M))(\beta, z)$.

We give algorithm $AdjacenciesOverNullifying02$ in Fig. 14. Note that if $\{c^0, c^2\}$ is a nullifying $\{0, 2\}$ adjacency of D', then c^0 nullifies at least one element of B, but there may be other elements of B which are delineable, rather than nullified, on c^0. If s is a section of $S(c^2)$ in such a case, and if the unique element F of B, in whose variety s is contained, is not nullified on c^0, then $\partial s \cap Z^*(c^0)$ consists of a unique section of $S^*(c^0)$, which can be determined by the method of Section 5. In fact, $AdjacenciesOverNullifying02$ handles such sections s in just this way.

7   Main algorithm

We summarize the preceding Sections with our main algorithm CADA3, given in Fig. 15.

THEOREM **7.1** *A cad of* E^3 *constructed by algorithm CADA3 has the boundary property.*

PROOF. Let D be the cad. It is clear from the defintion of algorithm $ExtendCellToStack$ of Section 2 that D is a basis-determined cad. By our discussion in Section 1, the induced cad D' of E^2 constructed by algorithm CADA2 of Arnon *et al.* (1984b) has the boundary property. If $\{c, d\}$ is an adjacency of D', with $dim(d) < dim(c)$, then by Theorems 3.1, 4.1, 4.3, 5.1, and 6.4, $S^*(c)$ has either the USBP or the SBP in $S^*(d)$. It follows that $S^*(c)$ has the boundary property in $S^*(d)$. Hence D has the boundary property. $\square$

8   Example

Let $F(x, y, z) = y^3 z + xy^2 - x^3$, and set $A \leftarrow \{F\}$. $\{F\}$ is a basis B for $prim(A)$. $cont(A)$ is trivial; $P = PROJ(B) = \{y^3, xy^2 - x^3\}$. Calling CADA2 with input P, we obtain the induced cad D' of E^2 shown in Fig. 16.

Continuing with step (1) of CADA3, we have $P^*(x, y) = y^5 - x^2 y^3$, and $B' = \{P^*\}$. Let c^0 denote cell (2,2) of D', i.e. the point $< 0, 0 >$. It is not hard to see that c^0 is a nullifying 0-cell, that F has no sections over the two 1-cells (1,4) and (3,4) (on which $x \neq 0$ and $y = 0$), and that F has one section over every other cell of D'. Thus in step (2) of CADA3, it is only for cell c^0 of D' that the call to $ExtendCellToStack$, *i.e.* the call to $CellExtensionPolynomial$, is interesting. In step (1) of this latter call, since the unique element F of B is nullified on $c = c^0$, we get $B_c = 1$, and so $\Gamma = \{1\}$. In step (2), we get $B_N = \{F\}$, so we will go through steps (2.1) - (2.3) just once. In step (2.1), we get $\hat{B}(x, y) = y^5 - x^2 y^3$, $G(x, z) = z^2$, and so we add $G(0, z) = z^2$ to Γ. In step (2.2), we get $H(x) = x$, $G(y, z) = z$, and so we add $G(0, z) = z$ to Γ. In step (2.3), we get a first $T(x, y)$ of $pp(Res_z(F, F_y)) = -y^4 + 3x^2 y^2$, we have $T(0, 0) = 0$, and we then get $G(x, z) = 27z^2 - 4$, and so we add $G(0, z) = 27z^2 - 4$ to Γ. Continuing in step (2.3), we get a second $T(x, y)$ of $pp(Res_z(F, F_z)) = 0$, and since this T is not of positive

$L \leftarrow$ **AdjacenciesOverNullifying02** (c^0, c^2, B', B, L')

Inputs: c^0 is a 0-cell, and c^2 a 2-cell, in a cad D' of E^2 induced by a basis-determined cad D of E^3, such that $\{c^0, c^2\}$ is a nullifying $\{0, 2\}$ adjacency of D'. B' is a basis for D', and B is a basis for D. Let c_1^1 and c_2^1 be the boundary 1-cells of c^2 with respect to c^0. L' is a collection of adjacencies of D that includes the interstack adjacencies over the adjacencies $\{c^0, c_1^1\}$ and $\{c^0, c_2^1\}$ of D', and the interstack adjacencies over the adjacencies $\{c_1^1, c^2\}$ and $\{c_2^1, c^2\}$ of D' (if L' does not contain these adjacencies, then algorithms *BoundaryOneCells*, *AdjacenciesOver01* and *AdjacenciesOver12* may be used to add them to it).

Output: L is a list of all interstack adjacencies between $S^*(c^0)$ and $S^*(c^2)$.

(1) [Initialize. Step through sections of $S(c^2)$ from bottom to top.] Set L to L'. Set $h(z) \leftarrow$ *CellExtensionPolynomial*(c^0, B', B). If c^2 is to the left of c^0, then set $leftOrBelow \leftarrow true$, else set $leftOrBelow \leftarrow false$. For each section s of $S(c^2)$, do the following steps (1.1) - (1.4):

(1.1) [Initialize for this section s of $S(c^2)$.] Initialize a set Σ_s to the empty set. Let $F(x, y, z)$ be the unique element of B such that $s \subset V(F)$.

(1.2) [Boundary 0-cells in $S^*(c^0)$ of s over the boundary 1-cells of its base.] $(c_1^1, c_2^1) \leftarrow$ *BoundaryOneCells*(c^0, c^2). Find the boundary section t_1^1 of s in $S^*(c_1^1)$ by querying L'; find the boundary section t_1^0 of t_1^1 in $S^*(c^0)$ by querying L'; add t_1^0 to Σ_s. If F is not nullified on c^0, then go to step 1.4. Find the boundary section t_2^1 of s in $S^*(c_2^1)$ by querying L'; find the boundary section t_2^0 of t_2^1 in $S^*(c^0)$ by querying L'; add t_2^0 to Σ_s.

(1.3) [Boundary 0-cells in $S^*(c^0)$ of s over interior sections of partial derivative projections.] Suppose that s is real root m of F over c^2. For each element $T(x, y)$ of the set $\{\ pp(Res_z(F, F_y)),\ pp(Res_z(F, F_z))\ \}$, do the following loop: set $(i, j) \leftarrow$ *InteriorSections*$(c^0, c^2, T(x, y))$; if $i > 0$, then for $k = i, i+1, ..., i+j$ do the following loop: set $L' \leftarrow SSADJ3(c^0, h(z), T(x, y), k, leftOrBelow, F(x, y, z))$, and then knowing that s is real root m of F over c^2, determine from L' the unique section t^0 of $S^*(c^0)$ which is contained in the boundary of s, and add t^0 to Σ_s.

(1.4) [Infer complete boundary of s in $S^*(c^2)$.] s is adjacent to all elements of $S^*(c^0)$ between the "lowest" and the "highest" sections of $S^*(c^0)$ that are elements of Σ_s, inclusive; add all such adjacencies to L.

(2) [Infer remaining interstack adjacencies.] Use the current contents of L to infer the remaining interstack adjacencies between $S^*(c^0)$ and $S(c^2)$, and add them to L $\square$

Figure 14: Algorithm AdjacenciesOverNullifying02.

CADA3(A; I, L, S)

Input: A is a subset of I_3.

Outputs: I is a list of the cell indices of the cells of a basis-determined A-invariant cad D of E^3. L is a list of all adjacencies of D, plus additional adjacencies involving infinite sections. S is a list of sample points for D. The boundary of each cell of D, and each cell of the cad's of E^2 and E^1 induced by D, is a disjoint union of lower-dimensional cells.

(1) [Construct induced cad D' of E^2.] Set $B \leftarrow$ a basis for $prim(A)$. Set $P \leftarrow PROJ(B) \cup cont(A)$. Call algorithm CADA2 of Arnon *et al.* (1984b) with input P to obtain outputs I', L', and S'. Let $P^*(x,y)$ be $\prod prim(P)$, and set $B' \leftarrow \{P^*\}$. (it is not hard to see that D' is a proper cad with defining polynomial P^*, and also a basis-determined cad with basis B').

(2) [Cell indices, intrastack adjacencies, and sample points for stacks of D.] Initialize I, L, and S to be the empty list. For each cell c of D', do the following: $ExtendCellToStack(c, B', B; g, J^*, I^*, L^*)$; add the elements of I^* to I; add the elements of L^* to L; use the sample points for D', the polynomial g, and the isolating intervals of J^* for real roots of g, to construct sample points for the cells of the stack $S(c)$, and add these sample points to S.

(3) [Adjacencies over $\{1,2\}$ adjacencies of D'.] For each $\{1,2\}$ adjacency $\{c^1, c^2\}$ of D', do $L \leftarrow L \cup AdjacenciesOver12(c^1, c^2, B)$.

(4) [Adjacencies over $\{0,1\}$ adjacencies of D'.] For each $\{0,1\}$ adjacency $\{c^0, c^1\}$ of D', do $L \leftarrow L \cup AdjacenciesOver01(c^0, c^1, B', B)$.

(5) [Adjacencies over nonnullifying $\{0,2\}$ adjacencies of D'.] For each nonnullifying $\{0,2\}$ adjacency $\{c^0, c^2\}$ of D', do $L \leftarrow L \cup AdjacenciesOverNonNullifying02(c^0, c^2, B', B, L)$

(6) [Adjacencies over nullifying $\{0,2\}$ adjacencies of D'.] For each nullifying $\{0,2\}$ adjacency $\{c^0, c^2\}$ of D', do $L \leftarrow L \cup AdjacenciesOverNullifying02(c^0, c^2, B', B, L)$ □

Figure 15: Main algorithm.

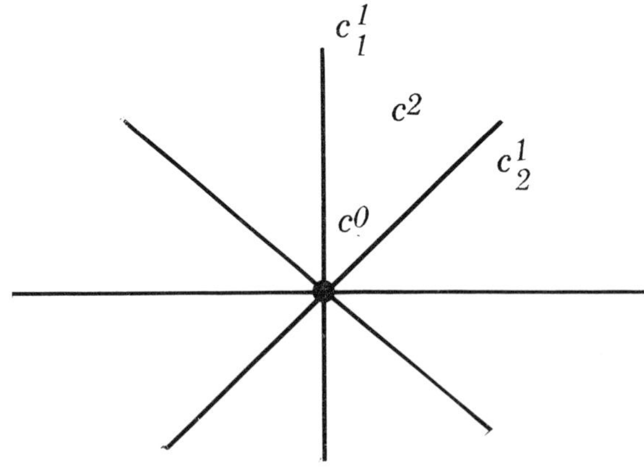

Figure 16: Induced decomposition of the plane for $F(x, y, z) = y^3 z + xy^2 - x^3$.

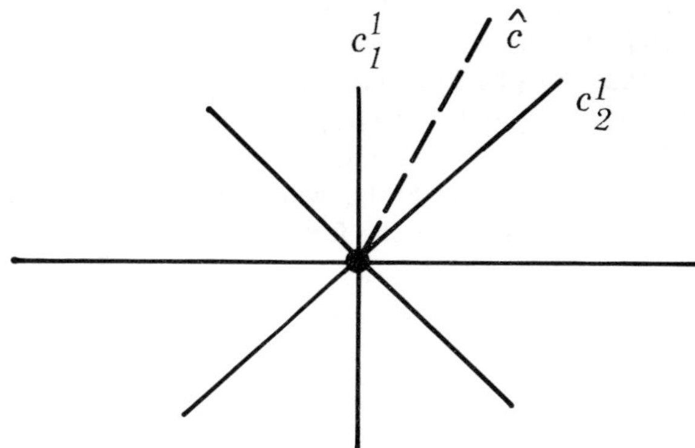

Figure 17: Interior section in the plane for $F(x, y, z) = y^3 z + xy^2 - x^3$.

degree in y, we exit step (2.3). Given the Γ we have created, it is clear that in step (3) of $CellExtensionPolynomial$ we will obtain $g(z) = z(27z^2 - 4)$. Thus returning to step (1) of $ExtendCellToStack$ where we isolate the real roots of $g(z)$, the data returned by $ExtendCellToStack$ will correspond to a stack in E^3 (over cell c^0) that has three sections: the 0-cells $< 0, 0, -2/3\sqrt{3} >$, $< 0, 0, 0 >$, and $< 0, 0, 2/3\sqrt{3} >$.

Let c^2 be the 2-cell (3,7) of D', let s be the unique section of D over c^2, and let us consider the boundary of s in $Z^*(c^0)$. Thus we are interested in the adjacencies found by that call to $AdjacenciesOverNullifying02$ in step (6) of CADA3, in which the first two inputs to $AdjacenciesOverNullifying02$ are c^0 and c^2. In step (1) of $AdjacenciesOverNullifying02$ we will set $h(z)$ to the same $g(z) = 27z^3 - 4z$ that we had above, we will have $leftOrBelow = false$, and since s is the unique section of $S(c^2)$, we will go through steps (1.1) - (1.4) just once. Clearly the two boundary 1-cells of c^2 are cell (2,3), which we will call c^1_1, and (3,6), which we will call c^1_2. In step (1.2) of $AdjacenciesOverNullifying02$, we will find that the boundary section of s in $S^*(c^1_1)$ is the unique finite section t^1_1 of this stack, and that the boundary section t^0_1 of t^1_1 in $S^*(c^0)$ is the section $< 0, 0, 0 >$ of $S(c^0)$. Also, the boundary section of s in $S^*(c^1_2)$ is the unique finite section t^1_2 of this stack, and the boundary section t^0_2 of t^1_2 in $S^*(c^0)$ is the section $< 0, 0, 0 >$ of $S(c^0)$. Hence at the end of step (1.2), we will have $\Sigma_s = \{< 0, 0, 0 >\}$.

In step (1.3) of $AdjacenciesOverNullifying02$, we will get the same $T(x, y)$'s that we had in our trace of $CellExtensionPolynomial$ above; thus only the first $T(x, y) = -y^4 + 3x^2 y^2$ may lead to a nontrivial computation and result in $InteriorSections$, and indeed for this $T(x, y)$, we get back a result of $(3, 0)$ from $InteriorSections$, telling us that the third real root of $T(\epsilon, y)$ corresponds to a (1-dimensional) "section" of $V(T)$ that lies between the two boundary 1-cells of c^2 and is adjacent to c^0. It is not hard to see that this "section" of $V(T)$ is the 1-cell $\hat{c}$, defined by the formula $x > 0 \ AND \ y^2 - 3x^2 = 0$, and indicated by the dotted line in Fig. 17.

Continuing in step (1.3) of $AdjacenciesOverNullifying02$, we make one call to SSADJ3, and from it learn that section $< 0, 0, -2/3\sqrt{3} >$ of $S(c^0)$ is contained in the boundary of s. Hence at the end of step (1.3), we have $\Sigma_s = \{< 0, 0, 0 >, < 0, 0, -2/3\sqrt{3} >\}$. In step (1.4), we infer that s is adjacent to all cells of $S(c^0)$ between $< 0, 0, 0 >$ and $< 0, 0, -2/3\sqrt{3} >$ inclusive, hence since $< 0, 0, 0 >$ is cell (2,2,4) of this stack, and $< 0, 0, -2/3\sqrt{3} >$ is cell (2,2,2), this means that s (which is cell (3,7,2)) is adjacent to cells (2,2,2), (2,2,3), and (2,2,4) of $S(c^0)$. In step (2) of *AdjacenciesOverNullifying02*, we infer the remaining interstack adjacencies between $S^*(c^0)$ and $S(c^2)$, i.e. that cell (3,7,1) is adjacent to cell (2,2,1), and that cell (3,7,3) is adjacent to cells (2,2,5), (2,2,6), and (2,2,7).

The reader may find that considering cross-sections of $V(F)$ by planes of the form $y = $ constant, for small positive constants, will help to understand the behavior of CADA3 for this example. Writing F in the form $z = k(x) = (x^3 - y^2 x) / y^3$, the local maxima and minima of k occur when $k'(x) = 3x^2 - y^2 / y^3 = 0$, i.e. $y = \pm \sqrt{3}x$. Thus at a local maximum of k, $z = (3x^3 - x^3) / (3\sqrt{3}x^3) = 2 / 3\sqrt{3}$, and at a local minimum of k, $z = -2 / 3\sqrt{3}$, independent of the (positive) value of y. The local minima occur over points $< x, y >$ in E^2 of the form $< x, \sqrt{3}x >$, which are points in c^2 when $x > 0$. Thus the point $< 0, 0, -2 / 3\sqrt{3} >$ in E^3 must be a boundary point of s. It is examples such as this $F(x, y, z)$ that make necessary the machinery of algorithm *AdjacenciesOverNullifying02* (and algorithm *CellExtensionPolynomial*), rather than just the simpler algorithm *AdjacenciesOverNonNullifying02*

9 References

Arnon, D. S., Collins, G. E., McCallum, S. (1984a). Cylindrical algebraic decomposition I: the basic algorithm, *SIAM J. Comp.* **13/4**, 865–877.

Arnon, D. S., Collins, G. E., McCallum, S. (1984b). Cylindrical algebraic decomposition II: an adjacency algorithm for the plane, *SIAM J. Comp.* **13/4**, 878–889.

Arnon, D. S. (1979). A cellular decomposition algorithm for semi-algebraic sets. Proceedings of an International Symposium on Symbolic and Algebraic Manipulation (EUROSAM '79). *Springer Lec. Notes Comp. Sci.* **72**, 301-315.

Arnon, D. S. (1981). *Algorithms for the Geometry of Semi-Algebraic Sets.* PhD thesis, Tech. Rept. #436, Comp. Sci. Dept., Univ. Wisconsin–Madison.

•Arnon, D. S. (1988). A cluster-based cylindrical algebraic decomposition algorithm. *J. Symb. Comp.* **5**, (this issue).

Collins, G. E. (1971). The calculation of multivariate polynomial resultants. *J. Assoc. Comp. Mach.***18/4**, 515–532.

Collins, G. E. (1975). Quantifier elimination for real closed fields by cylindrical algebraic decomposition. Proceedings of the Second GI Conference on Automata Theory and Formal Languages. *Springer Lec. Notes Comp. Sci.* **33**, 515–532.

Collins, G. E., Loos, R. G. K. (1982). Real zeros of polynomials. In (Buchberger, B., Loos, R., Collins, G. E., eds.) *Computer Algebra - Symbolic and Algebraic Computation* (Computing Supplementum 4), pp. 83-94. Vienna and New York: Springer-Verlag.

Kaltofen, E. (1982). Polynomial factorization. In (Buchberger, B., Loos, R., Collins, G. E., eds.) *Computer Algebra - Symbolic and Algebraic Computation* (Computing Supplementum 4), pp. 95-113. Vienna and New York: Springer-Verlag.

Kozen, D., Yap., C. K. (1985). Algebraic cell decomposition in NC. *Proc. IEEE Conf. on Foundations of Comp. Sci. (FOCS)*, 515–521.

Massey W. S. (1978). *Homology and Cohomology Theory.* New York, NY: Marcel Dekker.

McCallum, S. (1979). *Constructive Triangulation of Real Curves and Surfaces.* MSc thesis, Math. Dept., Univ. of Sydney, Australia.

McCallum, S. (1988). An improved projection operation for cylindrical algebraic decomposition of three-dimensional space. *J. Symb. Comp.* **5**, (this issue).

Prill, D. (1986). On approximation and incidence in cylindrical algebraic decompositions. *SIAM J. Comp.* **15**, 972–993.

Schwartz, J. T., Sharir, M. (1983). On the 'piano movers' problem II. General techniques for computing topological properties of real algebraic manifolds. *Adv. Applied Math.* **4**, 298–351.

Walker R. J. (1978). *Algebraic Curves,* New York, NY: Springer-Verlag.

J. Symbolic Computation (1988) **5**, 189–212

A Cluster-Based Cylindrical Algebraic Decomposition Algorithm*

DENNIS S. ARNON

Xerox PARC, 3333 Coyote Hill Road, Palo Alto, California 94304, U.S.A.

(Received 2 April 1985, and in revised form 15 November 1987)

Let $A \subset \mathbf{Z}\,[x_1, \ldots, x_r]$ be a finite set. An *A-invariant cylindrical algebraic decomposition (cad)* is a certain partition of r-dimensional euclidean space E^r into semi-algebraic cells such that the value of each $A_i \in A$ has constant sign (positive, negative, or zero) throughout each cell. Two cells are *adjacent* if their union is connected. Recently a number of methods have been given for augmenting Collins' cad construction algorithm (1975), so that in addition to specifying the cells that comprise a cad, it identifies the pairs of adjacent cells. Assuming the availability of such an adjacency algorithm, in this paper we give a modified cad construction algorithm based on the utilization of clusters of cells in a cad (a *cluster* is a collection of cells whose union is connected). Preliminary observations indicate that the new algorithm can be significantly more efficient in some cases than the original, although in other examples it is somewhat less efficient.

1   Introduction

Recently a number of methods have been given for augmenting the cad construction algorithm (Collins, 1975), so that in addition to specifying the cells that comprise a cad, it identifies the pairs of adjacent cells (see e.g. Arnon *et al.*, 1988, Prill, 1986, Kozen & Yap 1985, Schwartz & Sharir 1983). A *cluster* of cells in a cad is a collection of cells whose union is connected. Assuming the availability of an adjacency algorithm, in this paper we give a modified cad construction algorithm based on the utilization of clusters. The key idea is that, as a cad of E^{i-1} is extended to a cad of E^i, certain (possibly expensive) computations are performed only once for each cluster, rather than once for each cell as in the original algorithm. Offsetting this saving is the extra cost of adjacency computation. Preliminary observations indicate that the new algorithm can be significantly more efficient in some cases than the original, although in other examples it is somewhat less efficient. In this paper we give both a general framework for cluster-based cad construction, within which any available adjacency algorithm can be used, and a specific cluster-based cad algorithm that uses the 2-space and 3-space adjacency algorithms of Arnon *et al.* (1984b, 1988). The specific algorithm we give has the following properties: (1) it requires no coordinate changes, and (2) in any cad of E^1, E^2,

*This work was supported by the National Science Foundation (Grant MCS-8009357 to the University of Wisconsin-Madison), the Purdue Research Foundation, and the Xerox Corporation. This paper was typeset at Xerox PARC using TeX in the Cedar environment.

0747–1717/88/010189 + 24 $03.00/0

or E^3 that it builds, the boundary of each cell is a (disjoint) union of lower-dimensional cells. The particular clusters that occur in cluster-based cad construction are of mathematical interest in their own right. For example, if A consists of a single element F, then the (unions of the) r-space clusters are typically the connected components of the hypersurface $F = 0$ and its complement.

In this Introduction we sketch the broad outlines of the clustering strategy for cad construction, give an outline of the paper, and review prior related work.

1.1 Cad graphs and clusters

Let us begin our discussion of clustering by recalling terminology from Arnon *et al.* (1984a, 1984b, 1988). We say that a connected subset of E^r is a *region*. If $A = (A_1, ..., A_n)$ is a subset of $I_r = \mathbf{Z}\,[x_1, ..., x_r]$, if R is an A-invariant region in E^r (i.e. the value of each $A_i \in A$ has constant sign (-1, 0, or $+1$) throughout R), and if σ_i is the sign of A_i on R, then we say that the ordered n-tuple $\sigma = (\sigma_1, ..., \sigma_n)$ is the *signature* of R with respect to A (and also, the signature of A on R). A *cell triple* for a cell c of an A-invariant cad is a triple (I, σ, S), where I is the cell index of c (cell indices are defined in Section 4 of Arnon *et al.*, 1984a), σ is the signature of the cell (with respect to the set A of input polynomials), and S is a sample point for c. We temporarily proceed as though sample points are represented as in Arnon *et al.* (1984a, 1988); we will have more to say about their representation later.

Given $A \subset I_r$, a *graph representation* for an A-invariant cad D of E^r, or *cad graph*, is a quintuple $G = (A, B, V, E, G')$, defined as follows. B is a basis (as defined in Arnon *et al.*, 1988) for $prim(A)$, such that D is a basis-determined cad with basis B. (Recall that $prim(A)$ the set of primitive parts of those elements of A which have positive degree). V is a set of cell triples for the cells that comprise D. E is a set of unordered pairs of (distinct) elements of V, obeying the following condition: if (c_1, c_2) is an element of E, then cells c_1 and c_2 of the cad D are adjacent (thus (V, E) is a certain undirected graph). For any given pair of cells c_1 and c_2 of D, the converse may or may not hold. If for every pair of cells c_1 and c_2 of D the converse does hold, i.e. $(c_1, c_2) \in E$ if and only if c_1 and c_2 are adjacent in D, then we say that G is a *full* graph for D; otherwise, G is *partial*. The reader will notice a certain abuse of notation here: we freely identify a cell c with the triple that represents it. If $r > 1$, then G' is a graph representation for the cad D' of E^{r-1} induced by D, and $G' = \emptyset$ when $r = 1$. Typically the cad graph representations we work with are partial. In case G is full, the undirected graph (V, E) has been called the *connectivity graph* of D (Schwartz & Sharir, 1983, p. 320). We assume the availability of standard graph algorithms, e.g. depth-first search for connected components; see e.g. Aho *et al.* (1974).

It is appropriate to check that a graph representation for a cad supplies the information about that cad called for at the beginning of Section 4 of Arnon *et al.* (1984a). It was stated there that a description of a cad must inform one of the number of cells in the cad, how they are arranged into stacks, and the signature of each cell with respect to the set of input polynomials. Obviously a cad graph gives one the number of cells and each cell's signature. As detailed in Arnon *et al.* (1984a), the indices of the cells comprising a cad tell one how those cells are arranged into stacks.

Given $A \subset I_r$, let G denote a full graph for an A-invariant cad D. It is easy to show that the vertices of any connected subgraph of G correspond to a collection of cells of D whose union is a region in E^r. Turning this around, we say that a collection C of cells of D is a *cluster* (of D) if the subgraph of G induced by C is connected. Clearly C is a cluster if and only if the union of C is a region. The *dimension* of a cluster is the dimension of the largest cell in it. A partition of (the set of cells of) a cad D into clusters is called a *clustering* of D. Obviously any D can be clustered in many ways.

Assume now that G is either partial or full, and suppose given an equivalence relation R on the cells of D. Then R induces a clustering of D, which can be made explicit by computing the connected components of G subject to the constraint that we only "notice" an edge during the computation if the cells it joins belong to R. In this paper, we are exclusively interested in one particular equivalence relation, namely the relation to which a pair of cells belongs if and only if the two cells have the same signature (with respect to the set A of input polynomials). We call this the *sign-invariance* relation. (We will henceforth be using the term "sign-invariant" quite often in place of "A-invariant", to denote the condition that "each input polynomial is sign-invariant", without mentioning the particular set A of input polynomials). We call a clustering induced by the sign-invariance relation in a graph representation for a cad D a *sign-invariant clustering* of D, and the clusters which comprise it *sign-invariant clusters*.

Given two clusterings Γ_1 and Γ_2 of a cad D, we say that Γ_1 is *finer* than Γ_2, if each cluster of Γ_1 is a subset of some cluster of Γ_2. Equivalently, we say that Γ_1 is a *refinement* of Γ_2, and that Γ_2 is *coarser* than Γ_1. We say that a sign-invariant clustering of D is *maximal* if it is the coarsest possible sign-invariant clustering of D; its elements are then *maximal sign-invariant clusters*. Given A, we call the maximal connected A-invariant subsets of E^r the *A-components* of E^r, or in general the *sign-invariant components* of E^r (with respect to this A, of course). Note that this last definition is independent of any particular cad of E^r. Clearly a sign-invariant clustering of D is maximal if and only if the union of each of its clusters is a sign-invariant component of E^r. If G is a full graph for D, then clearly the sign-invariant connected components of G correspond to the maximal sign-invariant clusters of D, however if G is partial, this need not be the case.

If G is partial, then an equivalence relation on the cells of D still induces a clustering of D when (just as above) we compute connected components in the cad graph under the constraint that we only notice edges between equivalent cells. Such a clustering is in general finer than the clustering we get with the same relation applied to a full G, since the edges in a partial G are a subset of the edges in a full G. This observation is important, because in general we will build clusterings using partial graphs, and we will be interested in how closely these clusterings correspond to the clusterings that the same equivalence relation induces in a full graph.

Let us now look at some examples of the notions we have introduced. Consider the sample cad D from Section 5 of Arnon *et al.* (1984b), which we show in Fig. 1. Fig. 2 shows a full graph for D. The figure uses the convention that 0-cells are indicated as solid vertices, 1-cells as half-filled vertices, and 2-cells as unfilled vertices. Edges satisfiying the sign-invariance relation are shown as solid lines, and those not are shown as dotted lines. We see that there are 15 maximal sign-invariant clusters. Fig. 3 shows a partial graph for D. Again, edges satisfiying the sign-invariance relation are drawn as solid lines, and those not satisfiying it are drawn as dotted lines. For this graph, we get 19 sign-invariant clusters, many of which are not maximal.

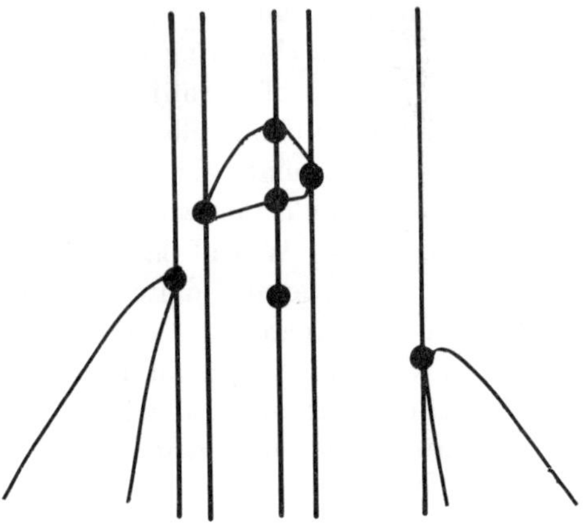

Figure 1: Sample Cad.

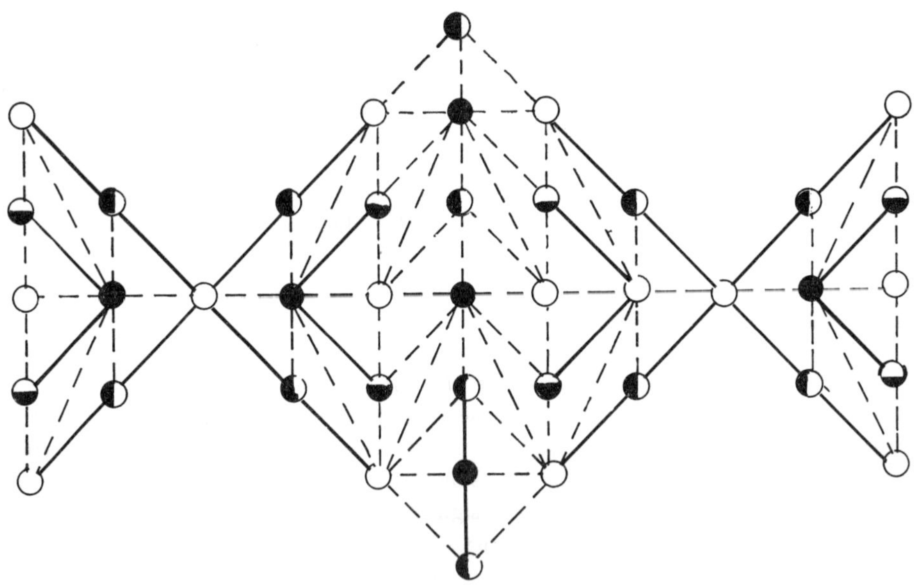

Figure 2: Full graph.

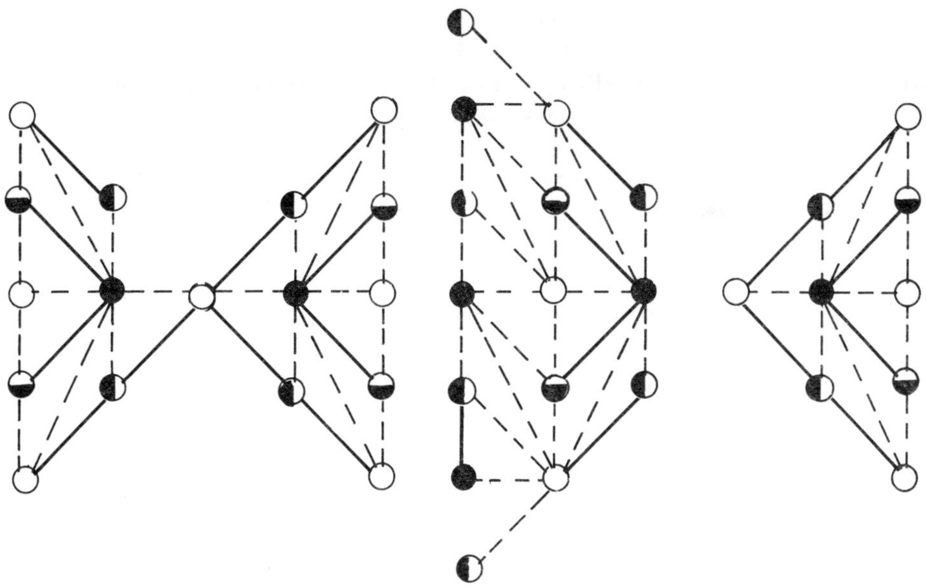

Figure 3: Partial graph.

1.2   Cluster-based cad construction

Let us now describe the basic idea of cluster-based cad construction. Essentially what we do is make the extension of a cad of E^{i-1} to E^i more efficient, by building stacks over sign-invariant clusters in E^{i-1}, rather than over individual cells in E^{i-1}. This general strategy requires the availability of adjacency algorithms, but does not require the use of any particular adjacency algorithm. We now explain in detail the formal basis for the strategy.

Assume given $A \subset I_r$. In Section 3 of Arnon *et al.* (1984a), a map $PROJ$, which takes a subset of I_r to a subset of I_{r-1}, is defined, and it is proved (Theorem 3.4) that over any $PROJ(A)$-invariant region in E^{r-1}, there exists an A-invariant stack. In applying this Theorem 3.4 to extend a cad of E^{r-1} to a cad of E^r in algorithm CAD of Arnon *et al.* (1984a), the $PROJ(A)$-invariant regions in E^{r-1} are the cells of the induced cad of E^{r-1}. However, given an arbitrary $PROJ(A)$-invariant decomposition $\hat{D}$ of E^{r-1}, Theorem 3.4 tells us that if we have a sample point for each region of $\hat{D}$, then we can extend it to a decomposition D^* of E^r consisting of the stacks over regions of $\hat{D}$, by exactly the steps used in CAD for extension over a single cell. Note that D^* is not necessarily cylindrical in the sense of Arnon *et al.* (1984a), *i.e.* it may not be the case that $\hat{D}$ consists of stacks over the regions of some decomposition of E^{r-2}. However, it is the case that if $\hat{D}$ is algebraic, *i.e.* its regions are semi-algebraic sets, then so is D^*.

Suppose now that for some $A \subset I_r$, we have (a graph for) a $PROJ(A)$-invariant cad D' of E^{r-1}, and a clustering of it into $PROJ(A)$-invariant clusters. Then forming the regions we get by taking the union of each cluster, we get a $PROJ(A)$-invariant, decomposition $\hat{D}$ of E^{r-1}. As above, let us extend $\hat{D}$ to a decomposition D^* of E^r by building stacks over $\hat{D}$'s regions, and let D denote the (A-invariant) cad of E^r that we get by extending D'. Then it is not hard to see that each element of D^* is the union of certain elements of D. In particular, if C is a cluster of D', and if R is the union of C, then for any $i \geq 1$, the ith element of the stack over R (this stack is part of D^*) is the union of the ith elements of the stacks over the cells of C (these stacks are each part of D). Furthermore, if cells c_1 and c_2 of C are adjacent elements of D', then for any $i \geq 1$, the ith element of the stack over c_1, and the ith element of the stack over c_2, are adjacent elements of D. We call such clusters and adjacencies in E^r *induced clusters* and *induced adjacencies*, because they are "induced" in a cad D of E^r by a cluster or adjacency in the cad D' of E^{r-1}.

Given D', a $PROJ(A)$-invariant clustering of D', and a sample point for (one cell in) each cluster, our observations above imply that we can build D as follows. For each cluster, we determine a stack over its union R using its sample point, as just discussed. Having determined the number of elements, i.e. sections and sectors, of this stack, and assuming that we have determined the signature of each element of the stack with respect to A, we next look to see which cells of D' comprise C (i.e. what their cell indices are), and we know immediately (i.e. we can write down the cell indices and signatures for) the cells of D which comprise each element of the stack over R. When we have processed all clusters of D' in this way, we will have compiled the indices and signatures of all cells of D. Furthermore, for each cluster C of D', each adjacency $\{c, d\}$ of elements of C induces an adjacency between the ith elements of the stacks over c and d. Clearly a graph for D which contains exactly these induced adjacencies gives rise to an induced clustering of D (into induced clusters).

Clearly the induced adjacencies of D are sign-invariant adjacencies. They are likely to be only a proper subset of the set of all of D's sign-invariant adjacencies, however. In particular, if we do a sign-invariant components computation in the graph for D in the form that it has after the steps we have described, the sign-invariant clusters we construct are likely not maximal. To get the most benefit from the use of clusters, we would like to have the largest possible sign-invariant clusters. Hence, the next step of our general strategy is to build larger sign-invariant clusters than those given to us by the induced adjacencies. As one might guess, we do this by computing further adjacencies in E^r using the adjacency algorithms that we assume are available to us.

Let us consider a simple example of these ideas. Let $A = \{u^2 + z^2 + y^2 + x^2 - 1\}$, i.e. A consists of the polynomial which defines the (three-dimensional) unit sphere in 4-space. We have $PROJ(A) = \{z^2 + y^2 + x^2 - 1\}$, $PROJ^2(A) = \{y^2 + x^2 - 1\}$, and $PROJ^3(A) = \{x^2 - 1\}$. The cad of 1-space clearly has five cells; recall from Section 4 of Arnon *et al.* (1984a) that we write the indices for these cells as (1), (2), (3), (4), (5). The maximal sign-invariant clusters for this cad of E^1 are just the five singleton sets. In 2-space, we have 13 cells, which can be partitioned into three maximal sign-invariant clusters: the unit circle (consisting of four cells, with indices (2,2), (3,2), (3,4), and (4,2)), its interior (consisting of one cell, with index (3,3)), and its exterior (consisting of eight cells, with indices (1,1), (2,1), (2,3), (3,1), (3,5), (4,1), (4,3), and (5,1)). The discussion of the previous paragraphs tells us that to determine the cad of 3-space, it suffices to have a sample point for each of these three 2-space clusters. When we extend over the unit circle, for example, we get a stack in E^3 consisting of three elements (i.e. two sectors and one sections), that corresponds to four stacks in the cad of 3-space that each have three elements. The adjacencies among the cells in E^2 that comprise the unit circle induce certain adjacencies (and three clusters) among the cells of these four stacks in 3-space. Similarly, extending into E^3 over the interior of the unit circle in E^2, we get a stack with five elements, and extending over the exterior of the circle, a stack in E^3 with one element. The latter stack corresponds to eight one-element stacks, with the obvious induced adjacencies and a single induced cluster, of the cad of 3-space. Clearly, the induced clusters in 3-space that we have described are not maximal sign-invariant clusters. Using a 3-space adjacency algorithm, we can compute additional adjacencies among the 3-space cells that enable us to obtain the three maximal sign-invariant clusters that there clearly are in 3-space (which correspond to the unit sphere, its interior, and its exterior). We can then build stacks in 4-space over these three 3-space clusters to determine a sign-invariant cad of 4-space.

Altogether the three steps of the cluster-based cad algorithm are: (1) If $r > 1$, call the algorithm recursively to build a graph for the induced cad of $r-1$ space, (2) If $r > 1$, extend, over the maximal sign-invariant clusters of the induced cad, to a graph for the cad of r-space, or if $r = 1$, build a graph directly, (3) Construct additional adjacencies in r-space. The simplest, trivial, case of cluster-based cad construction is the "original" cad algorithm, i.e. no adjacency computation at all, which means that we generally have just singleton clusters in the cad's of E^1, E^2, ... that we build.

1.3   Outline of the paper and prior work

Sections 2-4 fill in the details of the cluster-based cad construction strategy, by partitioning it into algorithms of four kinds: basis (Section 2), projection (Section 2), extension (Section 3), and adjacency (Section 4). Section 3 begins by defining several possible representations for sample points in cad graphs. This is fundamental material for this paper: careful management of sample point representations is an important reason why cluster-based cad construction is more efficient than previous cad algorithms in those cases that it is. Section 4 presents the particular adjacency algorithms for E^2 and E^3 that we currently use; these rely on certain adjacency subalgorithms from Arnon *et al.* (1984b) and Arnon *et al.* (1988). Section 5 presents a main algorithm CLCAD for cluster-based cad construction, which has procedure parameters for the four key subalgorithms. Also in Section 5 we specify the values of these procedure parameters that we use in our current implementation of CLCAD. Section 6 reports the comparative performance of algorithms CLCAD and CAD on a number of examples.

The work we report in this paper was done between 1979 and 1981. The use of adjacencies and clusters in cad construction was presaged by Arnon (1979), where it was shown that incidence of cells is decidable. A first version of CLCAD was presented, and some examples of its use and comparative performance with CAD given, in Arnon (1981). Applications of cluster-based cad construction can be found in Arnon & McCallum (1988), and Arnon (1988).

Defining formula construction is an important part of the cad algorithm, especially for applications to quantifier elimination (see e.g. Arnon & Mignotte, 1988). Constructing a defining formula for each cell of a cad is easily accomplished in cluster-based cad construction, by constructing such formulas for certain cells, and then inferring formulas for the remaining cells by much the same inference process as used for induced clusters and adjacencies in Section 1.2. See Arnon (1981) or Collins (1975) for details of Collins' original algorithm for cell defining formula construction.

As mentioned above, in the present paper we only make use of one equivalence relation of cells, namely the sign-invariance relation. The order-invariance relation of McCallum (1988) is another equivalence relation of cells in a cad whose use for cluster-based cad construction is attractive.

2   Basis, Projection, and Base steps

In this Section we discuss the first few steps of the cluster-based cad construction algorithm. We have relatively little to say about them. The reader may wish either to look ahead to algorithm CLCAD in Section 5, or skip this Section for the moment.

In general, it doesn't matter what type of basis (e.g. coarsest or finest squarefree basis) our basis procedure computes (see Arnon *et. al.*, 1988, and Collins, 1975, for basis-related definitions). The actual projection operator we currently use is determined by the adjacency algorithms of Arnon *et al.* (1984b, 1988) that we use (cf. Section 4). We want to build the same cad's as these algorithms do. Hence rather than use $PROJ(A)$ as a projection operator, as we did in Section 1, we henceforth assume that we have computed a basis B for $prim(A)$, and that we use $PROJ(B) \cup cont(A)$ as our projection operator ($cont(A)$ is the set of non-zero non-unit contents of elements of A; see Arnon *et al.*, 1988, for futher discussion). For $r \leq 3$, we could use McCallum (1988) projection instead; it would then be necessary that our basis procedure compute a finest squarefree basis. The projection operator would then be the P operator as defined in McCallum (1988). The resulting cad's of E^r, $1 \leq r \leq 3$, would still have the boundary property, i.e. the boundary of each cell would be a (disjoint) union of lower-dimensional cells, and if $r > 1$, then the induced cad of E^{r-1} would also have the boundary property.

The base step of our cad algorithm, i.e. the algorithm for construction of cad's of E^1, is essentially that of Arnon *et al.* (1984a). It is easy to make the graph for the cad of E^1 full, since we trivially know what its adjacencies are. The reader may consult algorithm CLCAD in Section 5 for details of the base step.

3   Extension step

Our task in this section is to develop the method (algorithm *ExtendCadClusters* of Fig. 6) that we use for the extension step of cluster-based cad construction. We begin by considering the issue of sample point representation. Assume throughout this section that our cad input polynomials have $r \geq 1$ variables.

So far we have assumed that cell sample points are represented as in Arnon *et al.* (1984a, 1984b, 1988). In fact, cell sample points in the cluster-based cad algorithm may have one of three representations: (a) *null* (no information), or (b) *extended*, consisting of a real algebraic number α, an $r - 1$ tuple of elements of $Q(\alpha)$, a nonzero squarefree polynomial $g(x) \in Q(\alpha)[x]$, and an isolating interval for a (real) root of $g(x)$ (this root is the rth coordinate of the sample point), or (c) *primitive*, consisting of a real algebraic number α (the primitive element) and an r-tuple of elements of $Q(\alpha)$.

In fact, this extended representation is present in passing in the extension step of the cad algorithms in Arnon *et al.* (1984a, 1984b, 1988), although ultimately all cell sample points become primitive in these algorithms. To be specific, when we have a sample point for the base of a stack, and we isolate the real roots of a squarefree univariate algebraic polynomial to determine the sections of the stack, the base sample point, the algebraic polynomial, and each isolating interval for one of its roots give us an extended sample point representation for a section of the stack. As described in Section

5 of Arnon *et al.* (1984a), we can use the NORMAL and SIMPLE algorithms of Loos (1982) to convert an extended representation to a primitive one. This conversion process has been observed to often be expensive, and avoiding it whenever possible is a major goal of the cluster-based cad algorithm. Working with extended rather than primitive representations whenever possible is one step towards that goal; another such step is to make do with null sample points, whenever possible, which we also will do.

As we have noted, however, it is a required of a cad algorithm to construct input polynomial signatures for each of its cells. Previous cad algorithms (such as those in Arnon *et al.*, 1984a, 1984b, 1988) have done so by evaluating the input polynomials at primitive cell sample points. We now show that it is possible to compute the signature (with respect to the input polynomials) of a cell in a basis-determined cad of E^r given an extended representation for the cell's sample point. In fact, the method we give could be used in the original as well as the cluster-based cad algorithm, to avoid extended-to-primitive conversions of section sample points in dimension r, i.e. the highest dimension.

We proceed in two steps. First, we show how to get the signatures of cells in E^r with respect to the basis polynomials from extended representations for their sample points. Second, we infer input polynomial signatures from these basis signatures plus signatures for the contents of the input polynomials. Here is a sketch of the first step. Suppose that $r \geq 2$, that s is a cell of a cad D of E^r, and that c is the unique cell of the induced cad D' of E^{r-1} for which $s \in Z(c)$. Suppose that we have already determined (i.e. found the number of sections of) the stack $S(c) \subset D$, by isolating the real roots of some suitable $g(x_r) \in Q(\alpha)[x_r]$. Thus, about each real root of $g(x_r)$, we have an open isolating interval with rational number endpoints. For each basis polynomial B_i, we compute the greatest squarefree divisor $d(x_r)$ of $B_i(\alpha, x_r)$. $d(x_r)$ has the same roots as $B_i(\alpha, x_r)$, but only simple roots; see Kaltofen (1982) for more information on greatest squarefree divisors. Since we are assuming $S(c)$ to be B-invariant, any root of $B_i(\alpha, x_r)$ is a root of $g(x_r)$. Hence for any section s of $S(c)$, B_i vanishes on s if and only if d has opposite signs at the endpoints of the isolating interval for the unique root of $g(x_r)$ that corresponds to s. If B_i doesn't vanish on s, then it has the same sign on s, on the sector immediately above s, and on the sector immediately below s. We can determine the sign of B_i on sectors of $S(c)$ as follows. The endpoints of the isolating intervals for the roots of $g(x_r)$ give us sample points of the form $< \alpha, b >$, b rational, for the sectors of $S(c)$ (much as we got sample points for the sectors of stacks in Section 5 of Arnon *et al.*, 1984a). By evaluating each $B_i(\alpha, b)$, we determine the sign of B_i on each sector of $S(c)$. Fig. 4 gives the algorithm *BasisSignaturesOverCell* that embodies this strategy. The map *gsfd* in the algorithm is "greatest squarefree divisor".

To infer input polynomial signatures from basis signatures, we need only a few more observations. Suppose $C(x) = content(A_i)$. If $C(\alpha) = 0$ then A_i vanishes on every element of $S(c)$, and we are done. If not, we use the sign of $content(A_i)$, and the factorization of $pp(A_i)$ ($pp(A_i)$ denotes the primitive part of A_i) as a power product of basis polynomials, to "infer" the sign of A_i on each element of the stack. Algorithm *InputSignaturesOverCell* in Fig. 5 has the details.

With algorithm *InputSignaturesOverCell* available to us, we have the following situation. We have no need to convert any extended sample point representations to primitive form in the cad of r-space. In dimensions less than r, we need primitive sample points for any cell or cluster that we are going to "extend over", i.e. build a stack over, as we extend our cad to the next higher-dimensional space. Thus the first step of *ExtendCadClusters* is to compute the sign-invariant connected components of

---

$$\Sigma \leftarrow \textbf{BasisSignaturesOverCell } (B, p, J)$$

Inputs: Given $A = (A_1, ..., A_n) \subset I_r$, $r \geq 1$, $B = (B_1, ..., B_m)$ is a basis for $prim(A)$. $p = (p_1, ..., p_{r-1})$ is a primitive sample point for a $PROJ(B)$-invariant and $cont(A)$-invariant cell c in a cad of E^{r-1}, i.e. each p_i is an element of $Q(\gamma)$ for some real algebraic number γ. If $r = 1$, then $p = \emptyset$ and $c = E^0$. $J = (J_1, J_2, ..., J_k)$, $k \geq 0$, is a list of open isolating intervals for the k real roots $\lambda_1 < ... < \lambda_k$ of some nonzero univariate real polynomial $g = g(x_r)$, such that for each j, $1 \leq j \leq k$, the point $(p_1, ..., p_{r-1}, \lambda_j)$ lies on the jth section of a B-invariant (and hence also A-invariant) stack $S(c)$ over c.

Output: $\Sigma = (\sigma_1, ..., \sigma_{2k+1})$, such that $\sigma_j = (\sigma_{1,j}, ..., \sigma_{m,j})$ is the signature of the jth element of $S(c)$ with respect to B.

(1) [Do it.] For $i = 1, ..., m$, do: set $h(x_r) \leftarrow B_i(p_1, ..., p_{r-1}, x_r)$; set $d(x_r) \leftarrow gsfd(h(x_r))$; set $d = 0$ if $h = 0$; set $\rho_{i,2k+1} \leftarrow sign(d) = sign(leadingCoefficient(d))$; set $\sigma_{i,2k+1} \leftarrow sign(h)$; for $j = k, k-1, ..., 1$ do: Let $J_j = (u_j, v_j)$; set $\rho_{i,2j-1} \leftarrow sign(d(u_j))$; if $\rho_{i,2j-1} \neq \rho_{i,2j+1}$, then $\sigma_{i,2j} \leftarrow 0$, and $\sigma_{i,2j-1} \leftarrow sign(h(u_j))$, else $\sigma_{i,2j-1} \leftarrow \sigma_{i,2j} \leftarrow \sigma_{i,2j+1}$ $\square$

Figure 4: Algorithm BasisSignaturesOverCell.

$$T \leftarrow \textbf{InputSignaturesOverCell } (A, B, p, J)$$

Inputs: $A = (A_1, ..., A_n) \subset I_r$, $r \geq 1$, and the remaining inputs are as for algorithm *BasisSignaturesOverCell*.

Output: $T = (\tau_1, ..., \tau_{2k+1})$, such that $\tau_j = (\tau_{1,j}, ..., \tau_{n,j})$ is the signature of the jth element of $S(c)$ with respect to A.

(1) [Get basis signatures.] $\Sigma \leftarrow BasisSignaturesOverCell(B, p, J)$.

(2) [Infer input polynomial signatures.] Recall that we follow the convention that $sign(content(F))$ is chosen to be $sign(F)$, for any $F \in I_r$. For each $A_i \in A$, there exist nonnegative integers $e_{i,1}, ..., e_{i,m}$ such that $A_i = content(A_i) \prod_{u=1}^{m} B_u^{e_{i,u}}$. For $i = 1, ..., n$, and for $j = 1, ..., 2k+1$ do: $\tau_{i,j} \leftarrow sign(content(A_i)) \prod_{u=1}^{m} \sigma_{u,j}^{e_{i,u}}$ $\square$

Figure 5: Algorithm InputSignaturesOverCell.

G', and for each (i.e. for each sign-invariant cluster of D'), insure that at least one of its constituent cells has a primitive sample point. Needing a primitive sample point only for one cell in each sign-invariant cluster, rather than for each cell of the cad, is a key reason why the cluster-based cad algorithm is faster than the original cad algorithm in those cases that it is. For cad's of E^3, however, the saving realized here in the extension step are somewhat offset by the fact that our current E^3 adjacency algorithm (given in Section 4) needs primitive sample points for certain additional cells of the induced cad of E^2. In general we should expect that adjacency algorithms may require us to perform certain addition extended-to-primitive conversions of sample point representations.

Fig. 6 gives the complete algorithm *ExtendCadClusters*. The reader will see that it is essentially a formalization of our discussion in Section 1.2. We say that a cad graph is *initial* if the "initial" adjacencies are present in it, where these are (1) the intrastack adjacencies of each stack of that cad, and (2) the induced adjacencies as defined in Section 1. Note that even though we use a basis B for $prim(A)$ to determine the stacks of our cad's, *ExtendCadClusters* constructs A-invariant clusters prior to extending. In general, A-invariant clusters will be coarser than B-invariant clusters, and we want the extend over the coarsest possible clusters to minimize the number of primitive sample points that we need.

Suppose $r = 2$. Since maximal sign-invariant clusters in a cad of E^1 each consist of a single cell, by Step (2.1) of *ExtendCadClusters*, we see that we get primitive sample points for all 1-cells and all 2-cells, and an extended or primitive sample point for each 0-cell, of the cad of E^2 that we build.

4   Adjacency step

For each i, the role of what we call the "adjacency" subalgorithm of cluster-based cad construction is to add non-initial (interstack) adjacencies to the graph for the cad of E^i. It is not necessary to actually have such an adjacency algorithm for each i. If we wish, we need compute no adjacencies beyond initial adjacencies, for any value(s) of $i \leq r$. For example, since at present we only have implemented adjacency algorithms for $i = 1, 2, 3$, the adjacency step in our implementation of the cluster-based cad algorithm is currently null for $i \geq 4$. As might be expected, if we have a null adjacency algorithm for the cad of E^i, then our graph for that cad is almost certainly partial, and the sign-invariant components of that graph give us a clustering of the cad that is almost certainly not maximal.

In Fig. 7 we give our 2-space adjacency algorithm, which is an adaptation of algorithm CADA2 of Arnon *et al.* (1984b). Note that when $r = 2$, the maximal sign-invariant clusters in the induced cad (of 1-space) are just singleton clusters, i.e. the individual cells. Hence when $r = 2$ we construct all adjacencies of the cad of E^2 that we build, and so clearly the sign-invariant components of the graph for this cad correspond to maximal sign-invariant clusters of D.

As for cells, we say that two (distinct) clusters (of a given clustering of a given cad) are *adjacent* if their union is connected. It is not hard to show that clusters C_1 and C_2 are adjacent if and only if there is a cell c_1 of C_1, and a cell c_2 of C_2, such that c_1 and c_2 are adjacent. We call an adjacency of cells belonging to different clusters an *intercluster* adjacency, whereas an adjacency of cells in the same cluster is an *intracluster* adjacency.

One possible adjacency algorithm for E^3 would be to simply build (all interstack) adjacencies in E^3 over all intercluster adjacencies of the induced cad of the plane, using

$$G \leftarrow \textbf{ExtendCadClusters } (A, B, G', ExtendCellToStack)$$

Inputs: $A \subset I_r$. $B \subset I_r$ is a basis for $prim(A)$. $G' = (A', B', V', E', G'')$ is a graph for a cad D' of E^{r-1} such that an A-invariant stack exists over each cell c of D'. $ExtendCellToStack(c, B', B; g, J, I, L)$ is a procedure with the following specifications. For inputs: c is a cell in a basis-determined cad D' of E^{r-1}, $r \geq 2$. $B' \subset I_{r-1}$ is a basis for D'. $B \subset I_r$ is a basis, such that each element of B is either delineable or nullified on c. For outputs: Let p be the sample point for c, and suppose the real algebraic number γ is a primitive element for p (see Arnon *et al.*, 1984a, for this terminology). g is a nonzero squarefree univariate polynomial $g(x_r)$ with coefficients in the field $Q(\gamma)$ whose real roots are in one-one correspondence with the sections of a B-invariant stack S over c. J is a list of isolating intervals for the real roots of g. I is a list of cell indices for the elements of S (since we know the cell index of c, we know the indices of elements of S). L is a list of the intrastack adjacencies of S.

Output: $G = (A, B, V, E, G')$ is an initial graph for an A-invariant cad of E^r.

(1) [Get sign-invariant clusters.] Do a sign-invariant connected components computation in G', to get a certain sign-invariant clustering of D'.

(2) [Process each cluster.] Initialize V and E to the empty set. For each sign-invariant cluster C of D', do the following steps (2.1) - (2.3):

(2.1) [Build a stack over the representative cell of the cluster.] Find a primitive sample point p for an (arbitrary) "representative" cell c of C; if none currently exists, construct one (by extended-to-primitive conversion) for some cell c of C, of highest possible dimension. Call $ExtendCellToStack(c, B', B; g, J, I, L)$. Set $T \leftarrow InputSignaturesOverCell(A, B, p, J)$. Using I, T, p, g, and J, we make a cell triple for each cell of the stack, as follows. We know the indices of all cells in the stack (from I), and their signatures (from T). Make a primitive sample point for each sector of the stack, and an extended sample point for each section (using p, g, and J). However, if the primitive element for p is of degree one, i.e. a rational number, then g has rational number coefficients, and so make a primitive rather than an extended sample point for each section of the stack. Add all cell triples to V. Add the intrastack adjacencies of L to E.

(2.2) [Infer stacks over the remaining cells of the cluster.] For all cells of C other than the one just used, do the following three steps: (1) make up cell triples for a stack over it, each triple consisting of an index inferred from the triple for the corresponding cell of the stack over c, a signature copied from the triple for the corresponding cell of the stack over c, and a null sample point; (2) add the triples for this stack to V; and (3) add the intrastack adjacencies for this stack to E.

(2.3) [Induced adjacencies of each induced cluster.] Let $2k + 1$ be the number of elements of the stack over the representative cell of C (thus each stack over an element of C also has $2k + 1$ elements). For each intracluster adjacency $\{d, e\}$ of C, and for $i = 1, ..., 2k + 1$, record in E that the ith element of the stack over d is adjacent to the ith element of the stack over e $\square$

Figure 6: Algorithm ExtendCadClusters.

AdjacenciesTwoSpace (G)

Inputs: $G = (A, B, V, E, G')$ is a graph for a basis-determined A-invariant cad D of E^2 (with basis B).

Output: G is modified so that it contains additional adjacencies among cells of D; in particular, if G is intial at input, then it is a full graph for D at output.

(1) [Interstack adjacencies.] Set $B^* \leftarrow \prod B$. Let $a_1 < a_2 < \cdots < a_{2k} < a_{2k+1}$, $k \geq 0$, be the sample points for the cells of D' (Each a_{2i+1} is a rational sample point for a 1-cell; each a_{2i} is an algebraic sample point for a 0-cell). For $i = 1, \ldots, k$, call algorithm SSADJ2 of Arnon *et. al.* (1984b) with inputs B^*, a_{2i}, a_{2i-1}, and a_{2i+1}, add the contents of its outputs L_1 and L_2 to G, i.e. to E. Note that the section numbers which occur in the adjacencies returned by SSADJ2 must first be converted into the indices of the corresponding cells of D; for example, if the list L_1 returned by the ith call to SSADJ2 contains the adjacency $\{3, 2\}$, it must be converted to $\{(2i, 6), (2i - 1, 4)\}$ before being added to L. Infer the remaining interstack adjacencies between $S(c_{2i})$ and $S(c_{2i-1})$, and between $S(c_{2i})$ and $S(c_{2i+1})$, as described at the end of Section 2 of Arnon *et al.* (1984b), and add them to G $\square$

Figure 7: Algorithm AdjacenciesTwoSpace.

the algorithms of Arnon *et al.* (1988). Assuming that we started with an initial graph for the cad of E^3, we clearly would end up with a full graph for this cad. A sign-invariant components computation in this graph would then obviously yield maximal sign-invariant clusters of the cad D of E^3. We now show that there is a proper subset Λ of the set of all intercluster adjacencies of the induced cad of E^2, such that given an initial graph for D, if we then build (all interstack) adjacencies over each element of Λ, then a sign-invariant connected components computation in the resulting graph yields maximal sign-invariant clusters of D. Besides the obvious reduction in the amount of adjacency determination we have to do in E^3, it turns out that the particular Λ that we show is sufficient allows us to often avoid the most costly extended-to-primitive sample point conversions in E^2.

Let us first determine what kinds of intercluster adjacencies can occur among maximal sign-invariant clusters of a cad of the plane. We assume that this cad has the boundary property. Clearly there can be no intercluster adjacency between 0-clusters. It is also clear that there can be no intercluster adjacencies between two maximal 2-clusters, since by the Intermediate Value Theorem, all 2-clusters have the same signature with respect to the input polynomials. So that leaves us with the possibility of adjacencies between 0- and 1-clusters, 0- and 2-clusters, 1- and 1-clusters, and 1- and 2-clusters. For the first of these cases, clearly all intercluster adjacencies are $\{0, 1\}$, i.e. involving a 0-cell of the 0-cluster and a 1-cell of the 1-cluster. In the second case, there can be both $\{0, 1\}$ and $\{0, 2\}$ intercluster adjacencies. In the third case, there can only be $\{0, 1\}$ intercluster adjacencies, since (by the boundary property) adjacent cells in the cad of the plane have different dimensions. In the fourth case, by the boundary property, there can be $\{0, 1\}$, $\{0, 2\}$, and $\{1, 2\}$ intercluster adjacencies.

In point of fact, adjacencies of two 1-clusters seem to occur rarely, for certain "peculiar" sorts of inputs. For example, let $F(x, y) = y + x$, $G(x, y) = y - x$, and $H(x, y) = y$, and let $A = \{FGH, FG, FH\}$. Fig. 8 shows an A-invariant cad of the plane that illustrates both intercluster adjacencies of a 1-cluster and a 1-cluster, and an adjacent

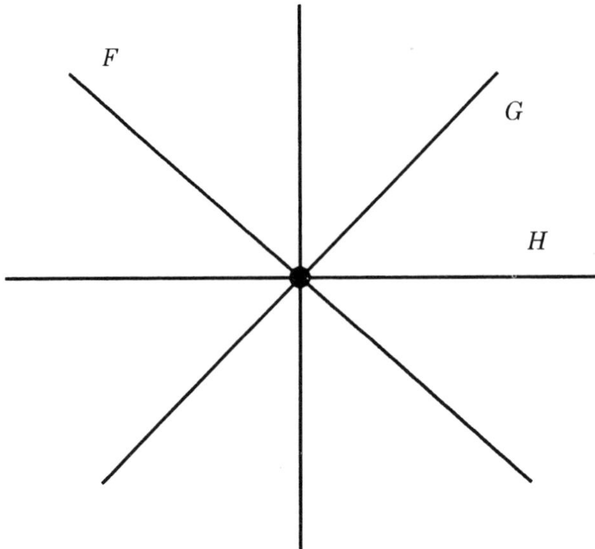

Figure 8: Sample cylindrical algebraic decomposition of the plane.

1-cluster and 2-cluster for which their only intercluster adjacency is a $\{0, 2\}$ adjacency of cells.

As for cells, the *boundary* of a cluster C is the set of all limit points of $R = \cup C$ which are not contained in R. It is not hard to see that clusters "do not have the boundary property", i.e. if two clusters are adjacent, then it is not necessarily the case that one is contained in the boundary of the other. For example, the tacnode curve is defined by the equation:

$$F(x, y) = y^4 - 2y^3 + y^2 - 3x^2y + 2x^4 = 0.$$

Fig. 9 shows an F-invariant cad of the plane. The curve itself is a maximal sign-invariant 1-cluster of this cad, and clearly, for any sign-invariant 2-cluster C, neither C nor the curve is contained in the boundary of the other.

We now prove a theorem that points the way to our actual 3-space adjacencies algorithm. The basic idea, given that clusters in the plane may fail to have the boundary property, is that for a pair of adjacent clusters of a cad of the plane, we find subclusters of each that are as large as possible while still having the property that one is contained in the boundary of the other. It then follows that it is sufficient to build adjacencies in E^3 over just one of the intercluster adjacencies between each such pair of subclusters.

We now define the central notion for our theorem. Given adjacent clusters C_1 and C_2 of some cad, we say that subclusters $Q_1 \subset C_1$ and $Q_2 \subset C_2$ are *cobounding subclusters* for C_1 and C_2 if

1. Each cell of Q_1 is in the boundary of one or more cells of Q_2, and

2. For each cell of Q_2, there are one or more cells of Q_1 contained in its boundary, and

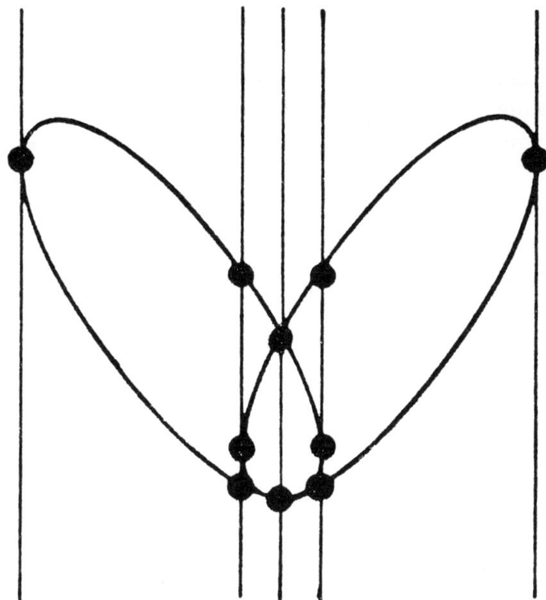

Figure 9: Cylindrical algebraic decomposition of tacnode curve.

3. If cells c_1 and c_2 of Q_2 are adjacent, then there are cells d_1 and d_2 of Q_1 such that $d_1 \subset \partial c_1$, $d_2 \subset \partial c_2$, and either $d_1 = d_2$, or d_1 and d_2 are adjacent.

Clearly if Q_1 and Q_2 are cobounding subclusters for C_1 and C_2, if $R_1 = \cup Q_1$, and if $R_2 = \cup Q_2$, then $R_1 \subset \partial R_2$. If Q_1 and Q_2 are cobounding subclusters for C_1 and C_2, and if for any other cobounding subclusters O_1 of C_1 and O_2 of C_2, it is the case that either $O_1 \cap Q_1 = \emptyset$, or $O_2 \cap Q_2 = \emptyset$, or $O_1 \subset Q_1$ and $O_2 \subset Q_2$, then we say that Q_1 and Q_2 are *maximal* cobounding subclusters for C_1 and C_2.

Let C_1 and C_2 be adjacent sign-invariant clusters of D' such that $R_1 \subset \partial R_2$, where $R_1 = \cup C_1$ and $R_2 = \cup C_2$. Let $S(R_1)$ be a stack over R_1 and $S(R_2)$ a stack over R_2. We say that $S(R_1)$ and $S(R_2)$ are *adjacent*. If for any section s of $S(R_2)$, $\partial s \cap Z^*(R_1)$ is a section t of $S^*(R_1)$, then we say that $S^*(R_2)$ has the *unique section boundary property (USBP)* in $S^*(R_1)$.

THEOREM **4.1** *Let D be a basis-determined cad of E^r, such that the induced cad D' of E^{r-1} has the boundary property. Let C_1 and C_2 be adjacent sign-invariant clusters of D', and suppose that $Q_1 \subset C_1$ and $Q_2 \subset C_2$ are cobounding subclusters for C_1 and C_2. Let R_1 and R_2 be the respective unions of Q_1 and Q_2, and suppose that for any cells c, d of $Q_1 \cup Q_2$, if $d \subset \partial c$, then $S^*(c)$ has the unique section boundary property in $S^*(d)$. Let $S(R_1)$ and $S(R_2)$ denote the unique stacks over R_1 and R_2 with which D is compatible, in the sense that each element of one of these stacks is a union of elements of D. Then $S^*(R_2)$ has the unique section boundary property in $S^*(R_1)$.*

PROOF. Suppose the assertion to be false, and let s be some section of $S^*(R_2)$ whose boundary points in $Z^*(R_1)$ are not a section of $S^*(R_1)$. Then there exist cells t_k and t_l

of D such that: t_k and t_l are contained in s, $t_k \in S(c_k)$ and $t_l \in S(c_l)$ for cells c_k and c_l of Q_2, there are cells d_k and d_l of Q_1 such that $d_k \subset \partial c_k$ and $d_l \subset \partial c_l$, and where $u_k \in S(d_k)$ and $u_l \in S(d_l)$ are the respective boundary sections of t_k and t_l, u_k is section n_k of its stack, u_l is section n_l of its stack, and $n_k \neq n_l$. There is a sequence (chain) of adjacent cells in s joining t_k and t_l. Each t_i in this chain is a section of $S(c_i)$ for some $c_i \in Q_2$, and for each such c_i, there is a $d_i \in Q_1$ such that $d_i \subset \partial c_i$, and either $d_i = d_{i+1}$ or d_i and d_{i+1} are adjacent. Since $S^*(c_i)$ has the USBP in $S^*(d_i)$, t_i has a boundary section u_i in $S^*(d_i)$. Then there exists a j for which u_j is section n_j of its stack, u_{j+1} is section n_{j+1} of its stack, and $n_j \neq n_{j+1}$. Suppose without loss of generality that $c_j \subset \partial c_{j+1}$, hence $t_j \subset \partial t_{j+1}$, hence $u_j \subset \partial t_{j+1}$. Since d_i and d_{i+1} are identical or adjacent, clearly section n_{j+1} of $S^*(d_j)$ is also contained in ∂t_{j+1}, hence both sections n_j and n_{j+1} of $S^*(d_j)$ are contained in ∂t_{j+1}, contradicting the USBP of $S^*(c_{j+1})$ in $S^*(c_j)$ $\square$

By the results of Arnon *et. al.* (1988), the hypotheses of Theorem 4.1 are satisfied for each pair of cobounding subclusters for each pair of adjacent maximal sign-invariant clusters of the induced cad of E^2. Hence, for each such pair of clusters in the plane, it is sufficient to build adjacencies in E^3 over just one of the intercluster adjacencies between each of their pairs of maximal cobounding subclusters, and this is what we will do.

Let us now consider the task of finding the pairs of maximal cobounding subclusters for a pair of adjacent clusters of the induced cad D' of E^2. Life is made easier with the following concept. Suppose for adjacent clusters C_1 and C_2 of D', that whenever cells $c_1 \in C_1$ and $c_2 \in C_2$ are adjacent, $c_1 \subset \partial c_2 \subset \partial C_2$. Then we say that C_1 and C_2 have *one-way boundary inclusions*. The next theorem tells us that clusters in E^2 have one-way boundary inclusions.

THEOREM **4.2** *Suppose given a maximal sign-invariant clustering of a cad with the boundary property, such that some cell of a cluster C_1 is contained in the boundary of (one or more cells of) a cluster C_2. Then for any cells $c_1 \in C_1$ and $c_2 \in C_2$, if c_1 and c_2 are adjacent, then $c_1 \subset \partial c_2 \subset \partial C_2$.*

PROOF. Suppose cell d_1 of cluster C_1 is contained in boundary of C_2; then clearly d_1 is in the boundary of some cell d_2 of C_2. Since C_1 and C_2 are different maximal sign-invariant clusters, there is some input polynomial F which vanishes on one but not the other. Since real varieties are closed, F must vanish on d_1, i.e. on C_1, but not on d_2, i.e. not on C_2. Suppose now that cells $c_1 \in C_1$ and $c_2 \in C_2$ are adjacent. Then one contains a limit (boundary) point of other, hence by the boundary property, one is contained in boundary of the other. But then $c_1 \subset \partial C_2$, since $c_2 \subset \partial C_1$ would imply that F vanishes on c_2, a contradiction $\square$

Fig. 10 gives an algorithm to find all pairs of maximal cobounding subclusters for a given pair of adjacent clusters. In Fig. 11 we give our 3-space adjacency algorithm *AdjacenciesThreeSpace*. It assumes that the particular value of the procedure parameter *ExtendCellToStack* of algorithm CLCAD that is called for in Section 5, i.e. algorithm *ExtendCellToStack* of Arnon *et al.* (1988), has been used to determine the stacks of the cad D of E^3. The various adjacency subalgorithms (e.g. *AdjacenciesOver01*) that *AdjacenciesThreeSpace* calls are from Arnon *et al.* (1988). Each such subalgorithm takes the two cells of an adjacency as inputs, and we are required to have primitive sample points for both. We assume that extended-to-primitive conversion is done as needed for these calls. A *nullifying* 0-cluster is a 0-cluster on whose unique constituent 0-cell some element of B is nullified.

$$K \leftarrow \textbf{MaximalCoboundingSubclusters } (C_1, C_2)$$

Inputs: C_1 and C_2 are disjoint clusters of a cad of E^r which has the boundary property, such that C_1 and C_2 have one-way boundary inclusions if they are adjacent.

Output: If C_1 and C_2 are not adjacent, then K is the empty list. Otherwise K is a list ($((Q_{1,1}, Q_{1,2}), L_1), ((Q_{2,1}, Q_{2,2}), L_2), ..., ((Q_{n,1}, Q_{n,2}), L_n)$), such that $(Q_{1,1}, Q_{1,2})$, $(Q_{2,1}, Q_{2,2})$, ..., $(Q_{k,1}, Q_{k,2})$ are the maximal cobounding subclusters for C_1 and C_2, and for each $(Q_{i,1}, Q_{i,2})$, L_i is a list of all intercluster adjacencies between $Q_{i,1}$ and $Q_{i,2}$.

(1) [Do it.] For each intercluster adjacency $\{c_1, c_2\}$ between C_1 and C_2, create an initial element ($(\{c_1\}, \{c_2\}, \{\{c_1, c_2\}\})$) of K. Then until no more coalescing is possible, attempt to "paste together" pairs $(Q_{i,1}, Q_{i,2}), L_i)$ and $(Q_{j,1}, Q_{j,2}), L_j)$ of elements of K. We attempt to paste such a pair by first checking whether $Q_{k,1} \leftarrow Q_{i,1} \cup Q_{j,1}$ is a subcluster of C_1 and whether $Q_{k.2} \leftarrow Q_{i,2} \cup Q_{j,2}$ is a subcluster of C_2. If so, then we set L_k to be $L_i \cup L_j$ plus any other intercluster adjacencies of $Q_{k,1}$ and $Q_{k,2}$, and check whether $Q_{k,1}$ and $Q_{k,2}$ are cobounding subclusters of C_1 and C_2 □

Figure 10: Algorithm MaximalCoboundingSubclusters.

The reader should now be convinced of the following proposition.

THEOREM **4.3** *Let D be a basis-determined sign-invariant cad of E^3, and let D' denote the induced sign-invariant cad of E^2. Suppose we have a graph representation for D which contains D's initial adjacencies, and the other adjacencies of it that are added by AdjacenciesThreeSpace. Then the clusters of D that we obtain by a sign-invariant connected components computation in the graph for D are maximal (sign-invariant) clusters.*

From algorithm *AdjacenciesThreeSpace* we see that we do not avoid all extended-to-primitive conversions of 0-cell sample points in the induced cad of the plane: we are required to have a primitive sample point for each 0-dimensional maximal sign-invariant cluster in E^2, and possibly also for certain 0-cells in 1-clusters. We now indicate how it is that the particular such conversions that actually are done are typically not as expensive as the ones that are not done. Consider for example the sample points of 0-clusters. Such 0-clusters are usually "topologically significant", e.g. they are typically the intersection points of two curves in E^2. It has been our empirical observation that the sample points of such "topologically significant" 0-cells often do not require field extension (i.e. nontrivial primitive element computation) in the conversion of their extended representations to primitive. In other words, the algebraic polynomial which is part of their extended representation is typically linear. Some explanation of this phenomenon is provided by Müller's observation (Müller, 1978), that probably at most one intersection of two (random) algebraic plane curves lies on any particular line in the plane, and so for any $F, G \in I_2$, the curve defined by F and the curve defined by G probably only have one intersection on a line $x = \alpha$, where α is the sample point of a 0-cell in the induced cad of E^1. If so, then $gcd(F(\alpha, y), G(\alpha, y))$ is linear, since each of its roots corresponds to an intersection point of the two curves. Hence the y-coordinates β of intersection points are likely to have the property that $Q(\alpha, \beta) = Q(\alpha)$, i.e. the primitive element algorithm is trivial.

The tacnode provides an illustrative example of how we are often able to avoid

---

AdjacenciesThreeSpace (G)

Inputs: $G = (A, B, V, E, G')$ is a graph for a basis-determined A-invariant cad D of E^3 (with basis B), such that D has the boundary property, and such that if cell d of D' is contained in ∂c for a cell c of D' on which no element of B is nullified, then $S^*(c)$ has the unique section boundary property in $S^*(d)$, and such that G' contains all adjacencies of D', and has a primitive or extended sample point for each of its cells.

Output: G is modified so that it contains additional adjacencies among cells of D; in particular, if G is initial at input, then the sign-invariant connected components of G correspond to maximal sign-invariant clusters of D.

(1) [Construct maximal sign-invariant clusters of induced cad.] Do a sign-invariant connected components computation in the G' graph, to get (maximal) sign-invariant clusters of D'.

(2) [Process (1-cluster, 1-cluster) adjacencies.] For each pair C_1, C_2 of adjacent 1-clusters of D', and for each of their $(0,1)$ intercluster adjacencies $\{c^0, c^1\}$, set $L \leftarrow AdjacenciesOver01(c^0, c^1, B', B)$ and add the adjacencies of L to E.

(3) [Process (1-cluster, 2-cluster) adjacencies.] For each pair C_1, C_2 of adjacent 1-cluster and 2-cluster of D', do $K \leftarrow MaximalCoboundingSubclusters(C_1, C_2)$. For each L_i of K, do the following loop. If L_i contains $(1,2)$ adjacencies, then let $\{c_i^1, c_i^2\}$ be one of them, set $L \leftarrow AdjacenciesOver12(c_i^1, c_i^2, B)$, add the adjacencies of L to E, and exit this loop iteration. Otherwise, if L_i contains $(0,1)$ adjacencies, then let $\{c_i^0, c_i^1\}$ be one of them, set $L \leftarrow AdjacenciesOver01(c_i^0, c_i^1, B', B)$, add the adjacencies of L to E, and exit this loop iteration. Otherwise, let $\{c_i^0, c_i^2\}$ be a $(0,2)$ adjacency of L_i, set $L \leftarrow AdjacenciesOverNonNullifying02(c_i^0, c_i^2, B', B)$, and add the adjacencies of L to E.

(4) [Process adjacencies of non-nullifying 0-clusters.] For each non-nullifying 0-cluster C, with unique constituent cell c^0, do the following two steps. First, for each 1-cluster C_1 which is adjacent to C, and for each of their $(0,1)$ intercluser adjacencies $\{c^0, c^1\}$, set $L \leftarrow AdjacenciesOver01(c^0, c^1, B', B)$ and add the adjacencies of L to E. Second, for each 2-cluster C_2 which is adjacent to C, do $K \leftarrow MaximalCoboundingSubclusters(C, C_2)$, and for each L_i of K, do the following loop. If L_i contains $(0,1)$ adjacencies, then let $\{c_i^0, c_i^1\}$ be one of them, set $L \leftarrow AdjacenciesOver01(c_i^0, c_i^1, B', B)$, add the adjacencies of L to E, and exit this loop iteration. Otherwise, let $\{c_i^0, c_i^2\}$ be a $(0,2)$ adjacency of L_i, set $L \leftarrow AdjacenciesOverNonNullifying02(c_i^0, c_i^2, B', B)$, and add the adjacencies of L to E.

(5) [Process adjacencies of nullifying 0-clusters.] For each nullifying 0-cluster C, with unique constituent cell c^0, do the following steps. For each $(0,1)$ adjacency $\{c^0, c^1\}$ of D', set $L \leftarrow AdjacenciesOver01(c^0, c^1, B', B)$ and add the adjacencies of L to E. For each $(0,2)$ adjacency $\{c^0, c^2\}$ of D', set $L \leftarrow AdjacenciesOverNullifying02(c^0, c^2, B', B)$, and add the adjacencies of L to E $\square$

Figure 11: Algorithm AdjacenciesThreeSpace.

$$G \leftarrow \textbf{CLCAD} \; (A, Basis, Projection, ExtendCellToStack, Adjacencies)$$

Inputs: A is a finite subset of I_r, for some $r \geq 1$. *Basis* is a procedure which, for any $i \geq 1$, given a subset U of I_i, computes a basis for $prim(U)$. *Projection* is a procedure which, for any $i \geq 2$, maps a subset of I_i to a subset of I_{i-1} having the expected properties (cf. Theorem 2.4 of Arnon *et al.*, 1988). *ExtendCellToStack*$(c, B', B; g, J, I, L)$ is a procedure with the same specifications as the input parameter of the same name to algorithm *ExtendCadClusters* of Section 3. *Adjacencies* is a procedure which, for any $i \geq 2$, given a graph for a cad of E^i, finds certain of its interstack adjacencies and adds them (i.e. adds the corresponding edges) to the graph.

Output: $G = (A, B, V, E, G')$ is a graph representation for an A-invariant cad D of E^r.

(1) $[r = 1$ (base case).$]$ Set $B \leftarrow Basis(A)$. If $r > 1$, then go to step (2). Construct a list J of open isolating intervals for the real roots of the elements of B, thus determining the cells of a cad D of E^1. Set $T \leftarrow InputSignaturesOverCell(A, B, \emptyset, J)$. Construct an index and a primitive sample point for each cell. From these and from T, create a triple for each cell, and set V to a list of all these triples. The adjacencies of D are obvious; collect them as the set E. Set G' to $\emptyset$, to complete the construction of a graph G for D. Return.

(2) $[r > 1$. Initial graph.$]$ Set $P \leftarrow Projection(A)$, and call CLCAD with inputs P, *Basis*, *Projection*, *ExtendCellToStack*, and *Adjacencies*, to obtain output G'. Call algorithm *ExtendCadClusters* of Section 3 with inputs A, B, G', and *ExtendCellToStack* to obtain an initial graph G for an A-invariant cad of E^r.

(3) $[r > 1$. Non-initial adjacencies among r-space cells.$]$ Apply *Adjacencies* to G $\square$

Figure 12: Algorithm CLCAD.

extended-to-primitive conversions of the sample points of 0-cells in 1-clusters. A sign-invariant cad of E^2 for the tacnode is shown in Fig. 9. Using an implementation of algorithm CAD (cf. Section 6), construction of this cad took 29 minutes, with 27 minutes of that spent in converting the extended representations of four 0-cell sample points to primitive: cells (4,2), (4,6), (8,2), and (8,6). Using CLCAD, we were able to construct the same cad of 2-space in 1 minutes; primitive sample points for cells (4,2), (4,6), (8,2), and (8,6) were not required, because they belong to a 1-dimensional sign-invariant cluster (the collection of all cells contained in the curve), which is adjacent only to 2-dimensional sign-invariant clusters, and for each such adjacent 2-cluster, each pair of maximal cobounding subclusters for the curve and the 2-cluster has an intercluster adjacency between a 2-cell of the 2-cluster and a 1-cell of the curve.

5   Main algorithm

In Fig. 12 we give our main algorithm CLCAD in a form which has various procedure parameters. Thus the exact version of the general cluster-based cad strategy that a user desires can be obtained by passing appropriate concrete procedures for these parameters, for example, one might use McCallum projection (McCallum, 1988) instead of the projection map assumed in Section 2, or one might use other adjacency algorithms than those we have given in Section 4.

Let us now list the particular concrete procedures that we pass for CLCAD's pro-

---

Adjacencies (G)

(1) If $r = 1$ or $r \geq 4$ then return. If $r = 2$ then $AdjacenciesTwoSpace(G)$. If $r = 3$ then $AdjacenciesThreeSpace(G)$ □

Figure 13: Algorithm Adjacencies.

cedure parameters in our current implementation; this information in effect summarizes Section 2-6. The default for calls to CLCAD is: we don't care what basis B for $prim(A)$ the procedure $Basis$ computes. We set $Projection(A) = PROJ(B) \cup cont(A)$, $2 \leq i \leq r$. We pass procedure $ExtendCellToStack$ of Arnon et al. (1988) as argument $ExtendCellToStack$. Finally, as one may expect from Sections 4, we pass the algorithm shown in Fig. 13 as argument $Adjacencies$ of CLCAD.

Given the these concrete procedures as values for CLCAD's procedure parameters, and given input polynomials $A \subset I_r$ with $1 \leq r \leq 3$, the sign-invariant connected components of the undirected graph (V, E) built by algorithm CLCAD correspond to maximal sign-invariant clusters of the cad D of E^r, and the boundary of each cell of D is a (disjoint) union of lower-dimensional cells, i.e. D has the boundary property.

6   Examples

6.1   General remarks.

We have not so far performed a detailed study of our implemented cluster-based cad algorithm's behavior, but preliminary experiments indicate that its performance is sometimes better, sometimes worse, than the "original" cad algorithm (i.e. algorithm CAD of Arnon et al., 1984a). Of course, the results of the comparisons we have carried out reflect the use of the particular adjacency algorithms given in Section 4. In any particular such comparison, the outcome seems to depend on the relative time of the extended-to-primitive sample point conversions that the original algorithm must do but which the cluster-based algorithm avoids, compared to the adjacency computations that the cluster-based algorithm must do but which do not occur in the original algorithm. Thus the Quartic and Ellipse examples below, for which the original algorithm was faster, most likely had easy sample point conversions relative to the cost of adjacency computations.

Fig. 14 contains a summary of the results of our comparisons. The times in it were obtained from algorithms CAD of Arnon et al. (1984a), and algorithm CLCAD of this paper, with both algorithms computing finest squarefree bases. Both algorithms were implemented in the SAC-2 computer algebra system (Collins, 1980), on a Vax 11/785 running Unix. The times given in the table are in minutes. $T_{original}$ is the time spent by the original cad algorithm, and $T_{clustered}$ the time spent by the cluster-based algorithm, for each example. A time of zero minutes means less than half a minute. The notation "$> n$ minutes" means that an algorithm ran for at least n minutes before either it was terminated or our computer went down. The column "Cells" gives the number of cells in the cad's built by both the original and cluster-based algorithms, and "Clusters" gives the number of maximal sign-invariant clusters in the cad built by the cluster-based

Name	$T_{original}$	$T_{clustered}$	Cells	Clusters
Tacnode	29	1	55	5
SIAM	1	1	41	15
Toptyp	> 120	4	37	9
Pair1	> 90	7	103	15
Pair2	> 83	9	127	27
Pair3	19	4	85	21
Pair4	1	1	63	15
Pair5	7	7	57	15
Quartic1	2	2	21	5
Quartic2	> 115	> 115	?	?
Quartic3	> 270	25	37	3
Quartic4	46	47	55	5
Quartic5	0	0	21	4
CADIII	0	1	51	3
Quartic	2	10	123	35
Implicit	> 300	89	855	9
SphereCatas	> 827	282	1393	9
Ellipse	9	74	2291	715

Figure 14: Sample comparisons of original and cluster-based cad algorithms.

algorithm. Sections 6.2 - 6.5 give the input polynomials for each example, and where applicable, cite a source for the example.

6.2   Miscellaneous bivariate examples.

6.2.1   Tacnode (Arnon *et al.*, 1984a)

$$y^4 \ - \ 2y^3 \ + \ y^2 \ - \ 3x^2 y \ + \ 2x^4$$

6.2.2   SIAM papers pair of polynomials (Arnon *et al.*, 1984a, 1984b)

$$144y^2 \ + \ 96x^2 y \ + \ 9x^4 \ + \ 105x^2 \ + \ 70x \ - \ 98$$
$$xy^2 \ + \ 6xy \ + \ x^3 \ + \ 9x$$

6.2.3   Toptyp algorithm example (Arnon & McCallum, 1988)

$$y^4 \ - \ 2xy^3 \ - \ x^2 y^2 \ + \ y^2 \ + \ 2x^3 y \ + \ x^2 \ - \ 1$$

6.3   Five randomly generated pairs of bivariate polynomials.

Each consisted of a quadratic and a cubic polynomial, with two-digit integer base coefficients.

6.3.1   first pair.

$$3y^2 - 2xy + 28x + 31$$
$$-8y^3 + 6x^2 y - 15xy - 7y - 7x^3 + 11x + 6.$$

6.3.2   second pair.

$$-9y^2 + 30xy - 22x^2 + 21$$
$$2y^3 - 12x^2y - 12xy - 8y + 11x^2 - 2x - 2.$$

6.3.3   third pair.

$$-2y - 13x + 22$$
$$-13y^3 + 5xy^2 + 12y^2 + 14x^2y + 11y - 10x^2 + 11.$$

6.3.4   fourth pair.

$$-12xy - 15y - 30x^2 + 4x + 21$$
$$-xy^2 + 15y^2 + 8x^2y - 12y + 12x^3 + 9.$$

6.3.5   fifth pair.

$$27xy + 9x^2 - 31x + 4$$
$$5y^3 - 14xy^2 + 15y^2 + 13x^2y + 2xy + 14y - 7x^3 - 3x.$$

6.4   Five randomly generated bivariate quartics.

Each had two-digit integer base coefficients.

6.4.1   first quartic.

$$44xy^3 + 57xy^2 + 25y + 37x^3 - 31x.$$

6.4.2   second quartic.

$$-62y^4 - 29x^2y^2 - 45y^2 + 45x^3y - 5x^2y + 26x^4 + 27x - 58.$$

6.4.3   third quartic.

$$-50y^4 + 48y^3 - 8y^2 - 34x^2y - 11x^3 - 5x.$$

6.4.4   fourth quartic.

$$60xy^3 + 59y^3 - 41y^2 - 55x^3y + 47xy + 45y + 22x^4 - 38x^3 + 3x^2 - 24.$$

6.4.5   fifth quartic.

$$52x^2y^2 + 30xy^2 + 49y^2 - 4x^2y + 62xy + 9x^4 + 33x^3.$$

6.5   Trivariate examples.

6.5.1   CADIII example surface (Arnon *et al.*, 1988)

$$y^3z + xy^2 - x^3$$

6.5.2   Positive definite canonical form quartic (Arnon & Mignotte, 1988)

$$p$$
$$8pr - 9q^2 - 2p^3$$
$$256r^3 - 128p^2 r^2 + 144pq^2 r + 16p^4 r - 27q^4 - 4p^3 q^2$$

6.5.3   Curve Implicitization (Arnon, 1988)

$$505t^3 - 864t^2 + 570t + x - 343$$
$$211t^3 - 276t^2 - 90t - y + 345$$

6.5.4   Unit sphere and catastrophe surfaces (McCallum, 1988)

$$z^2 + y^2 + x^2 - 1$$
$$z^3 + xz + y$$

6.5.5   Ellipse example (Arnon & Mignotte, 1988)

$$a$$
$$a - 1$$
$$b$$
$$b - 1$$
$$b - a$$
$$c$$
$$c - 1$$
$$c + 1$$
$$c + a + 1$$
$$c + a - 1$$
$$c - a + 1$$
$$c - a - 1$$
$$b^2 c^2 + b^4 - a^2 b^2 - b^2 + a^2$$

7   References

Aho, A. V., Hopcroft, J., Ullman, J. (1974). *The Design and Analysis of Computer Algorithms.* Reading, Massachusetts: Addison-Wesley.

Arnon, D. S. (1979). A cellular decomposition algorithm for semi-algebraic sets. Proceedings of an International Symposium on Symbolic and Algebraic Manipulation (EUROSAM '79). *Springer Lec. Notes Comp. Sci.* **72**, 301-315.

Arnon, D. S. (1981). *Algorithms for the Geometry of Semi-Algebraic Sets.* PhD thesis, Tech. Rept. #436, Comp. Sci. Dept., Univ. Wisconsin–Madison.

Arnon, D. S., Collins, G. E., McCallum, S. (1984a). Cylindrical algebraic decomposition I: the basic algorithm, *SIAM J. Comp.* **13/4**, 865–877.

Arnon, D. S., Collins, G. E., McCallum, S. (1984b). Cylindrical algebraic decomposition II: an adjacency algorithm for the plane, *SIAM J. Comp.* **13/4**, 878–889.

Arnon, D. S. (1988). Geometric reasoning with logic and algebra. *Artificial Intelligence* (special issue on Geometric Reasoning and Artificial Intelligence; to appear).

▸Arnon, D. S., McCallum, S. (1988). A polynomial-time algorithm for the topological type of a real algebraic curve. *J. Symb. Comp.* **5**, (this issue).

Arnon, D. S., Mignotte, M. (1988). On mechanical quantifier elimination for elementary algebra and geometry. *J. Symb. Comp.* **5**, (this issue).

Arnon, D. S., Collins, G. E., McCallum, S. (1988). An adjacency algorithm for cylindrical algebraic decompositions of three-dimensional space. *J. Symb. Comp.* **5**, (this issue).

Collins, G. E. (1975). Quantifier elimination for real closed fields by cylindrical algebraic decomposition. Proceedings of the Second GI Conference on Automata Theory and Formal Languages. *Springer Lec. Notes Comp. Sci.* **33**, 515–532.

Collins, G. E. (1980). SAC-2 and ALDES now available. *ACM SIGSAM Bull.* **14**, 19.

Kaltofen, E. (1982). Polynomial factorization. In (Buchberger, B., Loos, R., Collins, G. E., eds.) *Computer Algebra - Symbolic and Algebraic Computation* (Computing Supplementum 4), pp. 83-94. Vienna and New York: Springer-Verlag.

Kozen, D., Yap., C. K. (1985). Algebraic cell decomposition in NC. *Proc. IEEE Conf. on Foundations of Comp. Sci. (FOCS)*, 515–521.

Loos, R. (1982). Computing in algebraic extensions. In (Buchberger, B., Loos, R., Collins, G. E., eds.) *Computer Algebra - Symbolic and Algebraic Computation* (Computing Supplementum 4), pp. 173-187. Vienna and New York: Springer-Verlag.

McCallum, S. (1988). An improved projection operation for cylindrical algebraic decomposition of three-dimensional space. *J. Symb. Comp.* **5**, (this issue).

Müller, F. (1978). *Ein exakter Algorithmus zur nichtlinearen Optimierung für beliebige Polynome mit mehreren Veranderlichen*, Meisenheim am Glan: Verlag Anton Hain.

Prill, D. (1986). On approximation and incidence in cylindrical algebraic decompositions. *SIAM J. Comp.* **15**, 972–993.

Schwartz, J. T., Sharir, M. (1983). On the 'piano movers' problem II. General techniques for computing topological properties of real algebraic manifolds. *Adv. Applied Math.* **4**, 298–351.

J. Symbolic Computation (1988) **5**, 213–236

A Polynomial-time Algorithm for the Topological Type of a Real Algebraic Curve[*]

DENNIS S. ARNON   AND   SCOTT McCALLUM[†]

Xerox PARC, 3333 Coyote Hill Road, Palo Alto, California 94304, U.S.A.
† *Research School of Physical Science, Australian National University, Canberra ACT 2601,*
Australia

(Received 13 February 1985, and in revised form 15 November 1987)

It was proved over a century ago that an algebraic curve C in the real projective plane, of degree n, has at most $\frac{(n-1)(n-2)}{2} + 1$ connected components. If C is nonsingular, then each of its components is a topological circle. A circle in the projective plane either separates it into a disk (the *interior* of the circle) and a Möbius band (the circle's *exterior*), or does not separate it. In the former case, the circle is an *oval*. If C is nonsingular, then all its components are ovals if n is even, and all except one are ovals if n is odd. An oval is *included* in another if it lies in the other's interior. The topological type of (a nonsingular) C is completely determined by (1) the parity of n, (2) how many ovals it has, and (3) the partial ordering of its ovals by inclusion. We present an algorithm which, given a homogeneous polynomial $f(x, y, z)$ of degree n with integer coefficients, checks whether the curve defined by $f = 0$ is nonsingular, and if so, computes its topological type. The algorithm's maximum computing time is $O(n^{27} L(d)^3)$, where d is the sum of the absolute values of the integer coefficients of f, and $L(d)$ is the length of d.

1   Introduction

We begin with an example of what our algorithm does. Let $f(x, y, z)$ be the homogeneous polynomial

$$y^4 - 2xy^3 - x^2y^2 + y^2z^2 + 2x^3y + x^2z^2 - z^4 .$$

The equation $f = 0$ defines an algebraic curve C in the real projective plane. Let us draw a picture of C in two steps. Suppose that points in the projective plane have homogeneous xyz coordinates; then the points for which $z = 1$ constitute an affine xy-plane imbedded in the projective plane. The portion of C lying in this xy-plane is the locus of the equation

$$f(x, y, 1) = y^4 - 2xy^3 - x^2y^2 + y^2 + 2x^3y + x^2 - 1 = 0,$$

and is shown on the left in Fig. 1. Using the standard disk model for the projective plane, the full curve C is shown on the right in Fig. 1. C is of degree four, and happens

[*]This work was supported by the National Science Foundation (Grant MCS-8009357 to the University of Wisconsin-Madison), the Purdue Research Foundation, and the Xerox Corporation. This paper was typeset at Xerox PARC using TeX in the Cedar environment.

0747–1717/88/010213+24 \$03.00/0　　　　　　　　　　

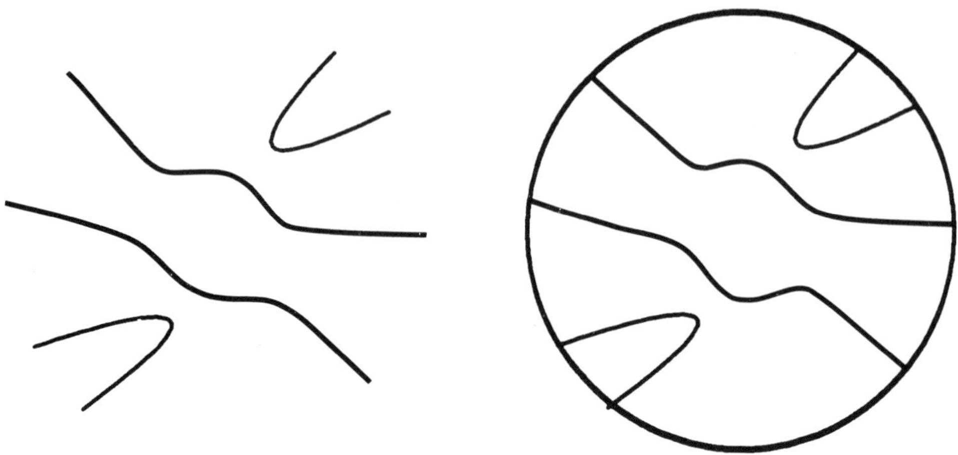

Figure 1: Sample algebraic curve.

to be nonsingular, so the facts cited in the Abstract tell us that C has at most four connected components, each of which is an oval. Recalling that antipodal boundary points are identified in the disk model of the projective plane, we can guess from Fig. 1 that C has two components, but it may not be obvious whether one includes the other or not. Given this particular $f(x, y, z)$ as input, the algorithm we present in this paper determines that C has two ovals, one included in the other.

Our algorithm divides naturally into two main steps. To describe them we use the notion of a cellular decomposition (cd) of a topological space. Let us recall the limited form of it we need here (cf. Massey, 1978, p. 54ff.). For any $i \geq 0$, an *i-cell* is essentially (to be precise, is homeomorphic to) an i-dimensional open ball. Thus a 0-cell is a point, a 1-cell is an "open arc", a 2-cell an "open region", etc. Let X be a subset of the projective plane RP^2; we can view X as a topological space with the topology it inherits from RP^2. A *cellular decomposition* of X is a nested sequence $X^0 \subset X^1 \subset X^2 = X$ of closed subspaces, such that X^0 consists of finitely many 0-cells, $X^1 - X^0$ consists of finitely many disjoint 1-cells, and $X^2 - X^1$ consists of finitely many disjoint 2-cells. Given a cd D of X, and a subset Y of X, we say that D is *compatible with* Y if Y is the union of certain cells of D. Fig. 2 shows a cd of the projective plane compatible with the curve that we looked at in Fig. 1. This cd consists of eleven 0-cells, twenty-three 1-cells, and thirteen 2-cells.

We will be much concerned with the precise arrangement of cells in the cd's we work with. Informally, two (distinct) cells of a cd are *adjacent* if they touch; formally, this is the condition that their union be connected. Clearly adjacency is a symmetric relation on a cd. Thus, we can represent it as an undirected graph which has a vertex for each cell of the cd, and an edge between every pair of adjacent cells. This is the *connectivity graph* of the cd, and is the basic data structure for our algorithm.

The first main step of our algorithm, the subject of Sections 3-6, starts with the input polynomial $f(x, y, z)$, determines whether the curve C defined by f is nonsingular, and if

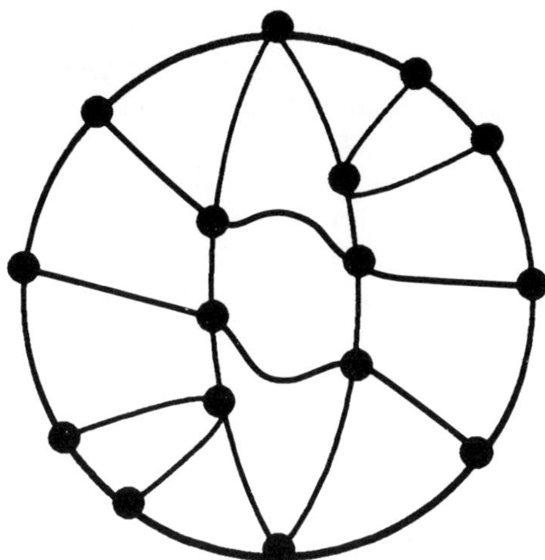

Figure 2: Cellular decomposition of projective plane compatible with sample algebraic curve.

so, constructs a (connectivity graph for a) cd of the projective plane. The key component here is construction of a certain *cylindrical algebraic decomposition* (cad) of the affine plane $z = 1$ (see e.g. Arnon *et al.*, 1984a, for information on cad's). This cad is a cd of the affine plane, and the cad algorithm from Arnon *et al.* (1984a, 1984b) that we use constructs its connectivity graph, so the only work that remains for us is to extend it to a cd of the projective plane and construct the enlarged connectivity graph. Fig. 2 shows the cd of the projective plane our algorithm produces for the sample $f(x, y, z)$ given at the beginning of this Introduction. Identification of the cells of this cd that belong to C is straightforward.

The second of the two main steps of our algorithm, discussed in Section 7, assumes that the input polynomial $f(x, y, z)$ defines a nonsingular curve C, that a cellular decomposition D^* of the projective plane compatible with C has been constructed, that D^*'s connectivity graph has been constructed, and that the cells (vertices) in the connectivity graph which belong to C are marked in some fashion. The topological type, i.e. the number and ordering of ovals, of C is then computed. The key idea is to reduce the determination of the ovals and their ordering to a series of connected components and Euler characteristic computations in appropriate subgraphs of the connectivity graph.

Section 8 summarizes our discussion with a main algorithm TOPTYP, and in Section 9 we trace TOPTYP for the sample $f(x, y, z)$ that we considered above. Section 10 contains an analysis of the TOPTYP's maximum computing time, which we show to be $O(n^{27} L(d)^3)$. This is not out of character with other algebraic algorithms, such as polynomial factorization (Kaltofen, 1982), and indeed the most costly parts of our algorithm are standard algebraic algorithms for such tasks as algebraic number computations, root isolation, and resultant computation. And as for other algebraic algorithms, despite a high worst-case computing time bound, our topological type algorithm has a useful range of application in practice. It has been implemented and applied to examples such as the one considered above.

We summarize basic properties of algebraic curves and homogeneous coordinates, and provide background on the facts cited in the Abstract, in Section 2. The purpose of this material is to establish connections between the conventions and viewpoints of algebraic geometry and those of computer algebra. The knowledgeable reader may skim or skip it.

We were led to seek a topological type algorithm by remarks of Gudkov (1974, p. 67). Polotovskii (1973) gave a topological type algorithm for so-called 'rough' curves of even degree, but did not discuss its feasibility or computing time. His approach is quite different from ours: he examines the curves $f(x, y, z) + \epsilon\, z^n$, $(n = degree(f))$, for various small values of ϵ. We have recently learned of an independently developed topological type algorithm by P. Gianni and C. Traverso (1983), which has some resemblance to our method, but does not make use of cellular decompositions. As noted by Fuks (1974), one could get a topological type algorithm directly from a decision procedure for elementary algebra and geometry (e.g. Tarski, 1951, Collins, 1975, or Ben-Or *et al.*, 1986), but it seems unlikely that such an algorithm would have a polynomial time bound.

The algorithm we give in this paper existed in rough form by summer 1982, and was presented in a seminar at Purdue University in February 1983. An expanded version of this paper has appeared as Arnon & McCallum (1983), and an abstract of it as Arnon & McCallum (1984). A graph data structure for cad's similar to the one we use in this paper is employed in the cluster-based cad algorithm (Arnon, 1988).

2   Algebraic curves

Kendig (1978) and Walker (1951) provide additional coverage of the material contained in this section. *Real affine space* is euclidean space without the notions of distance and angle (Walker, 1951). An important difference between euclidean and affine space is that the usual distance function gives euclidean space a (metric space) topology, which affine space lacks. The cylindrical algebraic decomposition algorithm makes essential use of this topological structure of euclidean space. In this paper, the term "affine space" is mainly useful to us as a means of referring to a certain distinguished subset of projective space. We view affine and euclidean space as essentially the same, i.e. we take for granted the existence of the usual topology on real affine space whenever we need it. We write E^i to denote i-dimensional euclidean space.

Curves in the affine plane defined by polynomial equations $g(x, y) = 0$ are called *affine algebraic curves*. Throughout this paper, when we say "curve", we mean what is usually referred to as a "real curve", i.e. the real (and not, for example, the complex) solutions to a polynomial equation such as $g(x, y) = 0$. It has long been the prevailing viewpoint in algebraic geometry that an affine curve is only a portion of some true algebraic curve. This point of view is motivated by consideration of the intersections of lines in the affine plane. Two lines in the affine plane have exactly one intersection, unless they are parallel, in which case they do not intersect. We can remove this exceptional behavior of parallel lines by extending affine lines to contain a "point at infinity", and specifying that lines which are parallel in the affine plane have the same point at infinity. Now we can say that any two (extended) lines, without exception, have exactly one intersection. Since we get to the same point at infinity no matter which direction we travel along an affine line towards infinity, the extended lines we have defined are topological circles.

In same spirit, what is considered to be a true algebraic curve will be obtained by adding certain of the points at infinity to an affine curve. First let us join the affine plane and the points at infinity into a new object: the real projective plane RP^2 is the real affine plane extended by the collection of all points at infinity. It is standard to model RP^2 as a closed disk (see e.g. Kendig, 1978, p. 6), with antipodal points on the boundary identified. The open interior of the disk corresponds to the affine plane, and its boundary circle to the points at infinity. Even though we are thinking of antipodal pairs of points as being the same point of the projective plane, the boundary circle is still topologically a circle in RP^2. It is customarily referred to as the *line at infinity* and denoted l_∞. Naturally any particular point of l_∞ is the point at infinity for the affine line through the origin which approaches it, and for all affine lines parallel to that line.

Since RP^2 has more points than the affine plane, to give its points coordinates we must somehow expand the coordinate system we used for the affine plane. We do so with homogeneous coordinates. We start with the convention that each point (x, y) in the affine plane is assigned the coordinate triple $[x, y, 1]$ as a point of the projective plane. We further adopt the convention that for any nonzero real number t, all triples $[tx, ty, t]$ correspond to exactly the same point of the projective plane (so a point in the projective plane has a whole equivalence class of coordiate triples, any one of which is its "homogeneous coordinates"). By definition, there is no point in the projective plane with coordinate triple $[0, 0, 0]$. As we saw, any point at infinity is the limit point of some affine line through the origin. Consider an affine line through the origin which also contains the affine point (a, b) different from the origin. All points on the line are of the form (ta, tb), for t real. Hence, as we travel out to infinity along the line, the projective coordinates of the points we pass over can be written in the form $[a, b, 1/t]$ for t steadily decreasing (in absolute value). Hence it is natural to assign the coordinate triple $[a, b, 0]$ to the point at infinity by which we extend this line. Note that with this coordinate assignment, we indeed approach the same point at infinity whether we travel out to infinity along the line by making t large positive or large negative.

In order to have algebraic curves in the projective plane, we must have polynomial equations which define them. Since points in the projective plane have coordinate triples $[x, y, z]$, we need trivariate polynomials $f(x, y, z)$. Furthermore, it must be the case that $f(x, y, z) = 0$ if and only if $f(tx, ty, tz) = 0$ for nonzero t. We achieve this by limiting ourselves to *homogeneous* polynomials, i.e. polynomials in which every monomial has the same total degree. For example, the polynomial $y^2 - x^2 + xz - z^2$ is homogeneous, but $y^2 - x^2 + x - 1$ is not. It is easy to see that any homogeneous f has the property that $f(x, y, z) = 0$ if and only if $f(tx, ty, tz) = 0$ for nonzero t.

Now suppose we have a defining polynomial $g(x, y)$ for an affine curve, say $y^2 - x^2 + x - 1$, and suppose we construct the "homogenization" of it, namely $f(x, y, z) = y^2 - x^2 + xz - z^2$. Clearly, for any point (a, b) of the affine plane, $g(a, b) = 0$ if and only if $f(a, b, 1) = 0$. Hence, in the affine plane, f defines the same curve as g. As one might expect, the manner in which we want to extend affine curves is so that they contain their limit points on l_∞. It turns out that all such limit points will be solutions of the equation $f(x, y, z) = 0$ which lie on l_∞. As we now show, this follows from reasoning which is essentially a repetition of our original derivation of homogenous coordinates. If we are approaching infinity on an affine curve, then in projective coordinates, we have points $[x, y, 1]$ in which either x, or y, or both, are approaching infinity. Suppose without loss of generality that y is. We can represent the points we are looking at as $[x/y, 1, 1/y]$. Hence as y approaches infinity, we approach the point $[x/y, 1, 0]$. By a straightforward continuity argument, the fact that all points $[x/y, 1, 1/y]$ satisfy $f = 0$, implies that $f(x/y, 1, 0) = 0$. Thus every point we want to add to our affine curve is indeed among the solutions of $f = 0$.

We define an algebraic curve in the projective plane (a *projective curve*) to be the point set (in the projective plane) defined by a homogeneous polynomial equation $f(x, y, z) = 0$. Obviously for any homogeneous $f(x, y, z)$, we can get a certain bivariate $g(x, y)$ (which in general is not homogeneous) by evaluating f at $z = 1$. g defines a certain affine curve, which we call an "affine representative" of the projective curve defined by f. We say that the curve $f = 0$ is the "projective completion" of the affine curve defined by g. The projective completions of affine lines are just the extended affine lines we began with above.

Let C be the projective curve defined by $f(x, y, z) = 0$ for some f. A point of the projective plane, with homogeneous coordinates $[x, y, z]$, is a *singular* point of C if $f(x, y, z) = f_x(x, y, z) = f_y(x, y, z) = f_z(x, y, z) = 0$ (f_w denotes the partial derivative of f with respect to w). C is *nonsingular* if it has no singular points. Walker (1951, p. 57) has pictures of some of the different kinds of singularities which can arise. Projective curves can have points on l_∞ which are not limit points of the curve's affine representative; these are always (isolated) singular points of the curve. Where $g(x, y) = f(x, y, 1)$, the singular points of f in the affine plane correspond to the common zeros of g, g_x, and g_y (see Walker, 1951, p. 54).

Pictures of singular curves such as Walker's may serve to make plausible the fact that a nonsingular projective curve is a compact one dimensional manifold (Wilson, 1978). This captures such observations as "nonsingular curves do not cross themsleves", and "nonsingular curves do not have isolated points", that the pictures suggest. A compact one-dimensional manifold is known to be homeomorphic to a disjoint union of topological circles (Milnor, 1965). A topological circle is known to have two possible imbeddings in the projective plane (Wilson, 1978). One is what we called an oval. The other is like the imbedding of a projective line in the projective plane (e.g. l_∞ has this second kind of imbedding in the projective plane). Harnack's theorem (Wilson, 1978), which we cited in the Abstract, established that a projective curve of degree n has at most $\frac{(n-1)(n-2)}{2} + 1$ connected components. It is known that for a nonsingular curve of even degree, each component is an oval (Wilson, 1978). For a nonsingular curve of odd degree, one component is like a projective line, and all the rest are ovals.

It is illustrative to consider nonsingular curves of degree two, i.e. conics. Recall that in the affine plane we have a variety of nonsingular conics, i.e. parabolas, hyperbolas, circles, ellipses. The results we have cited say that a nonsingular projective conic has at most $\frac{(2-1)(2-2)}{2} + 1 = 1$ components, and since degree two is even, this component is an oval. Thus the projective completion of any nonsingular affine conic is an oval in the projective plane. Consider, for example, the hyperbola $xy - 1 = 0$, which in the affine plane has two "branches": its projective completion is the nonsingular projective curve $xy - z = 0$. Setting $z = 0$, we see that it has the points $[1, 0, 0]$ and $[0, 1, 0]$ on l_∞. Thus the two branches "connect up" on l_∞, and so in the projective plane the curve consists of a single oval.

Projective curves C_1 and C_2 have the same *topological type* if there is a homeomorphism of the projective plane to itself which maps C_1 onto C_2. If C is a nonsingular curve defined by $f(x, y, z) = 0$, then the parity of the degree of f, the number of ovals that C has, and the partial ordering of its ovals induced by the inclusion relation, completely determine its topological type (this result depends upon elementary facts from two-dimensional topology, for example, the Schoenflies theorem; see Moise, 1977). Thus the algorithm we give in this paper produces information which completely determines the topological type of a nonsingular curve.

In 1900, Hilbert posed the problem (known since as Hilbert's Sixteenth Problem) of determining which logically possible topological types of a nonsingular curve of degree n are realized by an actual curve. It is known that the number which actually occur are a small fraction of the number which are logically possible, but the problem is still largely unsolved. Curves with the maximum possible number of ovals (so-called M-curves), of even degree, are of greatest interest. Obviously there is only one type of M-curve of degree two. It is easy to show that the only type of M-curve of degree four is four ovals all external to each other. Note that the sample $f(x, y, z)$ we looked at in Section 1 has degree four but only two ovals, and so is not an M-curve. Hilbert knew that there could not be more than three types of M-curves of degree six; the proof that there are exactly three was not completed until 1967. For degree eight and up the problem is unsolved. See Wilson (1978) and Gudkov (1974) for further information on Hilbert's Sixteenth Problem.

3   Preliminaries

Before taking up the essential ideas of the first main step of our algorithm we attend to some preliminaries. As before, C denotes the curve defined by $f = 0$, where $f(x, y, z)$ is the input polynomial to our algorithm. First, we will assume that $f(x, y, z)$ is squarefree (i.e. has no multiple factors). If it is not, we may replace it by its greatest squarefree divisor $h(x, y, z)$ (h is f divided by $gcd(f, f_z)$), since the curve defined by $h = 0$ is just C. See Kaltofen (1982, p. 98), or Collins & Loos, (1982, p. 84), for squarefree factorization algorithms. Second, assuming the convention that the affine plane is the set of points in RP^2 of the form $[x, y, 1]$ and that l_∞ is all points of the form $[x, y, 0]$, we want f to satisfy the following conditions:

(C1) C has only finitely many points on l_∞, each of which is simple (this is equivalent to the requirement that $f(x, y, 0)$ be nonzero and squarefree);

(C2) The point $[0, 1, 0]$ of l_∞ does not lie on C.

The purpose of these conditions is to simplify determination of adjacencies between cells in the affine plane and cells on l_∞, the subject of Section 6. Note also that if C satisfies (C1) and (C2), then it has no singularities on l_∞. We now show that if (C1) and (C2) are not satisfied initially, then we can perform a linear change of coordinates of RP^2 which transforms $f(x, y, z)$ to a polynomial $F(U, V, W)$, such that F is squarefree and homogeneous of the same degree as f, C is nonsingular if and only if the curve C' defined by $F = 0$ is nonsingular, C and C' have the same topological type, and C' satisfies conditions (C1) and (C2).

Our strategy is to find a line that has only simple intersections with C, and to change coordinates so that that line becomes the line at infinity. We need to work with a number of different homogeneous polynomials $p(x, y, z)$ in our derivation of this coordinate change. Thus, in this section only, we will write C_p to denote the real projective curve defined by $p(x, y, z)$. We may assume that the input polynomial $f(x, y, z)$ is squarefree. Let n be the degree of f, and let

$$f(x, y, z) \;=\; f_r(x, y,)z^{n-r} \;+\;\dots+\; f_n(x, y),$$

where $0 \le r \le n$, each $f_i(x, y)$ is homogeneous of degree i, and $f_r(x, y) \ne 0$. Suppose $f_n(x, y) = 0$. Then z divides $f(x, y, z)$, but z^2 doesn't divide $f(x, y, z)$, since $f(x, y, z)$ is squarefree. We can therefore write $f(x, y, z) = z\, \phi(x, y, z)$, where

$$\phi(x, y, z) \;=\; f_r(x, y)z^{n-r-1} \;+\;\dots+\; f_{n-1}(x, y)$$

and $f_{n-1}(x, y) \ne 0$. l_∞ is contained in the curve C_f, hence if C_ϕ has any point on l_∞ (that is, if either $f_{n-1}(0, 1) = 0$, or $f_{n-1}(1, y)$ has a real root), then C_f is singular, we report this fact, and exit from the algorithm. If C_ϕ does not meet l_∞, then C_f is nonsingular if and only if C_ϕ is nonsingular. Moreover, if C_ϕ is nonsingular, then C_f and C_ϕ have the same number and arrangement of ovals. Hence we can replace f by ϕ; since C_ϕ does not meet l_∞, conditions (C1) and (C2) are trivially satisfied.

Suppose now that $f_n(x, y) \ne 0$. Let us transform $f(x, y, z)$ to $F(X, Y, Z)$, such that $F(0, 1, 0) \ne 0$ (so that the point $[0,1,0]$ is not on C_F). We know $f_n(x, 1) \ne 0$, since otherwise $f_n(x, y) = 0$. Thus there is an integer λ such that $f_n(\lambda, 1) \ne 0$. Define $F(X, Y, Z)$ by

$$F(X, Y, Z) \;=\; f(X + \lambda Y, Y, Z)$$

Then $F(0, 1, 0) = f(\lambda, 1, 0) = f_n(\lambda, 1) \ne 0$. Let $G(X, Y) = F(X, Y, 1) \ne 0$ and let $D(X)$ be the discriminant of $G(X, Y)$. Then $D(X) \ne 0$, since $G(X, Y)$ is squarefree (and nonzero). Find an integer κ with $D(\kappa) \ne 0$. Change variables as follows: $X = W + \kappa U$, $Y = V$, $Z = U$. Since $W = X - \kappa Z$, the line $X = \kappa Z$ (which is the projective version of the affine line $X = \kappa$) corresponds to the line $W = 0$ (which is the line at infinity in U, V, W coordinates). Let

$$E(U, V, W) \;=\; F(W + \kappa U, \; V, \; U).$$

E is squarefree and homogeneous of the same degree as f. Observe that $E(0, 1, 0) = F(0, 1, 0) \ne 0$. We have $E(U, V, 0) = F(\kappa U, V, U)$, so that $E(1, V, 0) = F(\kappa, V, 1) = G(\kappa, V)$, a nonzero squarefree polynomial (since $D(\kappa) \ne 0$). Thus $E(U, V, 0)$ is nonzero and squarefree. Hence C_E satisfies conditions (C1) and (C2).

It remains to show that (i) C_f is nonsingular if and only if C_E is nonsingular; and (ii) C_f and C_E have the same topological type. Let $T(x, y, z) = (z, y, x - \kappa z - \lambda y)$. Then T is an invertible linear transformation of E^3 with inverse $T^{-1}(U, V, W) = (W + \kappa U + \lambda V, V, U)$. We have

$$E(U, V, W) \;=\; f(T^{-1}(U, V, W)) \,.$$

Applying the chain rule for differentiation, we obtain

$$\begin{pmatrix} E_U \\ E_V \\ E_W \end{pmatrix} = \begin{pmatrix} \kappa & 0 & 1 \\ \lambda & 1 & 0 \\ 1 & 0 & 0 \end{pmatrix} \begin{pmatrix} f_x \\ f_y \\ f_z \end{pmatrix}$$

The matrix on the right side of this equation is invertible. Hence by the two preceding equations, (U, V, W) is a singular point of C_E if and only if $T^{-1}(U, V, W)$ is a singular

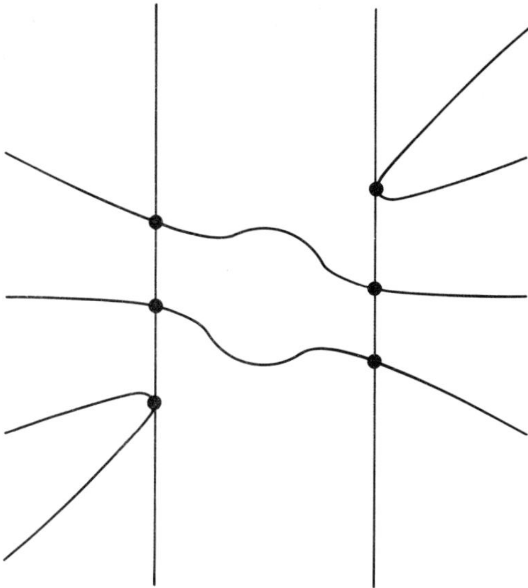

Figure 3: Sample cylindrical algebraic decomposition of E^2.

point of C_f. This proves (i). Since T is an invertible linear transformation of E^3, T induces a homeomorphism $\bar{T} : RP^2 \to RP^2$ given by $\bar{T}[x, y, z] = [T(x, y, z)]$. Clearly $\bar{T}$ carries C_f onto C_E. Thus C_f and C_E have the same topological type: so (ii) is proved.

The reader may wonder why we do not transform C_f to a curve which has no intersections with l_∞. Ragsdale (1906, p. 377 footnote) notes that there exist curves which, for any linear change of coordinates, will have points on l_∞.

4 Decomposition and adjacencies of the affine plane

We will discuss testing the curve C for the presence of singularities later in this section. Assume for the moment that we have determined that C is nonsingular, and we wish to construct an appropriate cd of the projective plane. As detailed in Section 2, we view the projective plane RP^2 as the disjoint union of an affine plane and a line at infinity. Our strategy is to separately construct decompositions and connectivity graphs for the affine plane and l_∞, and then to join the two connectivity graphs by adding edges that join a cell of one decomposition that is adjacent to a cell of the other. Thus we arrive at a connectivity graph for a decomposition of the projective plane. Both the decomposition of the affine plane and the decomposition of the line at infinity are constructed so as to be compatible with C, and so that the cells belonging to C are marked in some fashion. Thus, on completion of these steps we have a decomposition of the projective plane that is compatible with C, and in which the cells belonging to C are marked. In this section we discuss the decomposition of the affine plane. In Section 5 we discuss the decomposition of the line at infinity, and in Section 6, determination of adjacencies between cells in the affine plane and cells in l_∞.

Assume now that $f(x, y, z)$ is squarefree, homogeneous, and satisfies conditions (C1) and (C2) of Section 3. Let $g(x, y) = f(x, y, 1)$. As mentioned in Section 1, the decomposition we construct of the affine plane is a cylindrical algebraic decomposition (cad). Given input $g(x, y)$, we construct a cad that is g-invariant, i.e. for each cell c of the cad, either $g(x, y) < 0$ for all (x, y) in c, or $g(x, y) = 0$ for all (x, y) in c, or $g(x, y) > 0$ for all (x, y) in c. Fig. 3 shows the cad of E^2 that we construct for the example of Section 1, where $g(x, y) = y^4 - 2xy^3 - x^2y^2 + y^2 + 2x^3y + x^2 - 1$.

Algorithm $AffinePlaneDecomp$ given in Fig. 4 presents our complete algorithm for decomposition and adjacencies of the affine plane. It basically consists of portions of algorithm CADA2 of Arnon *et. al.* (1984b). It detects singularities of C in the affine plane as it builds the cad and halts if any are found. The method used for singularity detection is the same used for determination of basis polynomial signatures in the cluster-based cad algorithm (Arnon, 1988). As we have said, the cad that is constructed is a cd; $AffinePlaneDecomp$ produces the connectivity graph of this cd and marks the vertices (cells) in it that belong to C. The map $gsfd$ used in the algorithm denotes "greatest squarefree divisor", and the map $PROJ$ used in the algorithm is as defined in Arnon *et. al.* (1984a).

5 Decomposition and adjacencies of the line at infinity

We isolate the roots of $f(1, y, 0)$, a univariate polynomial with integer coefficients (Collins & Loos, 1982). For each root y, $[1, y, 0]$ is a point of C on l_∞, and we make it a 0-cell of our decomposition of l_∞. We make also the point $[0, 1, 0]$ (which is not on C by condition (C2) of Section 3) a 0-cell of the decomposition. We then take the complementary open intervals of these 0-cells to be the 1-cells of our decomposition of l_∞. The cells on the boundary circle of the disk in Fig. 2 illustrate these steps for the sample curve of Section 1. Recalling the identification of antipodal points in the disk model of RP^2, we see that l_∞ is decomposed into five 0-cells and five 1-cells in this example. In sum, once we know how many roots $f(1, y, 0)$ has, we know what the cells of the decomposition of l_∞ are, and what the connectivity graph of this decomposition is.

Let us introduce some notation for the cells of our decomposition of l_∞. Suppose that there are $k \geq 0$ points of C on l_∞. Since $[0, 1, 0]$ is not on C, these points can be written $[1, \gamma_1, 0] = P_1, ..., [1, \gamma_k, 0] = P_k$, where $\gamma_1 < ... < \gamma_k$ are the real roots of $f(1, y, 0)$. Our cellular decomposition of l_∞ consists of: $P_1, ..., P_k$, the point $[0, 1, 0] = P_0 = P_{k+1}$, and the 1-cells e_i, $0 \leq i \leq k$, where e_i is the open interval in l_∞ bounded by P_i and P_{i+1}. Fig. 5 illustrates this notation for the sample curve of Section 1. Let us assign to P_i the index $(0, 2i)$, and to e_i the index $(0, 2i + 1)$, for $0 \leq i \leq k$. This assignment of indices preserves the rule that the dimension of a cell is the sum of the parities of the components of its index (cell indices are defined in Section 6).

$G \leftarrow$ **AffinePlaneDecomp** ($g(x,y)$)

Input: $g(x,y)$ is a primitive bivariate polynomial with integer coefficients. Let C_g denote the curve in the real affine plane defined by $g = 0$.

Outputs: If C_g is nonsingular, then G is the connectivity graph for a g-invariant cad D of the euclidean plane in which the cells comprising C_g are marked. If C_g is singular, then T is the string "SINGULAR".

(1) [Base case.] Set $P \leftarrow PROJ(\{g\})$. Isolate the real roots of the irreducible factors of the nonzero elements of P to determine a cad D' of the real affine line. Construct a sample point for each cell of D' as in algorithm CADA2 of Arnon *et. al.* (1984b).

(2) [Extension and singularity check.] Let $a_1 < a_2 < \cdots < a_{2m} < a_{2m+1}$, $m \geq 0$, be the sample points for D' (each a_{2i+1} is a rational sample point for a 1-cell; each a_{2i} is an algebraic sample point for a 0-cell). For $i = 1, \ldots, 2m + 1$, let c_i denote the cell of D' whose sample point is a_i, and do the following five things: first, construct open isolating intervals for the real roots of $g(a_i, y)$ to determine the sections of a stack $S(c_i)$ in E^2; second, compare the signs of $gsfd(g_x(a_i, y))$ at the endpoints of each isolating interval, and the signs of $gsfd(g_y(a_i, y))$ at the endpoints of each isolating interval, and if the signs are different in either comparison, then set $G \leftarrow$ "SINGULAR" and exit; third, construct cell indices for the cells of $S(c_i)$ and add a vertex for each cell of $S(c_i)$ to G; fourth, mark each section of $S(c_i)$ as belonging to C_g; fifth, add the intrastack adjacencies for $S(c_i)$ to G. When we are done, we have built a g-invariant cad D of the affine plane that is *proper* in the sense of Arnon *et. al.* (1984b).

(3) [Adjacency computation.] For $i = 1, \ldots, m$, call algorithm SSADJ2 of Arnon *et. al.* (1984b) with inputs g, a_{2i}, a_{2i-1}, and a_{2i+1}, and add the contents of its outputs L_1 and L_2 to G. Note that the section numbers which occur in the adjacencies returned by SSADJ2 must first be converted into the indices of the corresponding cells of D; for example, if the list L_1 returned by the ith call to SSADJ2 contains the adjacency $\{3, 2\}$, it must be converted to $\{(2i, 6), (2i - 1, 4)\}$ before being added to G. Infer the remaining interstack adjacencies between $S(c_{2i})$ and $S(c_{2i-1})$, and between $S(c_{2i})$ and $S(c_{2i+1})$, as described at the end of Section 2 of Arnon *et al.* (1984b), and add them to G. All adjacencies of D are now recorded in G. $\square$

Figure 4: Algorithm AffinePlaneDecomp.

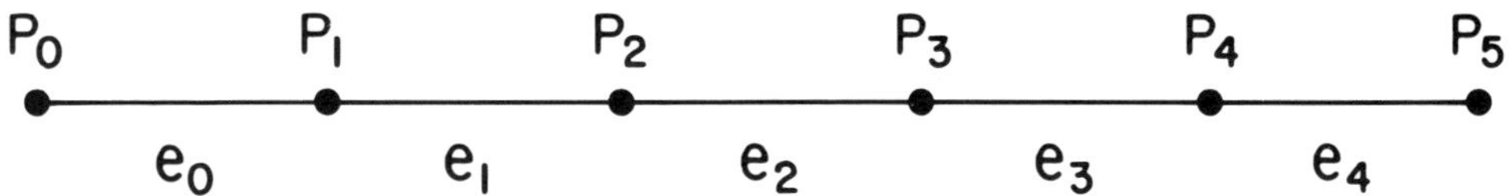

Figure 5: Cellular decomposition of l_∞ for sample curve.

6   Adjacencies between finite and infinite cells

Let us now consider the question of determining the adjacencies between a cell in our decomposition of the affine plane (which we call a *finite* cell) and a cell in our decomposition of the line at infinity (an *infinite* cell). Consider the sample curve C shown in Figs. 1 and 2. We see that for this example it is straightforward to describe all such adjacencies. The affine 1-cells which are on C, and which "go off to infinity" to the "right" or to the "left", are each adjacent to exactly one 0-cell on the line at infinity, and this 0-cell is not $[0, 1, 0]$, In fact, for this example, when we have either determined how many roots $f(x, 1, 0)$ has, or found out how many finite 1-cells that lie in C go off to infinity to the "right", or found out how many finite 1-cells that lie in C go off to infinity to the "left", we know exactly what the adjacencies among finite 1-cells that lie in C and infinite 0-cells are. Furthermore, all cells which are topmost or bottommost in some stack of the cad, and only those cells of the cad, are adjacent to the point $[0,1,0]$ on the line at infinity. This is precisely the situation that conditions (C1) and (C2) of Section 3 are designed to bring about. We prove with Theorem 6.1 below that they do.

We first review some definitions from Arnon *et al.* (1984a). Let X be a nonempty connected subset of E^1. The *cylinder over X*, written $Z(X)$, is $X \times E$. A *section* of $Z(X)$ is a set s of points $< x, \phi(x) >$, where x ranges over X, and ϕ is a continuous, real-valued function on X. s, in other words, is the graph of ϕ. The constant functions $\phi = -\infty$ and $\phi = +\infty$ are allowed; in these cases, s is an *infinite section*. A *sector* of $Z(X)$ is a set $\hat{s}$ of all points $< x, y >$, where x ranges over X, and $\phi(x) < y < \psi(x)$ for (continuous, real-valued) functions $\phi < \psi$. A *stack* over X is a collection of disjoint sections and sectors of $Z(X)$ whose union is $Z(X)$; X is the *base* of the stack. A *cylindrical decomposition D of E^2* is a finite cellular decomposition of E^2, such that for some cellular decomposition D' of the real line E^1, the cells of D comprise stacks over the cells of D'. The decomposition shown in Fig. 3 is a cylindrical decomposition D of E^2. For that example, D' consists of two 0-cells and the three complementary 1-cells.

Any cell c of a cylindrical decomposition D of E^2 can be assigned an index, consisting of an ordered pair of positive integers. The first component specifies the cell of D' which is the base of the stack containing c; the cells of D' are numbered 1, 2,... from left to right. The second component specifies the position of c within the stack; the cells of the stack are numbered 1, 2,... from bottom to top. The indices for the cells in Fig. 3 are shown in Fig. 6. It is easily seen that the dimension of a cell in a cylindrical decomposition is the sum of the parities (even $= 0$, odd $= 1$) of the components of its index, e.g. (1,9) is a 2-cell, (2,6) is a 0-cell.

In a cylindrical decomposition D of E^2, the cells over the leftmost cell of D' comprise the *leftmost stack* of D, and the cells over the rightmost cell of D' the *rightmost stack*. Thus in Fig. 6, the cells with indices $(1, j)$, $1 \leq j \leq 9$, are the leftmost stack, and the cells with indices $(5, j)$, $1 \leq j \leq 9$, are the rightmost stack. In general, the leftmost and rightmost stacks of D are distinct, however if D' consists of a single 1-cell (the entire real line), then the leftmost and rightmost stacks are identical. This concludes our review of definitions.

THEOREM **6.1** *Let $f(x, y, z)$ be a homogeneous trivariate polynomial with integer coefficients, which satisfies conditions (C1) and (C2) of Section 3. Suppose that $f(1, y, 0)$ has $k \geq 0$ real roots $\gamma_1 < ... < \gamma_k$. Let $g(x, y) = f(x, y, 1)$, and suppose that S and T are (respectively) the rightmost and leftmost stacks of a g-invariant cad of the affine plane. Then*

$$
\begin{array}{lllll}
(1,9) & & & & (5,9) \\
(1,8) & & & & (5,8) \\
(1,7) & (2,7) & & (4,7) & (5,7) \\
(1,6) & (2,6 & & (4,6) & (5,6) \\
(1,5) & (2,5) & (3,5) & (4,5) & (5,5) \\
(1,4) & (2,4) & (3,4) & (4,4) & (5,4) \\
(1,3) & (2,3) & (3,3) & (4,3) & (5,3) \\
(1,2) & (2,2) & (3,2) & (4,2) & (5,2) \\
(1,1) & (2,1) & (3,1) & (4,1) & (5,1)
\end{array}
$$

Figure 6: Cell indices for sample cylindrical decomposition.

(i) S has k sections, say $s_1 < ... < s_k$, and T has k sections, say $t_1 < ... < t_k$;

(ii) for $1 \leq i \leq k$, if s_i is the graph of the continuous real-valued function ϕ_i, and t_i is the graph of the continuous real-valued function ω_i, then

$$
\lim_{x \to +\infty} \frac{\phi_i(x)}{x} = \gamma_i ,
$$

$$
\lim_{x \to -\infty} \frac{\omega_i(x)}{x} = \gamma_{k-i+1} .
$$

PROOF. Let n be the degree of $f(x,y,z)$. By condition (C1) of Section 3, each γ_i is a simple root of $f(1,y,0)$. Let $G(X,Y) = f(1,Y,X)$. Since $f(0,1,0) \neq 0$ by condition (C2), $G(X,Y) = g_0 Y^n + g_1(X)Y^{n-1} + ... + g_n(X)$, for some constant $g_0 \neq 0$ and polynomials $g_1(X) , ..., g_n(X)$. Since $G(0,Y) = f(1,Y,0)$, $G(0,Y)$ has exactly k real roots $\gamma_1 < ... < \gamma_k$, each of them simple. Hence by root continuity, there is some $\delta > 0$ such that $|X| < \delta$ implies $G(X,Y)$ has exactly k real roots, each of them simple. The ith of these roots approaches γ_i as $|X| \to 0$. Since $g(x,y) = x^n G(1/x, y/x)$ for nonzero x, $g(x,y)$ has k real roots, each simple, for all sufficiently large positive x. Hence S has k sections. A similar argument shows that T has k sections.

For any x in the interval $(\alpha, +\infty)$ on which ϕ_i is defined, $\phi_i(x)$ is the ith real root of $g(x,y)$. Hence, for positive x greater than α, $\phi_i(x)/x$ is the ith real root of $G(1/x, Y)$. Hence, as $x \to +\infty$, $\phi_i(x)/x \to \gamma_i$. For any x in the interval $(-\infty, \beta)$ in which ω_i is defined, $\omega_i(x)$ is the ith real root of $g(x,y)$. Hence, for negative x less than β, $\omega_i(x)/x$ is the $(k-i+1)^{st}$ real root of $G(1/x, Y)$. Hence as $x \to -\infty$, $\omega_i(x)/x \to \gamma_{k-i+1}$ $\square$

As we have indicated, Figs. 1-3 illustrate Theorem 6.1. In this example, the stacks S and T of the theorem each have four sections. One sees that the asymptotic slope of s_i, namely γ_i, is equal to the asymptotic slope of t_{k-i+1}.

Let us now give a general procedure for determination of adjacencies between finite and infinite cells. First we relate adjacency of cells to the topological notion of boundary. It is easily shown that two cells are adjacent if and only if one contains a limit point of the other. For a subset X of a topological space, the *boundary* of X, written ∂X, is $\bar{X} - X$, where $\bar{X}$ denotes the closure of X. It is not difficult to show that ∂X is the set of all limit points of X which do not belong to X. Thus two cells are adjacent if and only if one meets the boundary of the other.

Now, for some given f, let S and T be as in Theorem 6.1. Let P_i and e_i be as in Section 5. Consider first adjacencies between sections of S and T and cells in l_∞. Suppose $S \neq T$. We claim that P_i is a limit point of the ith section s_i of S. As in the Theorem, let s_i be the graph of a function ϕ_i. Let $[x_i, \phi(x_i), 1]$ be a sequence of points in s_i, with x_i approaching $+\infty$. Then $\lim[x_i, \phi(x_i), 1] = \lim[1, \phi(x_i)/x_i, 1/x_i] = [1, \gamma_i, 0] = P_i$. It can be shown that P_i is in fact the unique limit point of s_i on l_∞. Similarly, P_{k-i+1} is the unique limit point of t_i on l_∞. If $S = T$, then $s_i = t_i$ has the limit points P_i and P_{k-i+1} in l_∞ (P_i and P_{k-i+1} may coincide).

Consider now sectors of S and T. Suppose the 1-cell c in the induced cad of the line is the base of S, and let $s_0 = c \times \{-\infty\}$ and $s_{k+1} = c \times \{+\infty\}$ denote the infinite sections of $Z(c)$. For $0 \leq i \leq k$, let $\hat{s}_i$ denote the sector of S between s_i and s_{i+1} (similar definitions hold for T by replacing s by t throughout). Note that $\bar{e}_i = e_i \cup \{P_i , P_{i+1}\}$. If $S \neq T$, it is evident that the portion of the boundary of $\hat{s}_i$ contained in l_∞ is $\bar{e}_i$, while the portion of the boundary of $\hat{t}_i$ contained in l_∞ is $\bar{e}_{k-i}$, for $0 \leq i \leq k$ (see example in Fig. 7). If $S = T$ is the only stack of D, the portion of the boundary of $\hat{s}_i$ contained in l_∞ is $\bar{e}_i \cup \bar{e}_{k-i}$ (see example in Fig. 8).

Now let R be any stack of D besides S and T. Let $r_1 < ... < r_l$ be the finite sections of R. Let $\hat{r}_0$ and $\hat{r}_l$ be defined as were $\hat{s}_0$ and $\hat{s}_k$. From the disk model for RP^2, it is evident that $\hat{r}_0$ and $\hat{r}_l$ are the only cells of R to have limit points in l_∞, and each in fact does have the unique limit point P_0 in l_∞ (Fig. 2 illustrates this discussion). This completes the determination of adjacencies between finite and infinite cells.

We omit the proof, in the present paper, that the decomposition D^* of the projective plane we construct is actually a cellular decomposition in the sense of Massey (1978) . A stronger result can in fact be shown: D^* gives RP^2 the structure of a cell complex. The proof is given in Arnon & McCallum (1983).

7   Topological type from cellular decomposition

In this section, we will think of the cd D^* of RP^2 as a certain collection of cells. Thus we write the connectivity graph as $G^* = (D^*, E)$, where E, the set of edges of G^*, is the set of all pairs of adjacent cells of D^*. We write D_C to denote the subset of D^* consisting of all cells contained in the curve C.

Our first task is to determine the components of C. This is accomplished by constructing the connected components of the subgraph of G^* induced by D_C. In the data structure for G^*, we mark each cell of D_C with an index identifying the particular component of C to which it belongs.

Now, for each component of C, we want to determine if it is an oval, i.e. if its complement with respect to the projective plane has two rather than one components. If so, we want to identify which component of its complement is its interior, and which its exterior. Let $D_1 \subset D_C$ be the cells which comprise a component C_1 of C. We compute the connected components of the subgraph of G^* induced by D^*-D_1 (call this subgraph G_1). If there is only one such component, than C_1 is not an oval, and we do no further processing for it.

The Euler characteristic χ, of a cd of a subspace of the projective plane, is $\alpha_0 - \alpha_1 + \alpha_2$, where α_i is the number of i-cells in the cd (cf. Massey, 1978, p. 61). Suppose

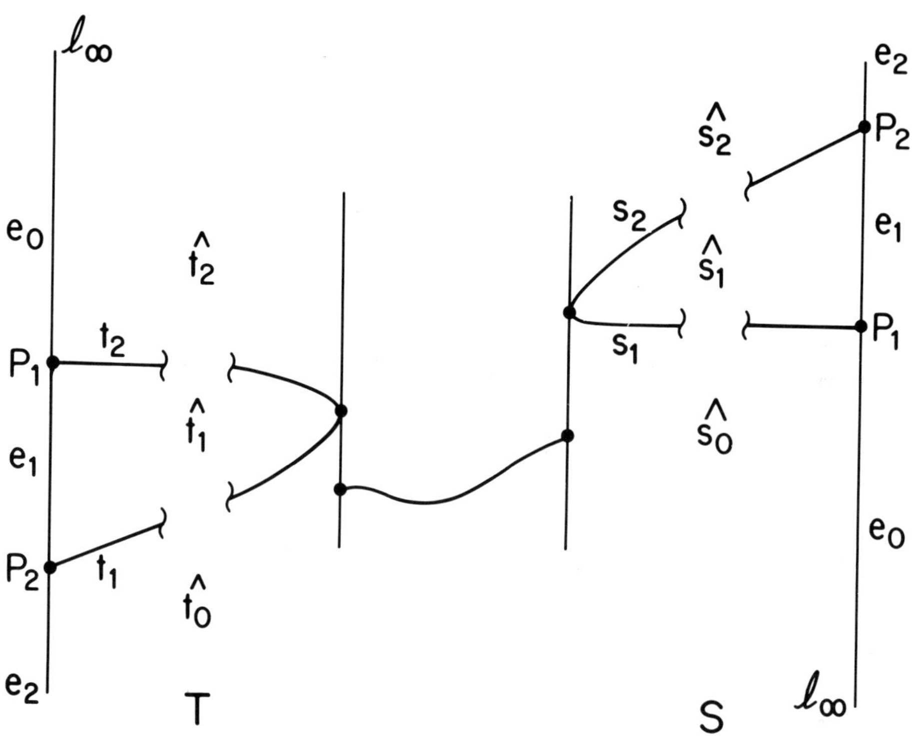

Figure 7: Example of adjacencies between S, T, and l_∞.

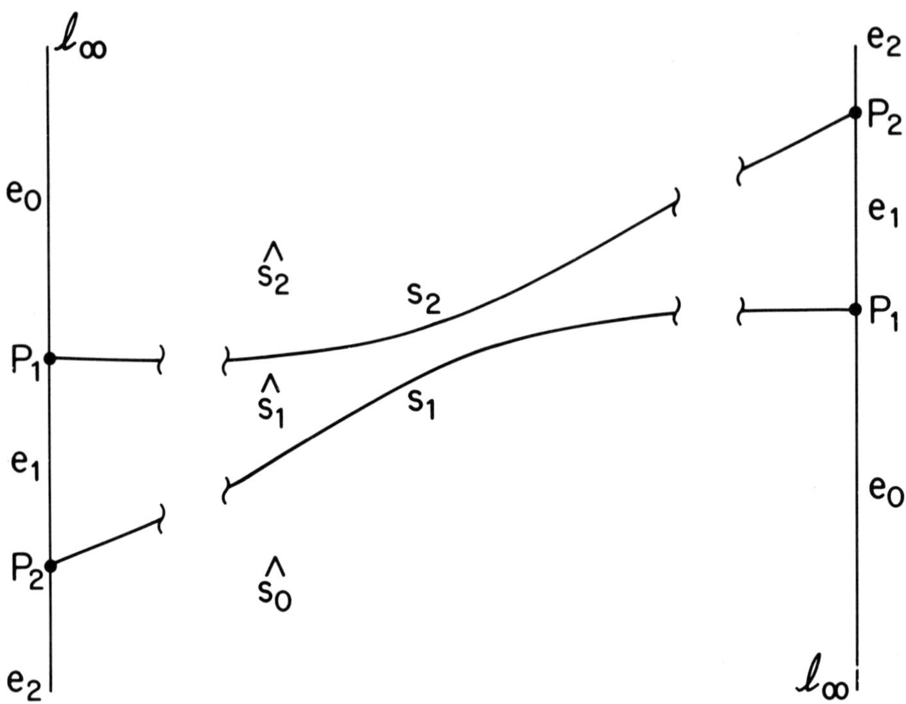

Figure 8: Example of adjacencies between $S = T$ and l_∞.

that C_1 is an oval, i.e. G_1 has two connected components. Since D^* is compatible with C, it follows that one of the components of G_1 is a cellular decomposition of the interior of C_1, and the other is a cellular decomposition of the exterior of C_1. As we have mentioned, the interior of C_1 is topologically a disk, and its exterior is topologically a Möbius band. It follows (e.g. from Exercise 2 of Massey, 1978, p. 61) that the Euler characteristics of the two components of G_1 must be different. In fact, $\chi(disk) = 1$, and $\chi(M\ddot{o}bius\ band) = 0$. Hence, by computing the Euler characteristics of the two components of G_1, we determine which is the interior and which the exterior of C_1. Finally, in the data structure for G^*, we record at each cell of $D^* - D_1$ whether it is in the interior or exterior of C_1.

When we have processed each component of C in the above-described fashion, it remains only to determine the partial ordering of C's ovals. We may do this in any number of ways, for example, by picking one cell in each oval and reading off the order information we have recorded with it in the data structure for G^*.

8   Main algorithm

We give a formal description of our main algorithm TOPTYP in Fig. 9. The map pp used in the algorithm denotes "primitive part" (see Arnon, 1988, for definition of primitive part).

9   Example

Let us now consider the example of Section 1 in more detail. Let $f(x, y, z)$ be as at the beginning of Section 1. f is irreducible, hence squarefree. $f(x, y, 0)$ has no multiple factors, and $[0,1,0]$ does not lie on C, so we need not change coordinates. Let $g(x, y) = f(x, y, 1)$. Recall that the cad of the affine plane $z = 1$ constructed by algorithm CADA2 with input $g(x, y)$ is shown in Fig. 3. We find that C is nonsingular. Continuing, we have

$$f(1, y, 0) \ = \ y\,(y - 1)\,(y + 1)\,(y - 2),$$

and so C has the four points $[1,0,0]$, $[1,1,0]$, $[1,-1,0]$, and $[1,2,0]$ on l_∞. Thus the cells of D^* on l_∞ are as shown in Figs. 2 and 5. In Fig. 10 we enlarge Fig. 2 and label each cell with its index. The connectivity graph of D^* is clear from Fig. 10 (or Fig. 2). We find that C has two components, composed respectively of the following collections of cells:

$$J_1 = \{\ (1, 2), (2, 2), (1, 4), (0, 6), (5, 6), (4, 6), (5, 8), (0, 8)\ \},$$

and

$$J_2 = \{\ (1, 6), (2, 4), (3, 2), (4, 2), (5, 2), (0, 2), (1, 8), (2, 6), (3, 4), (4, 4), (5, 4), (0, 4)\ \}.$$

Since f has even degree, both correspond to ovals.

Let O_1 denote the oval of C that comprises the cells of J_1. $complement(O_1)$ turns out to have two components, composed of $K_1 = \{\ (1, 3), (0, 7), (5, 7)\ \}$, and

$$L_1 \ = \ \{\ (0, 0), (0, 1), (0, 2), (0, 3), (0, 4), (0, 5), (0, 9), (1, 1), (1, 5), (1, 6), (1, 7),$$

$$T \leftarrow \textbf{TOPTYP} \; (\; f(x,y,z) \;)$$

Input: $f(x,y,z)$ is a homogeneous trivariate polynomial with integer coefficients, i.e. a homogeneous element of $\mathbf{Z}\,[x,y,z]$. Let C denote the curve in the real projective plane defined by $f = 0$.

Outputs: If C is nonsingular, then T is the number and partial ordering of the ovals of C. If C is singular, then T is the string "SINGULAR".

(1) [Transform f, if necessary.] If f is not squarefree, then replace f by $gsfd(f)$. Test whether $f(x,y,z)$ satisfies conditions (C1) and (C2) of Section 3. If not, then change coordinates as per Section 3 to get some new $f(x,y,z)$. Recall that we may detect during the coordinate change process that C is singular; in this event, set T to the string "SINGULAR" and exit. Set $g(x,y) \leftarrow f(x,y,1)$. By conditions (C1) and (C2), $V(content(g))$ is empty, hence $V(g) = V(pp(g))$, hence replace g by $pp(g)$.

(2) [Decomposition and adjacencies of the affine plane.] Set $G^* \leftarrow AffinePlaneDecomp(g)$. If G^* is the string "SINGULAR" then set T to the string "SINGULAR" and exit.

(3) [Decomposition and adjacencies of the line at infinity.] Determine the number of real roots of $f(1,y,0)$. From this information, place new vertices in the connectivity graph G^* for the corresponding 0- and 1-cells cells on the line at infinity, mark all 0-cells except $[0,1,0]$ as belonging to C, and add new edges to the connectivity graph corresponding to the adjacent pairs of cells within the line at infinity.

(4) [Adjacencies between finite and infinite cells.] As discussed in Section 6, all adjacencies between a 1-cell of the cad that is contained in C and a 0-cell on the line at infinity that is contained in C are now known; add them to the connectivity graph. From these adjacencies, infer the adjacencies of 2-cells of the cad with 0-cells and 1-cells on the line at infinity, and add them to the connectivity graph. Add an edge to the connectivity graph for each adjacency between the point $[0,1,0]$ on the line at infinity, and the topmost and bottommost cell in each stack of the cad of the affine plane. We now have a cellular decomposition D^* of the projective plane that is compatible with C, and all adjacencies of this decomposition are marked in the connnectivity graph.

(5) [Topological type from cellular decomposition.] Determine the components of C, determine which components are ovals, and for each oval, determine which cells of D^* comprise its interior, and which its exterior. From this information, determine the partial ordering of the ovals of C by inclusion. Set T to some representation of the ovals and their partial ordering, and exit $\square$

Figure 9: Algorithm TOPTYP.

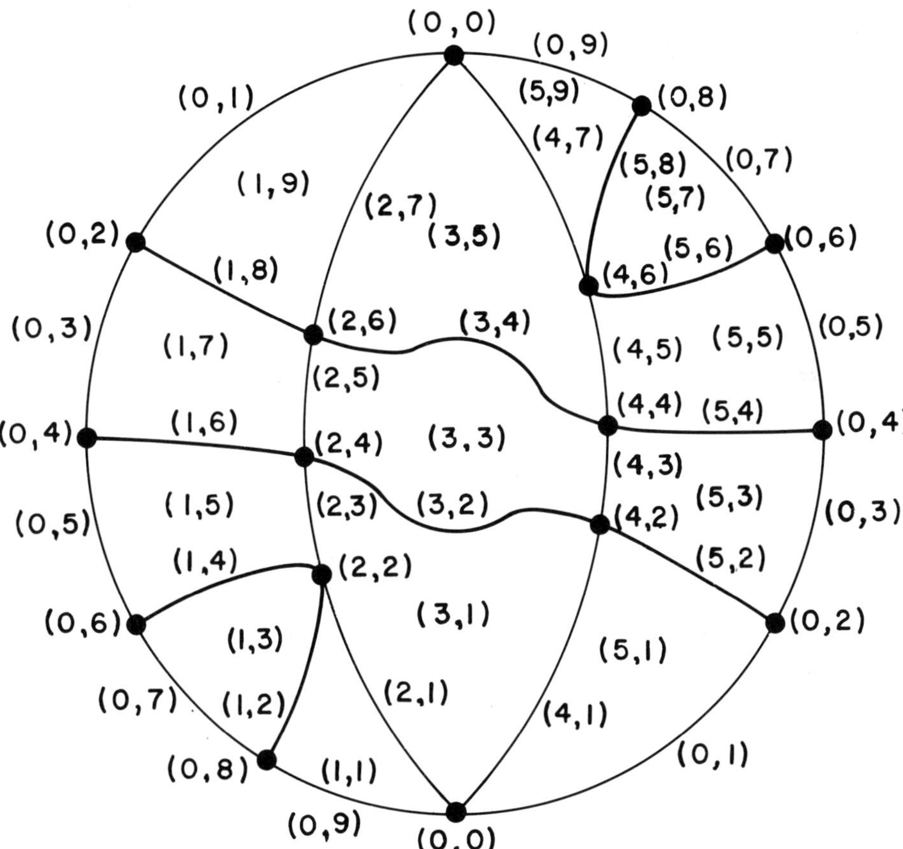

Figure 10: Cellular decomposition of projective plane compatible with sample curve.

$(1,8), (1,9), (2,1), (2,3), (2,4), (2,5), (2,6), (2,7), (3,1), (3,2), (3,3), (3,4),$

$(3,5), (4,1), (4,2), (4,3), (4,4), (4,5), (4,7), (5,1), (5,2), (5,3), (5,4), (5,5), (5,9)$ }.

Computing the dimension of each cell as the sum of the parities of the two components of its index, we find that the Euler characteristic of K_1 is $\chi = 0 - 1 + 2 = 1$. For L_1, we have $\chi = 7 - 18 + 11 = 0$. Hence K_1 corresponds to the interior, and L_1 to the exterior, of O_1.

Consider the second collection J_2 of cells comprising oval O_2 of C. We find two collections of cells corresponding to the components of $complement(O_2)$:

$$K_2 = \{ (1,7), (0,3), (5,3), (4,3), (3,3), (2,5) \}$$

and

$$L_2 = \{ (0,0), (0,1), (0,5), (0,6), (0,7), (0,8), (0,9), (1,1), (1,2), (1,3), (1,4), (1,5),$$

$$(1,9), (2,1), (2,2), (2,3), (2,7), (3,1), (3,5), (4,1), (4,5), (4,6), (4,7), (5,1), (5,5),$$

$$(5,6), (5,7), (5,8), (5,9) \}.$$

The Euler characteristic of K_2 is $\chi = 0 - 3 + 3 = 0$. For L_2, we have $\chi = 5 - 14 + 10 = 1$. Hence K_2 corresponds to the exterior, and L_2 to the interior, of O_2. We see that the cells comprising O_1 occur among the cells comprising the interior of O_2. Hence O_1 is included in O_2, and we have determined the topological type of C.

The time required for this example was approximately five minutes, using an implementation of TOPTYP in the SAC-2 computer algebra system (Collins, 1980) on a Vax 11/785.

10   Computing time analysis

We count the cost of the individual steps of algorithm TOPTYP.

10.1   TOPTYP Step (1)

In computing the greatest squarefree divisor of f, the gcd calculation takes $O(n^5 L(d)^2)$ steps (Collins & Loos, 1982, p. 84; Loos, 1982). This dominates the cost of the division. Using Mignotte's bound (Collins & Loos, 1982, p. 84), the maximum coefficient $\bar{d}$ of the greatest squarefree divisor satisfies $\bar{d} \leq 2^n d$. We do not consider this greatest squarefree divisor calculation to be really a part of our algorithm, and so we will ignore this potential coefficient growth in the remainder of our analysis. Note that the greatest squarefree divisor has lower degree, if different from f.

The significant operations in the coordinate transformation, if it is carried out, are two linear changes of coordinates and computation of the discriminant of a bivariate polynomial of degree n. The cost of the two coordinate changes is dominated by the discriminant computation, and since we will always do a discriminant computation on an input of the same or larger size in step (2), we ignore the cost of the coordinate transformation.

Examination of the two linear changes of coordinates shows that the transformed f may have a sum norm (i.e. sum of the absolute values of its integer coefficients) of dn^n, where d was the sum norm of the original polynomial (consider, for example, the use of a Horner type evaluation to actually do the changes of coordinates). We will assume that $L(dn^n) = L(d) + nL(n)$ is $O(nL(d))$, and so assume from now on that the length of the sum norm of our input polynomial is $O(nL(d))$. The cost of computing $pp(g)$ is less than the cost of the discriminant computation that we will count in Step (2).

10.2   TOPTYP Step (2)

In Step (1) of $AffinePlaneDecomp$, since $[0, 1, 0]$ is not on C, g has constant leading coefficient, hence $PROJ(\{g\}) = Discriminant(g)$. Discriminant computation is $Resultant(g, g')$, i.e. resultant of two bivariate polynomials of degree n or less and sum norm $n \cdot dn^n$. Note that the length of this new sum norm is still $O(nL(d))$. Computing the resultant of two bivariate polynomials of degree s and sum norm u takes time $O(s^5 L(u)^2)$ (Loos, 1982, p. 134), which gives us time $O(n^7 L(d)^2)$ altogether, and produces a polynomial of degree n^2 or less. If e is the sum norm of the discriminant, then $L(e) = O(n \cdot nL(d))$. The factorization of a univariate integral polynomial of degree s and sum norm u takes $O(s^{12} + s^9 L(u)^3)$ (Kaltofen, 1982, p. 111). Hence the factorization of $D(x)$ into irreducibles takes $O(n^{24} + n^{18} L(e)^3) = O(n^{24} + n^{24} L(d)^3) = O(n^{24} L(d)^3)$. For simplicity let us assume that the discriminant has only one irreducible factor; if it has more than one, that will complicate the analysis but will reduce the computing time. Root isolation applied to a squarefree univariate integral polynomial of degree s and sum norm u is $O(s^6 L(u)^2)$ (Collins & Loos, 1982, p. 93), so for the discriminant this gives us $O(n^{16} L(d)^2)$. The remaining actions of Step (1) of $AffinePlaneDecomp$ are not significant.

In Step (2) of $AffinePlaneDecomp$, let m be the number of roots of the discriminant $D(x)$ just computed; we have $m \le n^2$. The dominant cost of this step is the root isolations of the polynomials $g(a_{2j}, y)$, $1 \le j \le m$, since in these cases a_{2j} is an algebraic, rather than a rational, number. The only bound presently available to us for root isolation of algebraic polynomials is that given for the Collins-Loos algorithm (Collins & Loos, 1982) by Rump (1975). We get a better bound assuming the use of an alternative, somewhat roundabout, algebraic polynomial root isolation algorithm. The alternative algorithm is to compute the "normal polynomial" $V(y) = Resultant(g(x, y), D(x))$ (whose roots include those of $g(a_{2j}, y)$), isolate its roots, and use $gsfd(g(a_{2j}, y))$ to select the intervals that contain a root of $g(a_{2j}, y)$. To compute the normal polynomial, we compute a resultant of one polynomial of degree n^2 or less and sum norm e (given our assumption that $D(x)$ has only one factor) and one of degree n or less and sum norm d; so this is $O(n^{10} L(e)^2) = O(n^{14} L(d)^2)$. The resultant is a polynomial of degree n^3 or less with sum norm v with $L(v) = O(nL(e))$, so root isolation takes $O(n^{27} L(d)^2)$ (we assume that $V(y)$ is squarefree). Let α denote the algebraic number a_{2i}, and let $g_\alpha(y) = g(a_{2j}, y)$. We must compute $gsfd(g_\alpha) = g_\alpha / gcd(g_\alpha, g'_\alpha)$. Evaluation of g at α takes no time; we just interpret $g(x, y)$ as a polynomial in y over $Q(\alpha)$. Suppose that we compute $gcd(g_\alpha, g'_\alpha)$ by a natural polynomial polynomial remainder sequence (Loos, 1982). At each of the $O(n)$ steps we have to do a division with remainder of two polynomials in $Q(\alpha)[y]$, each of which has degree n or less, and each of which has sum norm of $O(nL(n^2 d))$ (Loos, 1982, p. 133). Thus, since the degree of the minimal polynomial of α is n^2 or less, the $O(n)$ arithmetic operations in $Q(\alpha)$ we do at each step, each have a cost of $O((n^2)^3 n^2 L(nd)^2)$, thus our total cost for this gcd calculation is $O(n^{10} L(nd)^2)$. This surely dominates the cost of $g_\alpha / gcd(g_\alpha, g'_\alpha)$, and so we will take it to be the cost of the entire $gsfd$ computation. We next must evaluate the signs of the $gsfd$ at the endpoints of at most n^3 isolating intervals for roots of the polynomial $V(y)$ we computed above. Let u/v be one of these endpoints. Given that the degree of V_j is $O(n^3)$, by Horner's rule, this costs

$$n^3 L(u)\{n^3 L(u) + n^3 L(v) + nL(nd)\},$$

(Collins & Loos, 1982, p. 84), where $L(u)$ and $L(v)$ are each $O(n^3 L(n^3 d))$ (Collins & Loos (1982, p. 84). Thus the total time for the evaluation is $O(n^9 L(n^3 d)^2)$. The resulting element of $Q(\alpha)$ is represented as a polynomial with rational number coefficients, each of which has length that is $O(n^3 L(n^3 d))$. By Rump (1976), the sign of an algebraic number whose minimal polynomial has degree s, and such that u is the largest coefficient occurring either in the minimal polynomial or in the representing polynomial for that algebraic number, can be found in time $O(s^5 L(u)^3)$. Hence since in our case the degree of the minimal polynomial is n^2 or less, and the length of u is $O(n^3 L(n^3 d))$, we get a time of $O(n^{19} L(n^3 d)^3)$. Since we have $O(n^3)$ of these sign determinations to do, this gives us a total time of $O(n^{22} L(n^3 d)^3)$.

We have to do two more similar $gsfd$ calculations in this step for g_x and g_y, and evaluate those at the endpoints of the isolating intervals we have found to contain roots of $gsfd(g(\alpha, y))$, but as the cost of these computations is dominated by the cost of what we've already done for g, we take the cost of the Step (2) of $AffinePlaneDecomp$ for this i to be $O(n^{22} L(n^3 d)^3)$. Since $m = O(n^2)$, the total cost for Step (2) of $AffinePlaneDecomp$ is $O(n^{24} L(n^3 d)^3)$.

Now let us go on to Step (3) of $AffinePlaneDecomp$, and consider the cost of a single call to SSADJ2. Step (1) of SSADJ2 calls for a root isolation that we already did in Step (2) of $AffinePlaneDecomp$; we may assume that it is not repeated. In Step (2), we start knowing that (b_1, b_2) is an isolating interval for α as a root of its minimal polynomial $M(x)$; we must shrink (bisect) (b_1, b_2) until no $g(x, s_j)$ has a root in $[b_1, b_2]$, and (b_1, b_2) must still contain α. We can think of this as having to isolate the roots of product polynomials $M(x) g(x, s_j)$, for successive j. The cost of these isolations depends on the minimum root separation of $M(x) g(x, s_j)$ versus the minimum root separation of $M(x)$. For simplicity we will assume that the coefficients of these two polynomials have the same maximum size. Then it follows from Collins & Loos (1982, p. 84), that since the degree of $M(x)$ is $O(n^2)$, and the degree of $g(x, s_j)$ is $O(n)$, we have to do at most n bisections for each j, and since there are $O(n)$ successive values of j, we obtain a cost of $O(n^2)$ so far for step (2) of SSADJ2. Clearly this is not a significant cost, even given the fact that our discussion ignored the cost of the (rational number) arithmetic for each bisection. In the remaining steps of SSADJ2, we see that we have $O(n)$ calls to a root isolation algorithm for an integral polynomial of degree n. Let us count $O(n^6 L(d)^2)$ for each such call; this gives us $O(n^7 L(d)^2)$ total for one call to SSADJ2. SSADJ2 is executed $m = O(n^2)$ times, so altogether we have time $O(n^9 L(d)^2)$ for step (3) of $AffinePlaneDecomp$.

10.3   TOPTYP Step (3)

The only significant cost in Step (3) of TOPTYP is the root isolation of $f(1, y, 0)$, which is $O(n^6 L(d)^2)$.

10.4   TOPTYP Step (4)

Negligible cost.

10.5   TOPTYP Step (5)

We consider the computing time of the steps described in Section 7. First, note that the decomposition of the projective plane we construct has $O(n^3)$ cells, and $O(n^4)$ adjacencies. Thus the connectivity graph for our decomposition has $O(n^3)$ vertices and $O(n^4)$ edges. To construct the components of C involves constructing the connected components of connectivity graph, hence, if we use depth first search, then the time is $O(n^3 + n^4) = O(n^4)$ (Aho *et. al.*, 1974). As we have mentioned (cf. Abstract and Section 2), C has $O(n^2)$ components. For each component, we must determine if it is an oval (which we do with a connected components computation in a subgraph of the connectivity graph), and if so, do two Euler characteristic computations. Thus for each component of C, we have a cost of $O(n^4)$ for the connected components computation, and a cost of $O(n^3)$ for the Euler characteristic computations, hence a cost of $O(n^6)$ for all components of C. The cost of the steps we have described dominates the cost of determining the partial ordering of ovals. Hence the total time for this step is $O(n^6)$.

10.6   Summary

We see that altogether, the maximum computing time of TOPTYP is $O(n^{27} L(d)^3)$.

11   Acknowledgements

We are indebted to G. Brumfiel for the observation that Euler characteristic suffices to distinguish the interior of an oval from its exterior (we had originally contemplated a homology calculation). R.H. Bing was kind enough to provide us with a detailed account of some fundamental facts of the topology of plane curves. The second author would like to acknowledge helpful and inspiring conversations on the subject of this paper with the following people: G. Collins, E. Fadell, T.-C. Kuo, E. Mansfield.

12   References

Aho, A. V., Hopcroft, J., Ullman, J. (1974). *The Design and Analysis of Computer Algorithms.* Reading, Massachusetts: Addison-Wesley.

Arnon, D. S., McCallum, S. (1983). *A Polynomial-Time Algorithm for the Topological Type of a Real Algebraic Curve.* Tech. Rept. #454, Comp. Sci. Dept., Purdue Univ.

Arnon, D. S., McCallum, S. (1984). A polynomial-time algorithm for the topological type of a real algebraic curve – extended abstract. *Rocky Mountain J. Math.*, **14**, 849-852.

Arnon, D. S., Collins, G. E., McCallum, S. (1984a). Cylindrical algebraic decomposition I: the basic algorithm, *SIAM J. Comp.* **13/4**, 865-877.

Arnon, D. S., Collins, G. E., McCallum, S. (1984b). Cylindrical algebraic decomposition II: an adjacency algorithm for the plane, *SIAM J. Comp.* **13/4**, 878-889.

Arnon, D. S. (1988). A cluster-based cylindrical algebraic decomposition algorithm. *J. Symb. Comp.* **5**, (this issue).

Ben-Or, M., Kozen, D., Reif, J. (1986). The complexity of elementary algebra and geometry. *J. Comp. Sys. Sci.*, **32**, 251-264.

Collins, G. E. (1975). Quantifier elimination for real closed fields by cylindrical algebraic decomposition. Proceedings of the Second GI Conference on Automata Theory and Formal Languages. *Springer Lec. Notes Comp. Sci.* **33**, 515–532.

Collins, G. E. (1980). SAC-2 and ALDES now available. *ACM SIGSAM Bull.* **14**, 19.

Collins, G. E., Loos, R. G. K. (1982). Real zeros of polynomials. In (Buchberger, B., Loos, R., Collins, G. E., eds.) *Computer Algebra - Symbolic and Algebraic Computation* (Computing Supplementum 4), pp. 83-94. Vienna and New York: Springer-Verlag.

Fuks, D. (1974). Review of Polotovskii, G.M., An algorithm for determining the topological type of a structurally stable plane curve of even degree. *Math. Reviews 58*, No. 28000.

Gianni, P., Traverso, C. (1983). *Shape Determination for Real Curves and Surfaces.* Tech. Rept. #23, Math. Dept., Univ. Pisa.

Gudkov, D. A. (1974). The topology of real projective algebraic varieties. *Russian Math. Surveys* **29**, 1-79.

Kaltofen, E. (1982). Polynomial factorization. In (Buchberger, B., Loos, R., Collins, G. E., eds.) *Computer Algebra - Symbolic and Algebraic Computation* (Computing Supplementum 4), pp. 95-113. Vienna and New York: Springer-Verlag.

Kendig, K. (1978) *Elementary Algebraic Geometry.* New York, NY: Springer-Verlag.

Loos, R. (1982). Generalized polynomial remainder sequences. In (Buchberger, B., Loos, R., Collins, G. E., eds.) *Computer Algebra - Symbolic and Algebraic Computation* (Computing Supplementum 4), pp. 115-137. Vienna and New York: Springer-Verlag.

Massey W. S. (1978). *Homology and Cohomology Theory.* New York, NY: Marcel Dekker.

Milnor, J. (1965). *Topology from the Differentiable Point of View.* Charlottesville, VA: University Press of Virginia.

Moise, E. E. (1977). *Geometric Topology in Dimensions Two and Three.* New York, NY: Springer-Verlag.

Polotovskii, G. M. (1973). Algorithm for determining the topological type of a rough plane algebraic curve of even degree. In: Qualititative methods of the theory of differential equations and their applications, *Gor'kov Gos. Univ. Ucen. Zap. Vyp.* **187**, pp. 143-187 (Russian).

Ragsdale, V. (1906). On the arrangement of the real branches of plane algebraic curves. *Amer. J. Math.* **28**, 377-404.

Rump, S. (1975). *Isolierung der reellen Nullstellen algebraischer Polynome.* Bericht der Arbeitsgruppe Computer Algebra No. 10, FB Informatik, University of Kaiserslautern (German) .

Rump, S. (1976). On the sign of a real algebraic number. *Proc. ACM SYMSAC '76.* Yorktown Heights, NY, (R.D. Jenks, ed.), 238-241.

Tarski, A. (1951). *A Decision Method for Elementary Algebra and Geometry.* Report R-109, second revised edn.. Santa Monica, CA: The Rand Corporation.

Walker, R. J. (1951). *Algebraic Curves.* Princeton, NJ: Princeton University Press.

Wilson, G. (1978). Hilbert's sixteenth problem. *Topology,* **17**, 53-73.

J. Symbolic Computation (1988) **5**, 237–259

On Mechanical Quantifier Elimination for Elementary Algebra and Geometry[*]

DENNIS S. ARNON   AND   MAURICE MIGNOTTE[†]

Xerox PARC, 3333 Coyote Hill Road, Palo Alto, California 94304, U.S.A.
† *Université Louis Pasteur, Centre de Calcul de l'Esplanade, 7 rue René Descartes, F-67084 Strasbourg Cédex, France*

(*Received 1 July 1986, and in revised form 15 November 1987*)

We give solutions to two problems of elementary algebra and geometry: (1) find conditions on real numbers p, q, and r so that the polynomial function $f(x) = x^4 + p\,x^2 + q\,x + r$ is nonnegative for all real x, and (2) find conditions on real numbers a, b, and c so that the ellipse $\frac{(x-c)^2}{a^2} + \frac{y^2}{b^2} - 1 = 0$ lies inside the unit circle $y^2 + x^2 - 1 = 0$. Our solutions are obtained by following the basic outline of the method of quantifier elimination by cylindrical algebraic decomposition (Collins, 1975), but we have developed, and have been considerably aided by, modified versions of certain of its steps. We have found three equally simple but not obviously equivalent solutions for the first problem, illustrating the difficulty of obtaining unique "simplest" solutions to quantifier elimination problems of elementary algebra and geometry.

1   Introduction

A. Tarski (1951) gave a quantifier elimination (q.e.) method for elementary algebra and geometry (EAG). G. Collins (1975) gave a new q.e. method for EAG, the first to be fully implemented on a computer (Arnon, 1981). In this paper we solve two problems of EAG by formulating them as q.e. problems and applying certain of the basic components of Collins' method, together with modified versions of certain others of its steps. The two sample problems are:

Quartic problem Find the conditions on real numbers p, q, and r so that the polynomial function $f(x) = x^4 + p\,x^2 + q\,x + r$ is positive semi-definite, i.e. nonnegative for all real x.

X-axis ellipse problem In the equation

$$\frac{(x - c)^2}{a^2} + \frac{y^2}{b^2} - 1 = 0,$$

let a and b be interpreted as the principal semi-axes of an ellipse, and $(c, 0)$ as its center; find the conditions on a, b, and c so that the ellipse is nondegenerate and

[*]This work was supported by the Xerox Corporation. This paper was typeset at Xerox PARC using TEX in the Cedar environment.

                © 1988 Academic Press Limited

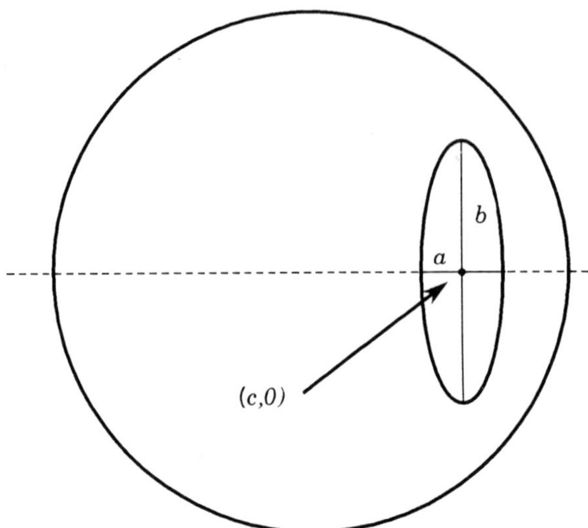

Figure 1: X-axis ellipse problem.

inside the unit circle $y^2 + x^2 - 1 = 0$. "Nondegenerate" means here $a \neq 0$ and $b \neq 0$, and "inside the circle" means either "strictly inside" (no contact), or "inside and touching" (the ellipse contacts the circle but has no points outside the circle), or "coincident" (the ellipse is identical with the circle). We call this problem the "x-axis" ellipse problem since the center of the ellipse is constrained to lie on the x-axis. Fig. 1 illustrates it.

We give three solutions to the quartic problem, and one to the x-axis ellipse problem:

Quartic problem: first solution

$$\delta \geq 0 \; AND \; [\, p \geq 0 \; OR \; L < 0 \; OR \; (L = 0 \; AND \; q = 0)\,],$$

where $\delta = 256r^3 - 128p^2r^2 + 144pq^2r + 16p^4r - 27q^4 - 4p^3q^2$ and $L = 8pr - 9q^2 - 2p^3$.

Quartic problem: second solution

$$[\, r = 0 \; AND \; q = 0 \; AND \; p \geq 0\,] \; OR$$

$$[\, r > 0 \; AND \; \{\, 9q^2 - 32pr < 0 \; OR \; (\, 32r^2 - 8p^2r + 3pq^2 \geq 0 \; AND \; \delta \geq 0\,)\,\}\,],$$

where δ is as in the first solution.

Quartic problem: third solution

$$[\, r = 0 \; AND \; q = 0 \; AND \; p \geq 0\,] \; OR$$

$$[\, r > 0 \; AND \; \{\, 3q^2 - 8pr < 0 \; OR \; (6q^2r - 16pr^2 - p^2q^2 + 4p^3r \geq 0 \; AND \; \delta \geq 0)\,\}\,],$$

where δ is as above.

X-axis ellipse problem solution

$$[\, c = 0 \ AND \ 0 < a \le 1 \ AND \ 0 < b \le 1 \,] \ OR$$

$$[\, 0 < a < 1 \ AND \ 0 < b < 1 \ AND \ c^2 < 1 \ AND \ (c + a - 1)(c - a + 1) \le 0 \ AND$$

$$[\, b^2 - a \le 0 \ OR \ b^2 c^2 + b^4 - a^2 b^2 - b^2 + a^2 \le 0 \,] \,].$$

An EAG quantifier elimination problem has the following general form: given a (well-formed) quantified formula $\Psi(x_1, \ldots, x_k)$, $k \ge 0$, of EAG, write down an equivalent formula $\Phi(x_1, \ldots, x_k)$ which is free of quantifiers. Φ is a *solution formula* for Ψ. We follow the convention that $\Psi(x_1, \ldots, x_k)$ denotes a formula in which all occurrences of $x_1, \ldots, x_k$ are free, each x_i may or may not occur in Ψ, and no variable besides $x_1, \ldots, x_k$ occurs free in Ψ. We call the set of all points in real $(x_1, \ldots, x_k)$-space that satisfy Ψ (or equivalently, Φ) the *solution set* of Ψ. Thus, solution sets of q.e. problems are certain subsets of "free-variable" space. If $k = 0$, i.e. Φ is a sentence, then its solution set is 0-dimensional space (a one-point space defined by the formula $0 = 0$) if Φ is true, or the empty set (defined by the formula $0 = 1$) if Φ is false. Arnon (1988b), Collins (1982), and Tarski (1951) provide more information on EAG and its q.e. and decision problems.

Our two sample problems can be formulated as q.e. problems as follows:

Quartic problem Eliminate the quantifier from the formula:

$$(\forall x) \ (x^4 + p \ x^2 + q \ x + r \ge 0). \qquad (I)$$

X-axis ellipse problem Eliminate the quantifiers from the following formula, which asserts that "for all points (x, y) in the plane, if (x, y) is on the ellipse, then (x, y) is on or inside the unit circle":

$$ab \ne 0 \ AND \ (\forall x)(\forall y) \ [b^2(x - c)^2 + a^2 y^2 - a^2 b^2 = 0 \ \Rightarrow \ y^2 + x^2 - 1 \le 0]. \qquad (II)$$

In Section 2 we review those parts of Collins' method that we use in this paper, and give an example of our overall q.e. methodology applied to a variant of the ellipse problem which has only one free variable. Our solution process for this sample problem exhibits the three main steps that comprise our q.e. methodology: decomposition of free-variable space, solution set determination, and solution formula construction. Sections 3-5 present these steps in greater detail and illustrate their application to the quartic problem (first solution) and the x-axis ellipse problem. Section 6 gives two additional solutions to the quartic problem, which when compared with the first solution, illustrate the difficulty of finding "simplest" solutions to EAG q.e. problems. Section 7 contains concluding remarks.

The quartic problem was posed by Delzell (1983, 1984) in connection with finding a solution of a certain kind to Hilbert's 17th problem for a form of degree four. The x-axis ellipse problem is a special case of a more general ellipse problem posed by Kahan (1975), in which the center of the ellipse may be located anywhere in the plane, rather than just on the x-axis. Our first solution to the quartic problem previously appeared as Arnon (1985). Lauer (1977) and Lazard (1988) have given mathematical derivations of solutions to the general ellipse problem, using a computer only for routine computations (such as resultants). For the x-axis ellipse problem, our solution is identical to one given by Lazard, while Lauer's solution evaluates to a formula which is longer than ours, and not obviously equivalent. Preliminary work on applying Collins' q.e. method to the x-axis ellipse problem was reported in Arnon & Smith (1983). Alternative derivations of solutions to the quartic and x-axis ellipse problems were given by Mignotte (1986).

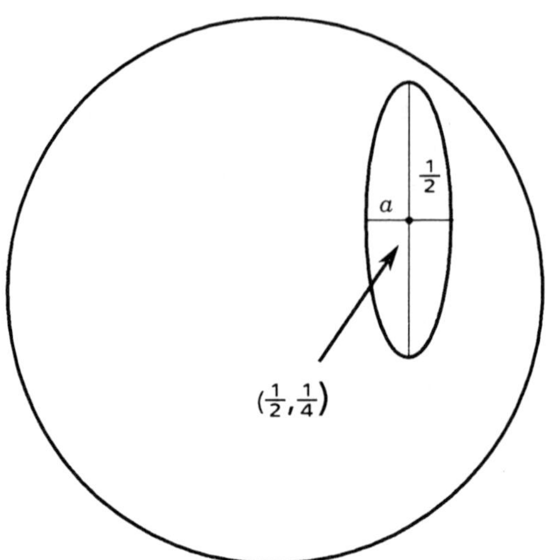

Figure 2: Off-axis ellipse problem.

2   Quantifier elimination by cylindrical algebraic decomposition

Consider the following variation of the x-axis ellipse problem of Section 1: the equation

$$\frac{(x - \frac{1}{2})^2}{a^2} + \frac{(y - \frac{1}{4})^2}{(\frac{1}{2})^2} - 1 = 0$$

defines an ellipse with center $(\frac{1}{2}, \frac{1}{4})$, height $\frac{1}{2}$, and width a. Let us find the conditions on a so that the ellipse is (nondegenerate and) inside the unit circle $y^2 + x^2 - 1 = 0$. "Inside the circle" in this case means either "strictly inside" or "inside and touching". Fig. 2 illustrates the problem, which we call the "off-axis ellipse problem", since the center of the ellipse is fixed at a point off the x-axis.

Let us now recall certain foundational results on which Collins' method depends. We assume the definitions of the terms region, cell, cell index, semi-algebraic set, positive polynomial, resultant, sector, section, stack, cylinder, decomposition, cylindrical decomposition, algebraic decomposition, cylindrical algebraic decomposition (c.a.d.), and induced c.a.d. given in Arnon *et. al.* (1984) or Arnon (1988b). We review some additional definitions from these references. Let X and Y be any sets, and let $T : X \to Y$ be a function. T is *invariant* on X, and X is *T-invariant*, if T is constant on X. For any $k \geq 0$, let E^k denote k-dimensional euclidean space, i.e. *Reals^k* with the usual topology. Let X be a subset of E^k, and let $\Psi(x_1, \dots, x_k)$ be a formula. We say that Ψ is invariant on X (and that X is Ψ-*invariant*), if the function $truthValue(\Psi)$ is invariant on X. Let $\phi(x_1, \dots, x_r)$ be a quantifier-free formula (q.f.f.). As we know, ϕ consists of a (finite) collection of atomic formulas, combined using the propositional connectives. Each atomic formula can be written in the form $[\, F(x_1, \dots, x_r) ~ 0 \,]$, where $F \in I_r$ (we write I_r to denote $\mathbf{Z}[x_1, \dots, x_r]$), and $~$ is one of $<, \leq, >, \geq, =, \neq$. When its atomic formulas are written in this form, we say that ϕ is a *standard* formula, and refer to the set of all F's which occur in at least one of its atomic formulas as its *(set of) polynomials*. For example, formulas (I) and (II) above are in standard form, with respective sets of polynomials $\{x^4 + p\,x^2 + q\,x + r\}$ and $\{ab,\ b^2(x-c)^2 + a^2y^2 - a^2b^2,\ y^2 + x^2 - 1\}$.

Let X be a subset of E^r, and let F be an element of I_r. We say that F is invariant on X (and that X is *sign-invariant (with respect to F)*, and *F-invariant*), if the function $signum(F)$ is invariant on X. Let $A = \{A_1, \ldots, A_n\}$, be a (finite) subset of I_r. We say that A is invariant on X (and that X is *sign-invariant (with respect to A)*, and *A-invariant*) if each A_i is invariant on X. A collection C of subsets of E^r is A-invariant if each element of the collection is. We say that C is sign-invariant if there exists some $A \subset I_r$ such that C is A-invariant. The reader will note that our definition of "F-invariant" (and "A-invariant") is an abuse of language: its most natural meaning would be that the *value* of the function $F(\alpha_1, ..., \alpha_r)$ is invariant as $\alpha = (\alpha_1, ..., \alpha_r)$ ranges over X, but instead, we define it to mean that the *sign* of $F(\alpha_1, ..., \alpha_r)$ is invariant on X. This is intentional: it is the sign of $F(\alpha_1, ..., \alpha_r)$, and not its value, which is crucial for Collins' method.

The following Lemma is rather obvious but should be carefully considered.

LEMMA **2.1** *Let $\phi(x_1, ..., x_r)$, $r \geq 0$, be a quantifier-free formula in standard form, and let $A \subset I_r$ be the polynomials of ϕ. An A-invariant decomposition of E^r is ϕ-invariant.*

A subset X of E^n is *definable* if for some formula $\phi(x_1, ..., x_n)$ of L, X is the set of points in E^n that satisfy ϕ. We say that X is *defined* by ϕ, and that ϕ is a *defining formula* for X. Given formula $\phi(x_1, ..., x_n)$, we write $Set(\phi)$ to denote the set defined by ϕ. Thus given quantifier-free ϕ, Lemma 2.1 tells us that if we construct an A-invariant c.a.d. D of E^r, then each cell of D is ϕ-invariant, and if we have a sample point for each cell, then we can determine the cells comprising $Set(\phi)$ by evaluating ϕ at these sample points. The following extension of Lemma 2.1 is basic for Collins' method, and points the way to the first step of the q.e. methodology we employ in this paper. It is implicit in Collins (1975), and is proved in Arnon (1988b). A *prenex* formula of L is a formula $\Psi(x_1, \ldots, x_k)$ of the form:

$$(Q_{k+1} x_{k+1})(Q_{k+2} x_{k+2})...(Q_r x_r)\ M(x_1, ..., x_r), \quad 0 \leq k \leq r,$$

where each Q_i is an existential or universal quantifier, and $M(x_1, ..., x_r)$ is a quantifier-free formula, the *matrix* of Ψ. Formulas (I) and (II) above are prenex; the matrix of (I) is $x^4 + p\,x^2 + q\,x + r \geq 0$.

THEOREM **2.2** *Let $\Phi(x_1, ..., x_k)$, $k \geq 0$, be a standard prenex formula whose matrix M involves $r \geq k$ variables, If D is an M-invariant c.a.d. of E^r, then the c.a.d. of E^k that is induced by D is Φ-invariant.*

Theorem 2.2 tells us that if $\Phi(x_1, ..., x_k)$ is a standard prenex formula with matrix M, if D is an M-invariant c.a.d. of E^r, and if D' is the c.a.d. of E^k that is induced by D, then $Set(\Phi)$ is the union of certain of the cells of D'. Thus, given Φ, a first step towards eliminating its quantifiers is to construct a c.a.d. of free-variable space E^k that is the c.a.d. of E^k that is induced by some M-invariant c.a.d. of E^r. Where A is the set of polynomials of M, by Lemma 2.1 it suffices to construct a c.a.d. of free-variable space that is induced by some A-invariant c.a.d. of E^r.

$$a$$

$$y^2 + x^2 - 1$$
$$16a^2y^2 - 8a^2y + 4x^2 - 4x - 3a^2 + 1$$

Figure 3: Input polynomials A for off-axis ellipse problem.

Arnon *et. al.* (1984) and Arnon (1988b) discuss the c.a.d. construction algorithm. As McCallum (1988) details, a key component of it is the projection operation, which maps a finite set of polynomials in n variables to a finite set of polynomials in $n-1$ variables. In this paper we use a projection operator $Proj$, defined as follows. Given $A \subset I_r$, $r \geq 2$, let $prim(A)$ denote the set of primitive parts of those elements of A which have positive degree, and let B be a finest squarefree basis for $prim(A)$. Let $cont(A)$ denote the set of non-zero non-unit contents of the elements of A. For any set S of elements of I_n, let $PPDIF(S)$ denote the set of all distinct positive irreducible factors of elements of S, subject to the further constraint that each element of $PPDIF(S)$ have positive degree in some variable. Then we define $Proj(A)$ to be $PPDIF(cont(A) \cup PROJ(B))$, where the map $PROJ$ is as defined in Arnon *et. al.* (1984). Hence given $A \subset I_r$, $r \geq 2$, each element of $Proj(A)$ is a positive polynomial, has positive degree in at least one of the $r-1$ variables of I_{r-1}, and is irreducible. From the results of Arnon *et. al.* (1984), it follows that any $Proj(A)$-invariant c.a.d. of E^{r-1} is induced by some A-invariant c.a.d. of E^r.

We write $Proj^i$ to denote the i-fold composition of the $Proj$ map, i.e. $Proj^i(A) = Proj(Proj^{i-1}(A))$. Thus, if $\Phi(x_1, ..., x_k)$ is a standard prenex formula with matrix M, and if A is the set of polynomials of M, then a $Proj^{r-k}(A)$-invariant c.a.d. of E^k is induced by some M-invariant c.a.d. of E^r, and so is Φ-invariant. Hence the first step of our q.e. methodology is to compute $Proj^{r-k}(A)$ and construct a $Proj^{r-k}(A)$-invariant c.a.d. of E^k. Let us see the details of these computations for the off-axis ellipse problem. Clearing denominators, we have an input formula of:

$$a \neq 0 \ AND \ (\forall x)(\forall y) \ [16a^2y^2 - 8a^2y + 4x^2 - 4x - 3a^2 + 1 = 0 \ \Rightarrow \ y^2 + x^2 - 1 \leq 0].$$

which gives us the set A of input polynomials shown in Fig. 3. We get a first projection set $P = Proj(A)$ shown in Fig. 4, and a second projection set $P^2 = Proj(P)$ shown in Fig. 5. We find that a P^2-invariant c.a.d. of 1-space has 51 cells, and takes 54 seconds to construct using an implementation of Collins' algorithm in the SAC-2 computer algebra system (Collins, 1980) on a Vax 11/785. (All execution times given in this paper are CPU, not elapsed, time). We know that the solution set is the union of certain of these cells. This completes the first step of our q.e. methodology.

Let us now continue on to solve the off-axis ellipse problem, and in so doing introduce the second and third steps of our methodology. Assume that we do not allow negative a, i.e. assume that ellipses have positive width. Then only those cells in 1-space that lie in the interval $0 < a \leq \frac{1}{2}$ may possibly be in the solution set. It turns out that the ten cells with indices (27), (28), ... (36) of the c.a.d. of 1-space, each of which represents successively larger values of a, comprise the interval $(0, \frac{1}{2}]$. We know that each of these cells either consists entirely of points in the solution set, or has no points in the solution set. Thus, for any given one of these cells c, by choosing a sample $a \in c$, substituting

$$a$$

$$x - 1$$

$$x + 1$$

$$2x - 2a - 1$$

$$2x + 2a - 1$$

$$4x^2 - 4x - 3a^2 + 1$$

$$16x^4 - 32x^3 + 64a^4 x^2 - 24a^2 x^2 + 24x^2 + 24a^2 x - 8x - 55a^4 - 6a^2 + 1$$

$$256a^4 x^4 - 128a^2 x^4 + 16x^4 + 128a^2 x^3 - 32x^3 - 352a^4 x^2 + 72a^2 x^2 + 24x^2 - 104a^2 x - 8x + 105a^4 + 26a^2 + 1$$

Figure 4: First projection set P for off-axis ellipse problem.

$$a$$

$$2a + 1$$

$$2a - 1$$

$$2a + 3$$

$$2a - 3$$

$$a^2 - 3$$

$$3a^2 - 1$$

$$5a^2 + 1$$

$$3a^2 + 6a + 2$$

$$3a^2 - 6a + 2$$

$$16a^2 + 16a - 11$$

$$16a^2 - 16a - 11$$

$$64a^2 + 64a - 47$$

$$64a^2 - 64a - 47$$

$$880a^8 - 1224a^6 + 3231a^4 - 1242a^2 + 351$$

$$26880a^8 - 70624a^6 + 30537a^4 - 5022a^2 + 297$$

Figure 5: Second projection set P^2 for off-axis ellipse problem.

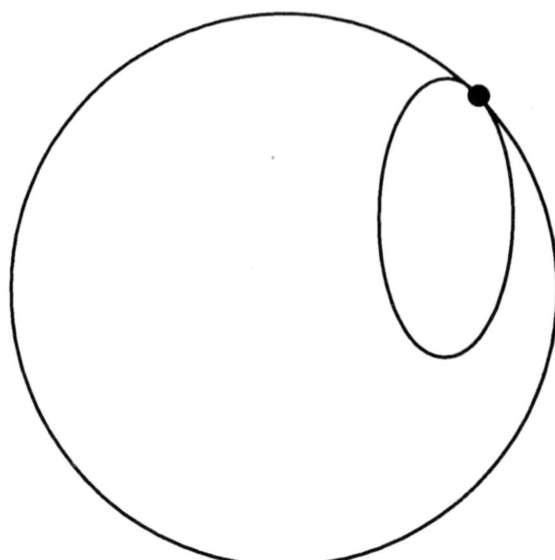

Figure 6: Boundary case of off-axis ellipse problem.

this a into the input formula above, and deciding the resulting sentence, we determine whether c is contained in or is disjoint from the solution set. A sample point for cell (27) is $a = \frac{107}{256}$; substituting this into the off-axis problem input formula we obtain:

$$(\forall x)(\forall y) \; [\frac{183184}{65536} y^2 - \frac{91592}{65536} y + 4x^2 - 4x + \frac{31189}{65536} = 0 \;\Rightarrow\; y^2 + x^2 - 1 \leq 0].$$

Applying the original Collins q.e. method, implemented as above in SAC-2 on a Vax 11/785, we are told in 14 seconds that the sentence is true. A sample point for cell (29) is $a = \frac{27}{64}$; substituting this into the input formula yields:

$$(\forall x)(\forall y) \; [\frac{11664}{4096} y^2 - \frac{5832}{4096} y + 4x^2 - 4x + \frac{1909}{4096} = 0 \;\Rightarrow\; y^2 + x^2 - 1 \leq 0].$$

The Collins algorithm tells us in 28 seconds that this sentence is false. Cell (28) must be in the solution set, since clearly the solution set has a "boundary point", as illustrated by Fig. 6. This reasoning is substantiated by the fact that all ellipses whose widths a belong to cell (27) in fact are strictly inside the unit circle, which we establish by applying the Collins algorithm to the variant of the sentence we had above for cell (27) that results from replacing "$y^2 + x^2 - 1 \leq 0$" by "$y^2 + x^2 - 1 < 0$". This variant sentence is found to be true in 11 seconds. We establish that cell (28) is in the solution set by this "indirect reasoning", rather than by the substitution and direct check that we did for cells (27) and (29), because its sample point is an algebraic number of degree 8 (see below), and so the direct check is likely to be quite time-consuming.

We have now completed the second step of our q.e. methodology: the determination of which cells of the c.a.d. of free-variable space comprise the solution set, i.e. cells (27) and (28) for the off-axis ellipse problem. The third step is to write down a solution formula. Cell (28) is the unique root α of the polynomial

$$f(a) = 26880a^8 - 70624a^6 + 30537a^4 - 5022a^2 + 297$$

$$p$$

$$q$$

$$9q^2 + 2p^3$$

$$27q^2 + 8p^3$$

$$27q^2 - p^3$$

Figure 7: Second projection set P^2 for quartic problem.

in the interval $(\frac{107}{214}, \frac{215}{512})$, and cell (27) is the open interval $(0, \alpha)$. Examining the real roots of $f(a)$ we find that α is its smallest positive real root, hence since $f(0)$ is positive, a defining formula for the union of cells (27) and (28), i.e. a solution formula for the off-axis ellipse problem, is:

$$0 < a < \frac{215}{512} \ AND \ f(a) \geq 0.$$

This completes the third step of our q.e. methodology, namely solution formula construction.

For this sample problem, Collins' original q.e. method would call for us to extend the c.a.d. of 1-space to a c.a.d. of 2-space, extend the c.a.d. of 2-space to a c.a.d. of 3-space, and then evaluate the matrix M of the input formula at sample points for the cells of the 3-space c.a.d. We attempted to carry out these c.a.d. extensions, and found that the induced c.a.d. of 2-space took 47 minutes of CPU time to construct and had 1081 cells, and that construction of the c.a.d. of 3-space was still incomplete after two hours of CPU time (these computations were done using algorithm CLCAD of Arnon, 1988a, implemented in SAC-2 on a Vax 11/785). Thus the solution process used above is advantageous for the off-axis ellipse problem.

3   Decomposition of free-variable space

Let us now apply our three-step q.e. methodology to the quartic and x-axis ellipse problems. In this section, we shall consider the first step: decomposition of free-variable space. For the quartic problem, the set A of input polynomials contains one element:

$$A = \{x^4 + p\,x^2 + q\,x + r\}.$$

$x^4 + p\,x^2 + q\,x + r$ is irreducible, hence we get a projection set $P = Proj(A) = \{\delta, L, p\}$, where δ and L are as in Section 1. Our c.a.d. algorithm (algorithm CLCAD of Arnon, 1988a) took 10 minutes to construct a P-invariant c.a.d. of E^3, which has 123 cells. $P^2 = Proj(P)$ is shown in Fig. 7. $P^3 = Proj(P^2) = \{p\}$. The P^3-invariant induced c.a.d. of 1-space obviously has three cells, and took 18 seconds to construct. The P^2-invariant induced c.a.d. of 2-space has 21 cells and took 24 seconds to construct.

For the ellipse problem, we will find a way of reducing the size of projection sets. Ours is a different projection set reduction than that given by McCallum (1988). Ideally, both his reduced projection set and the technique we give here should be used. The set

$$a$$

$$b$$

$$x - 1$$

$$x + 1$$

$$x - c + a$$

$$x - c - a$$

$$b^2 x^2 - a^2 x^2 - 2b^2 cx + b^2 c^2 - a^2 b^2 + a^2$$

Figure 8: First projection set P for x-axis ellipse problem.

of polynomials of formula (II) of Section 1, i.e. our set A of input polynomials, is:

$$A = \{ \ ab, \ b^2 x^2 - 2b^2 cx + b^2 c^2 + a^2 y^2 - a^2 b^2, \ \ y^2 + x^2 - 1 \ \},$$

$P = Proj(A)$ is as shown in Fig. 8. Now let us make some basic observations. Clearly the semi-axes of the ellipse must each be less than or equal to the radius of the circle, and the center of the ellipse must be (strictly) inside the circle. Whether to allow negative a and b seems to us to be a matter of choice; we choose not to. Hence we have the restrictions

$$(R1) \quad 0 < a \leq 1,$$

$$(R2) \quad 0 < b \leq 1,$$

$$(R3) \quad c^2 < 1.$$

Let us next observe that whenever $a = 1$, in any solution we must necessarily have $c = 0$ and $b \leq 1$. A similar argument applies whenever $b = 1$. Finally, whenever $c = 0$, then any choice of a and b in the range $(0, 1]$ gives us a solution. Hence we know the solutions of the problem whenever $a = 1$, $b = 1$, or $c = 0$, and need only be concerned about the portion of the solution set for which $0 < a < 1$ and $0 < b < 1$ and $0 < c^2 < 1$.

Returning now to our solution process, recall that the $PROJ$ operator defined in Section 3 of [ACM84a] performs certain computations once for each reductum of its argument polynomials (see McCallum, 1988, for the definition of reductum), to cover the possiblity that one or more of the leading coefficients of these argument polynomials vanish. We see that in set P above, only element $Q(x) = b^2 x^2 - a^2 x^2 - 2b^2 cx + b^2 c^2 - a^2 b^2 + a^2$ of $prim(P)$ has a potentially vanishing leading coefficient (i.e. $b^2 - a^2$). Note that since we have defined the elements of our projection sets to be irreducible polynomials, $prim(P)$ is a finest squarefree basis for itself, hence $Proj(P) = PPDIF(cont(P) \cup PROJ(prim(P)))$. Following ideas in Section 5 of Collins (1975), we can ignore the reducta of all elements of $prim(P)$ other than Q in constructing $PROJ(prim(P))$. The reducta of Q are

$$red^0(Q) = Q$$

$$red^1(Q) = -2b^2 cx + b^2 c^2 - a^2 b^2 + a^2$$

$$red^2(Q) = b^2 c^2 - a^2 b^2 + a^2$$

$$a$$

$$b$$

$$b + a$$

$$b - a$$

$$c + a + 1$$

$$c + a - 1$$

$$c - a + 1$$

$$c - a - 1$$

$$b^2 c^2 + b^4 - a^2 b^2 - b^2 + a^2$$

Figure 9: First partial second projection set P_0^2 for x-axis ellipse problem.

Let us consider separately the subsets of $PROJ(prim(P))$ that derive from each of these reducta, and for each, see if we can determine that certain of its elements can be withheld from $Proj(P)$. Again, this sort of pruning of projection sets is suggested in Section 5 of Collins (1975).

First, assuming that the leading coefficient of $Q(x)$ doesn't vanish, i.e. $b^2 \neq a^2$, we get the partial second projection set P_0^2 shown in Fig. 9. Suppose now that $b = a$, i.e. the leading coefficient of $Q(x)$ does vanish. Using $red(Q)$ in place of Q, we get another partial second projection set P_1^2 shown in Fig. 10. Let us now apply the fact that $b = a$ to the elements of P_1^2 (this amounts to simplification in the presence of the relation $b = a$). When $b = a$, $bc + ab + a = b(c + a + 1)$, and similarly for the other polynomials $bc \pm ab \pm a$. When $b = a$, $b^2 c^2 - 2b^2 c - a^2 b^2 + a^2 = b^2(c^2 - 2c - a^2 + 1) = b^2(c - a + 1)(c + a + 1)$, and a similar factorization occurs for $b^2 c^2 + 2b^2 c - a^2 b^2 + a^2$. Hence we see that the elements of P_1^2 which are relevant for us are the subset $\{b, c + a + 1, c + a - 1, c - a + 1, c - a + 1\}$ of P_0^2, plus the polynomial c, and so P_0^2 augmented by c is still satisfactory as our tentative second projection set.

If $b = a$ and $c = 0$, the first two leading coefficients of $Q(x)$ vanish, however we have already dealt with the case $c = 0$. Hence $P_0^2 \cup \{c\}$ suffices as a tentative projection set. To arrive at our final projection set, we add the polynomials $a - 1$, $b - 1$, and $c^2 - 1 = (c - 1)(c + 1)$ from (R1) - (R3). We can omit $b + a$ since (R1) - (R3) imply $b + a > 0$. Hence we take our second projection set P^2 to be as shown in Fig. 11.

The time required to construct a P^2-invariant c.a.d. of E^3 was 75 minutes; it has 2291 cells. $P^3 = Proj(P^2)$ is shown in Fig. 12. $P^4 = Proj(P^3)$ is shown in Fig. 13. The P^4-invariant induced c.a.d. of 1-space has 11 cells, and took 47 seconds to construct. The P^3-invariant induced c.a.d. of 2-space has 157 cells and took two minutes to construct.

4   Solution set determination

Let us determine the cells comprising the solution set of the quartic problem. Given a sample point (π, ζ, ρ) for a cell in a c.a.d. of E^3, our task is to determine whether the polynomial $g(x) = x^4 + \pi\, x^2 + \zeta\, x + \rho$ is positive semi-definite. We do this by isolating the real roots of g and checking the sign of g at sample rational number elements of each

$$b$$
$$c$$
$$c + a + 1$$
$$c + a - 1$$
$$c - a + 1$$
$$c - a - 1$$
$$bc + ab + a$$
$$bc + ab - a$$
$$bc - ab + a$$
$$bc - ab - a$$
$$b^2 c^2 - 2b^2 c - a^2 b^2 + a^2$$
$$b^2 c^2 + 2b^2 c - a^2 b^2 + a^2$$

Figure 10: Second partial second projection set P_1^2 for x-axis ellipse problem.

$$a$$
$$a - 1$$
$$b$$
$$b - 1$$
$$b - a$$
$$c$$
$$c - 1$$
$$c + 1$$
$$c + a + 1$$
$$c + a - 1$$
$$c - a + 1$$
$$c - a - 1$$
$$b^2 c^2 + b^4 - a^2 b^2 - b^2 + a^2$$

Figure 11: Second projection set P^2 for x-axis ellipse problem.

$$a$$

$$a + 1$$

$$a - 1$$

$$a + 2$$

$$a - 2$$

$$b$$

$$b + 1$$

$$b - 1$$

$$b + a$$

$$b - a$$

$$b^2 + a$$

$$b^2 - a$$

$$b^4 - a^2 b^2 + a^2$$

Figure 12: Third projection set P^3 for x-axis ellipse problem.

$$a$$

$$a + 1$$

$$a - 1$$

$$a + 2$$

$$a - 2$$

Figure 13: Fourth projection set P^4 for x-axis ellipse problem.

of the five or fewer open intervals into which g's roots divide the real line. Determining the solution set of the quartic problem by this method took our software four minutes.

For the ellipse problem, to determine which cells of the c.a.d. D of E^3 belong to the solution set, we first reduce the number of quantifiers to be eliminated from two to one. Let $E(a,b,c,x,y) = b^2(x-c)^2 + a^2y^2 - a^2b^2 = 0$ be the equation of the ellipse, and $C(a,b,c,x,y) = y^2 + x^2 - 1 = 0$ be the equation of the unit circle. Let $S(a,b,c,y) = Resultant_x(E,C)$, and $R(a,b,c,x) = Resultant_y(E,C)$. It is a special case of Theorem 1 of Lauer (1977), that for any particular triple (a,b,c) satisfying (R1) - (R3), the corresponding ellipse is inside the circle if and only if either $S(y)$ has no real roots, or $R(x)$ has no real roots, or $S(y)$ and $R(x)$ have only multiple real roots. Let $\varphi(a,b,c)$ denote the formula:

$$S(a,b,c,x) \ has \ no \ real \ roots \ or \ R(a,b,c,x) \ has \ no \ real \ roots$$

$$or \ S(a,b,c,x) \ and \ R(a,b,c,x) \ both \ have \ only \ multiple \ real \ roots.$$

Let Φ denote formula (II) of Section 1. By Lauer's Theorem 1, a cell of D that satisfies (R1) - (R3) is Φ-invariant if and only if it is φ-invariant, and the truth values of Φ and φ are the same on any such cell. Hence, to determine whether a cell in our P^2-invariant c.a.d. of E^3 that satisfies (R1) - (R3) is in the solution set we need only determine the truth value of $\varphi(a,b,c)$ on it. Given a sample point (α,β,γ) for such a cell in E^3, we do this by isolating the real roots of $S(\alpha,\beta,\gamma,x)$ and $R(\alpha,\beta,\gamma,x)$. Given that φ is concerned not only with the number of real roots of S and R, but also their multiplicities, we use a multiplicity-reporting real root isolation algorithm, such as the Collins-Loos algorithm [CLO82].

5   Solution formula construction

Our strategy for solution formula construction for our two sample problems is to display the solution set in a manner which enables us to read off a (hopefully simple) defining formula for it. Naturally this approach is most attractive for problems which have solution sets in E^1, E^2, and E^3. In Section 2, for example, we essentially "displayed" the solution set for the off-axis ellipse problem as a subset of E^1 and then wrote down a defining formula for it.

Let us now describe our display method in detail. Note that the quartic and x-axis ellipse problems each have three free variables. We compile a sequence of two-dimensional "slices" of the c.a.d. of 3-space by planes of the form $p = p_0$ for the quartic problem (of the form $a = a_0$ for the x-axis ellipse problem), for a sequence of increasing p_0's (a_0's) chosen from cells of the induced c.a.d. of 1-space. Each such slice yields a c.a.d. of 2-space, in which we mark the cells that belong to the solution set. We then examine all slices and attempt to write down a solution formula.

Our method of choosing the p_0's is to pick one sample p_0 from each cell d of the induced c.a.d. of 1-space for which there may possibly be cells in the c.a.d. of 3-space that lie "over" d and are contained in the solution set. Thus our p_0's are sample points for either 0-cells or 1-cells of the induced c.a.d. of 1-space. Consider any 1-cell d in the induced c.a.d. of 1-space, and let $S(d)$ denote the stack over this cell that is part of the induced c.a.d. of 2-space. $S(d)$ consists of 1-dimensional sections and 2-dimensional sectors. For each adjacency of a 1-section c^1 and a 2-sector c^2 of $S(d)$, we know (Arnon et. al., 1988) that where $S(c^1)$ and $S(c^2)$ denote the stacks over these cells that are part of the c.a.d. of 3-space, and $S^*(c^1)$ and $S^*(c^2)$ denote the extensions of these stacks, that $S^*(c^2)$ has the unique section boundary property in $S^*(c^1)$, i.e. for any section s of $S^*(c^2)$, the set of boundary points of s that lie "over" c^1 is a section of $S^*(c^1)$. (See Arnon *et. al.*, 1988, for the precise definition of the unique section boundary property). Hence, for any values of p_0 from d that we choose, the c.a.d.s of the plane that we get by slicing the c.a.d. of 3-space by the planes $p = p_0$ are all "topologically equivalent", in the sense that they consist of the same numbers of cells, arranged into stacks in the same way, and having the same adjacencies. Hence, a single slice for any such d suffices to show us the portion of the solution set that lies "over" that d. When d is a 0-cell, then $d = p_0$ for some real algebraic p_0, and we slice by the plane $p = p_0$.

If for any reason we are not able to produce a solution formula by examining the slices that we construct, then we can always fall back to Collins original algorithm for constructing solution formulas, namely to construct defining formulas for the individual cells comprising the solution set, and then form the disjunction of these formulas. This has not been necessary for the sample problems we consider in this paper.

For the quartic problem, we saw that the induced c.a.d. of E^1 consists of the three cells defined by $p < 0$, $p = 0$, and $p > 0$. Hence if α is a positive real number, then it suffices to "slice" the c.a.d. of E^3 by the planes $p + \alpha = 0$, $p = 0$, and $p - \alpha = 0$, to get an accurate picture of the solution set. Figs. 14 and 15 show the slice $p + \alpha = 0$, with the viewer standing at some large negative value of p and looking toward the origin. Fig. 14 shows the structure of the c.a.d., i.e. the adjacencies of its cells. We stress that our figures are not intended to show the actual shapes of curves and surfaces (which don't matter for q.e.), but only their topological structure. In Fig. 14, dots correspond to (cross-sections of) 1-cells in 3-space, arcs to (cross-sections of) 2-cells, and patches of white space to (cross-sections of) 3-cells. (Cross-sections of) 2-cells belonging to the zero set of δ are shown as dotted lines, and 2-cells belonging to the zero set of L are shown as solid lines. For clarity, we do not show all cells of the c.a.d. in these diagrams, but only those relevant to the determination of the solution set. Note that there are five "columns" of dots in Fig. 14, and recall that the polynomials that determine the induced cad of 2-space are shown in Fig. 7. The dots in the leftmost column in Fig. 14 correspond to 1-cells in 3-space that lie "over" a 1-cell in the induced cad of E^2 that is contained in the zero set of $9q^2 + 2p^3$. The dots in the next column to the right correspond to 1-cells in 3-space that lie "over" a 1-cell in the induced cad of E^2 that is contained in the zero set of $27q^2 + 8p^3$. The dots in the center column lie over a 1-cell in E^2 on which q vanishes, and so on. Fig. 15 indicates the cells in this slice that belong to the solution set by shading or crosshatching them. Similarly Figs. 16 and 17 show the slice $p = 0$, and Figs. 18 and 19 the slice $p - \alpha = 0$. In Fig. 16, dots correspond to 0-cells in 3-space, arcs to 1-cells, and patches of white space to 2-cells. In Figs. 16-19, the viewer is standing at some large positive value of p and looking toward the origin.

From Figs. 14-19, it is not hard to see that $\delta \geq 0$ on every cell in the solution set of the quartic problem. Also, when $p \geq 0$, "$\delta \geq 0$" defines precisely the cells in the solution set. When $p < 0$, all cells defined by "$\delta \geq 0$ *AND* $L < 0$" are in the solution set, but this formula fails to define one additional 1-cell in the solution set, which can be defined by "$p < 0$ *AND* $\delta \geq 0$ *AND* $L = 0$ *AND* $q = 0$". Altogether we obtain a solution formula of:

$$\delta \geq 0 \ AND \ [\, p \geq 0 \ OR \ L < 0 \ OR \ (L = 0 \ AND \ q = 0)\,].$$

Let us now display the portion of the x-axis ellipse problem solution set satisfying

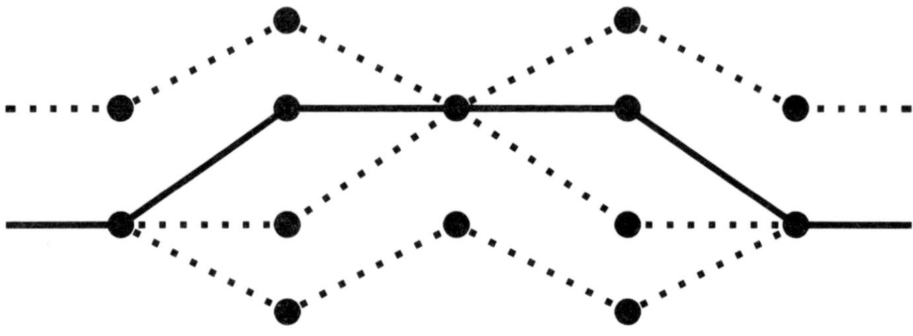

Figure 14: Quartic problem, slice for $p < 0$.

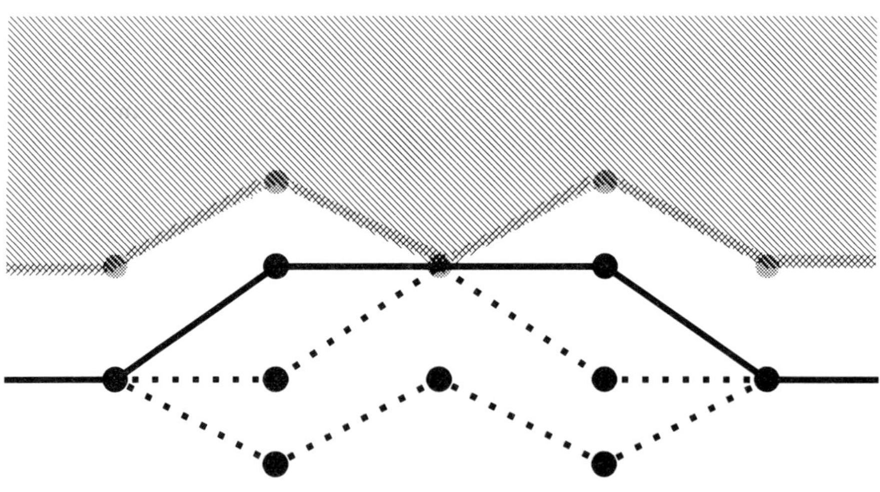

Figure 15: Quartic problem, slice for $p < 0$, solution set marked

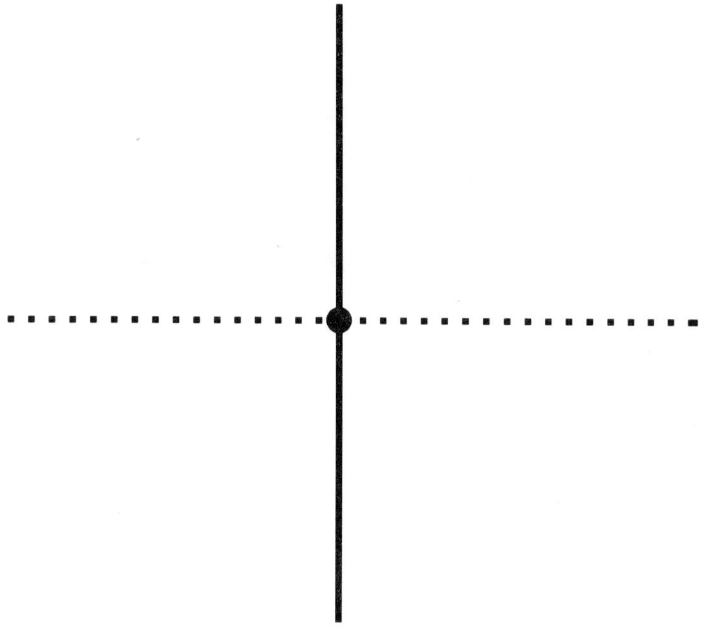

Figure 16: Quartic problem, slice for $p = 0$.

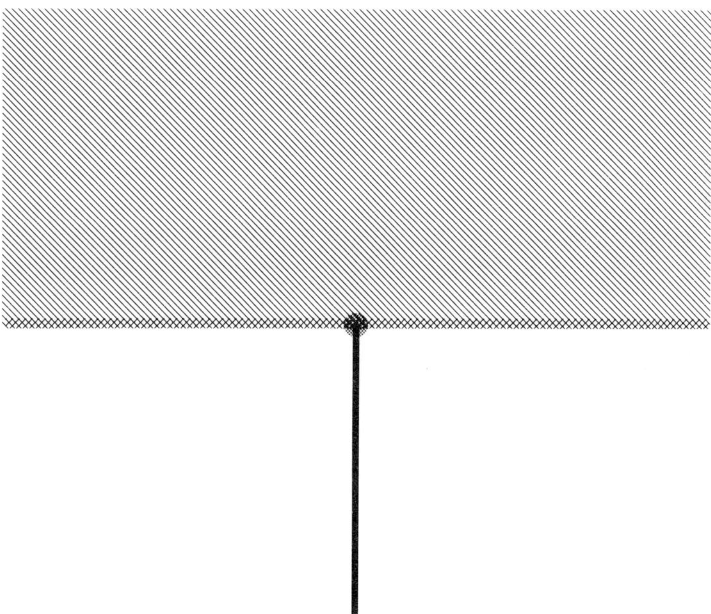

Figure 17: Quartic problem, slice for $p = 0$, solution set marked

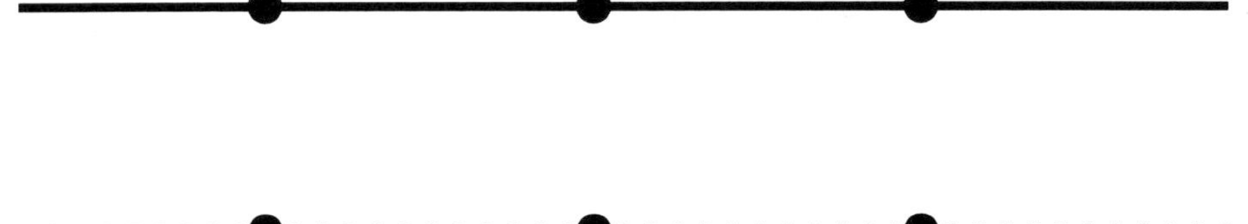

Figure 18: Quartic problem, slice for $p > 0$.

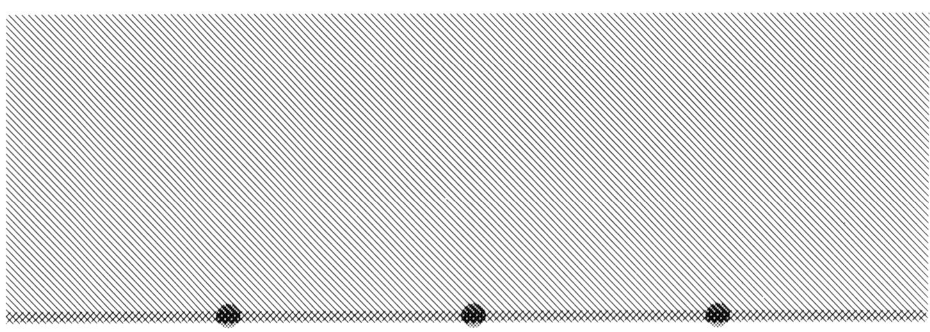

Figure 19: Quartic problem, slice for $p > 0$, solution set marked

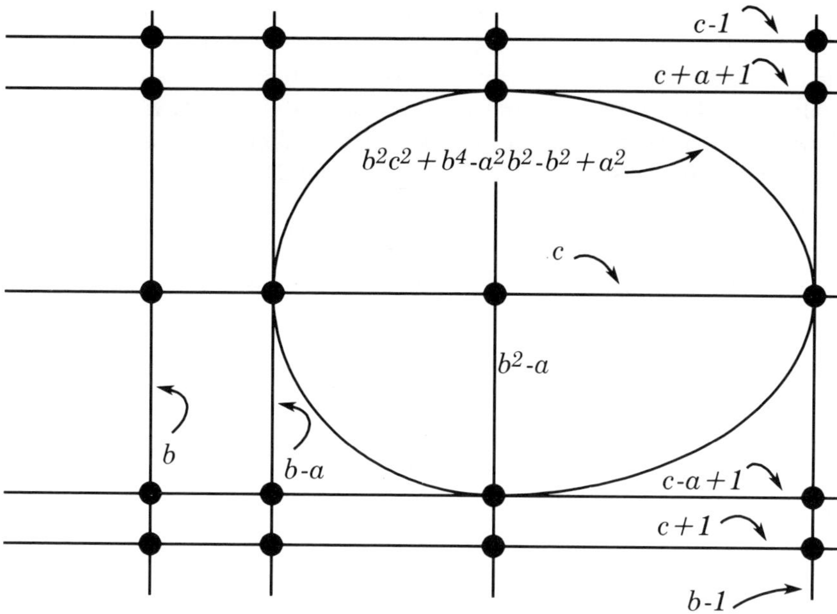

Figure 20: X-axis ellipse problem, slice for $0 < a < 1$.

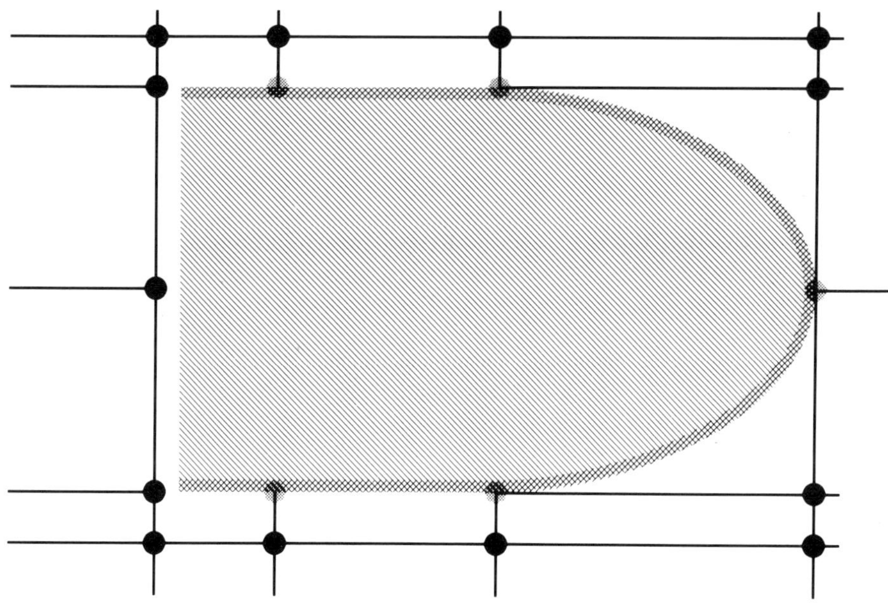

Figure 21: X-axis ellipse problem, slice for $0 < a < 1$, solution set marked

$0 < a < 1$ and $0 < b < 1$ and $c^2 < 1$. Clearly (cf. Fig. 13) the open interval $0 < a < 1$ is a 1-cell in the induced c.a.d. of E^1. Hence a single slice of three-dimensional space by a plane $a = \alpha$, $0 < \alpha < 1$, suffices to show the structure of those cells in E^3 which may possibly be in the solution set, and which we do not already have a description of. Fig. 20 shows such a slice, and Fig. 21 indicates with shading and cross-hatching which of these cells belong to the solution set. In Figs. 20 and 21 the viewer is standing at a large positive value of a and looking toward the origin.

We see that when $b^2 \leq a$, $(c + a - 1)(c - a + 1) \leq 0$ defines the cells in the solution set, and when $b^2 > a$, $b^2 c^2 + b^4 - a^2 b^2 - b^2 + a^2 \leq 0$ defines the cells in the solution set. Also, when $b^2 > a$, it is still the case that $(c + a - 1)(c - a + 1) \leq 0$ on all cells in the solution set. Hence our overall solution to the ellipse problem is:

$$[\, c = 0 \; AND \; 0 < a \leq 1 \; AND \; 0 < b \leq 1 \,] \; OR$$

$$[\, 0 < a < 1 \; AND \; 0 < b < 1 \; AND \; c^2 < 1 \; AND \; (c + a - 1)(c - a + 1) \leq 0 \; AND$$

$$[\, b^2 - a \leq 0 \; OR \; b^2 c^2 + b^4 - a^2 b^2 - b^2 + a^2 \leq 0 \,]\,].$$

6   Additional solutions for the quartic problem

6.1   Quartic problem, second solution

Let $f(x) = x^4 + p\, x^2 + q\, x + r$, and observe that since $f(x)$ tends to $+\infty$ with $|x|$, formula (I) of Section 1 is true if and only if $min(f(x)) \geq 0$. Furthermore, $min(f(x)) \geq 0$ if and only if $(\forall x)[\, f'(x) = 0 \; \Rightarrow \; f(x) \geq 0 \,]$.

To take advantage of this observation, we consider separately the cases $r = 0$ and $r \neq 0$. Suppose $r = 0$. Then "$r = 0 \; AND \; (I)$" is equivalent to the conjunction of:

$$r = 0 \; AND \; (\forall x)[\, x \leq 0 \; \Rightarrow \; x^3 + p\, x + q \leq 0 \,]$$

and

$$r = 0 \; AND \; (\forall x)[\, x \geq 0 \; \Rightarrow \; x^3 + p\, x + q \geq 0 \,].$$

It is not hard to see that the conjunction of these two conditions is equivalent to

$$r = q = 0 \; AND \; p \geq 0.$$

Suppose now that $r \neq 0$. From the requirement that $f(0) \geq 0$, we see that in any solution, $r \neq 0$ implies $r > 0$, hence we assume $r > 0$. Set $\tilde{r} = 1/r$, $\tilde{p} = p/r$, and $\tilde{q} = q/r$. Then $f(x)$ is never negative if and only if

$$Q(y) \equiv (y^4/r) \cdot f(1/y) = y^4 + \tilde{q}\, y^3 + \tilde{p}\, y^2 + \tilde{r} \qquad (2)$$

is never negative. Hence "$r > 0 \; AND \; (I)$" is equivalent to

$$\tilde{r} > 0 \; AND \; (\forall y)[\, Q'(y) = 0 \; \Rightarrow \; Q(y) \geq 0 \,]. \qquad (3)$$

We have

$$Q'(y) = 4y^3 + 3\tilde{q}\, y^2 + 2\tilde{p}\, y = y \cdot (4y^2 + 3\tilde{q}\, y + 2\tilde{p}) = y \cdot \hat{Q}(y),$$

$$\tilde{q}$$

$$9\tilde{q}^2 - 32\tilde{p}$$

$$\tilde{r}$$

$$B/2$$

$$R = 256\tilde{r}^2 - 27\tilde{q}^4\tilde{r} + 144\tilde{p}\tilde{q}^2\tilde{r} - 128\tilde{p}^2\tilde{r} - 4\tilde{p}^3\tilde{q}^2 + 16\tilde{p}^4$$

Figure 22: Projection set P for second solution of quartic problem.

where $\hat{Q}(y) \equiv 4y^2 + 3\tilde{q}\,y + 2\tilde{p}$. The root $y = 0$ of $Q'(y)$ implies that in any solution formula for (3), $\tilde{r} > 0$ (which we already know). By Euclidean division, we find that

$$64 \cdot Q(y) \equiv A\,y + B \quad [\, mod\ \hat{Q}(y) \,],$$

where

$$A = 9\tilde{q}^3 - 32\tilde{p}\tilde{q} = \tilde{q} \cdot (9\tilde{q}^2 - 32\tilde{p}),$$

and

$$B = 64\tilde{r} + 6\tilde{p}\tilde{q}^2 - 16\tilde{p}^2.$$

Hence (3) is equivalent to

$$\tilde{r} > 0 \ AND \ (\forall y)[\ \hat{Q}(y) = 0 \ \Rightarrow \ A\,y + B \geq 0 \,]. \qquad (4)$$

The set of polynomials of formula (4) is $A = \{\tilde{r}, \hat{Q}(y), A\,y + B\}$, and $P = Proj(A)$ is shown in Fig. 22. Construction of a P-invariant c.a.d. of E^3 took 3 minutes; there were 127 cells. After determination and display of the solution set, we obtain the following quantifier-free formula equivalent to (4):

$$\tilde{r} > 0 \ AND \ \{\ 9\tilde{q}^2 - 32\tilde{p} < 0 \ OR \ (\ \frac{B}{2} \geq 0 \ AND \ R \geq 0 \ ) \ \}.$$

Substituting back $r = 1/\tilde{r}$, $p = r \cdot \tilde{p}$, $q = r \cdot \tilde{q}$, we obtain the following quantifier-free formula equivalent to (I):

$$[r = 0 \ AND \ q = 0 \ AND \ p \geq 0] \ OR$$

$$[r > 0 \ AND \ \{\ 9q^2 - 32pr < 0 \ OR \ (\ 32r^2 - 8p^2r + 3pq^2 \geq 0 \ AND \ \delta \geq 0 \ ) \ \} \,],$$

where δ is as in Section 1. This is our second solution of the quartic problem.

6.2   Quartic problem, third solution

Another approach is to solve directly the formula

$$\tilde{r} > 0 \ AND \ (\forall y) \ [Q(y) \geq 0].$$

From this formula we obtain the set $A = \{\tilde{r}, \ y^4 + \tilde{q}\,y^3 + \tilde{p}\,y^2 + \tilde{r}\}$ of input polynomials. $P = Proj(A)$ is shown in Fig. 23; the polynomial R is as defined in Section 6.1.

$$3\tilde{q}^2 - 8\tilde{p}$$

$$\tilde{r}$$

$$6\tilde{q}^2\tilde{r} - 16\tilde{p}\tilde{r} - \tilde{p}^2\tilde{q}^2 + 4\tilde{p}^3$$

$$R$$

Figure 23: Projection set P for third solution of quartic problem.

Construction of a P-invariant c.a.d. of E^3 took 11 minutes; there were 123 cells. After determination and display of the solution set, we obtain the following solution formula:

$$\tilde{r} > 0 \; AND \; \{ \; 3\tilde{q}^2 - 8\tilde{p} < 0 \quad OR \; (\; 6\tilde{q}^2\tilde{r} - 16\tilde{p}\tilde{r} - \tilde{p}^2\tilde{q}^2 + 4\tilde{p}^3 \geq 0 \; AND \; R \geq 0 \;) \; \}.$$

Substituting back $r = 1/\tilde{r}$, $p = r \cdot \tilde{p}$, $q = r \cdot \tilde{q}$, we get the following formula equivalent to (I):

$$[r = 0 \; AND \; q = 0 \; AND \; p \geq 0] \; OR$$

$$[r > 0 \; AND \; \{ \; 3q^2 - 8pr < 0 \; OR \; (6q^2r - 16pr^2 - p^2q^2 + 4p^3r \geq 0 \; AND \; \delta \geq 0) \; \} \;],$$

where δ is as above. This is our third solution.

7   Concluding remarks

In fact, we were able to use the original Collins' algorithm to solve the quartic problem. The c.a.d. of E^4 took about 18 hours to construct and had 489 cells. The time then required to determine the solution set was about 19 minutes. The solution formula produced was 80 lines long: it is the disjunction of the defining formulas for the 56 cells in 3-space that comprise the solution set. This sort of experience has been one motivation for the development of the q.e. methodology presented in this paper.

We gave three solution formulas for the quartic problem. It is difficult to say that one of them is "simplest". And perhaps there exist yet other solution formulas which are even "simpler". In general, the solution set of a q.e. problem is a certain set of points in free-variable space, and our goal is to write down a "simplest" defining formula for it. A rigorous understanding of the problem of constructing such a "simplest" defining formula for a semi-algebraic set, given some (quantifier-free) defining formula for it, would enhance both the theory and practice of solving elementary algebra and geometry problems by quantifier elimination.

8   Acknowledgement

The work of the first author has benefited in numerous ways from the unique environment and stimulating colleagues of the Computer Science Laboratory at Xerox PARC.

9 References

Arnon, D. S. (1981). *Algorithms for the Geometry of Semi-Algebraic Sets*. PhD thesis, Tech. Rept. #436, Comp. Sci. Dept., Univ. Wisconsin–Madison.

Arnon, D. S., Smith, S. (1983). Towards a mechanical solution of the Kahan ellipse problem I. Proceedings of the European Computer Algebra Conference. *Springer Lec. Notes Comp. Sci.* **162**, 36-44.

Arnon, D. S., Collins, G. E., McCallum, S. (1984). Cylindrical algebraic decomposition I: the basic algorithm, *SIAM J. Comp.* **13/4**, 865–877.

Arnon, D. S. (1985). On mechanical quuantifier elimination for elementary algebra and geometry: automatic solution of a nontrivial problem. Proceedings of EUROCAL '85. *Springer Lec. Notes Comp. Sci.* **204**, 270-271.

Arnon, D. S. (1988a). A cluster-based cylindrical algebraic decomposition algorithm. *J. Symb. Comp.* **5**, (this issue).

Arnon, D. S. (1988b). Geometric reasoning with logic and algebra. *Artificial Intelligence J.* (special issue on Geometric Reasoning and Artificial Intelligence; to appear).

Arnon, D. S., Collins, G. E., McCallum, S. (1988). An adjacency algorithm for cylindrical algebraic decompositions of three-dimensional space. *J. Symb. Comp.* **5**, (this issue).

Collins, G. E. (1975). Quantifier elimination for real closed fields by cylindrical algebraic decomposition. Proceedings of the Second GI Conference on Automata Theory and Formal Languages. *Springer Lec. Notes Comp. Sci.* **33**, 515–532.

Collins, G. E. (1980). SAC-2 and ALDES now available. *ACM SIGSAM Bull.* **14**, 19.

Collins, G. E. (1982). Quantifier elimination for real closed fields: a guide to the literature. In (Buchberger, B., Loos, R., Collins, G. E., eds.) *Computer Algebra - Symbolic and Algebraic Computation* (Computing Supplementum 4), pp. 79-81. Vienna and New York: Springer-Verlag.

Collins, G. E., Loos, R. G. K. (1982). Real zeros of polynomials. In (Buchberger, B., Loos, R., Collins, G. E., eds.) *Computer Algebra - Symbolic and Algebraic Computation* (Computing Supplementum 4), pp. 83-94. Vienna and New York: Springer-Verlag.

Delzell, C. (1983). Private communication.

Delzell, C. (1984). A continuous, constructive solution to Hilbert's 17th problem. *Invent. math.* **76**,, 365-384.

Kahan, W. (1975). Problem No. 9: An ellipse problem *ACM SIGSAM Bull.* **9/35**, 11.

Lauer, M. (1977). A solution to Kahan's problem (SIGSAM problem no. 9). *ACM SIGSAM Bull.* **11**, 16-28.

Lazard, D. (1988). Quantifier elimination: optimal solution for two classical examples. *J. Symb. Comp.* **5**, (this issue).

McCallum, S. (1988). An improved projection operation for cylindrical algebraic decomposition of three-dimensional space. *J. Symb. Comp.* **5**, (this issue).

Mignotte, M. (1986). Computer versus paper and pencil. *Proc. CALSYF*, **4**, 63-69.

Tarski, A. (1951). *A Decision Method for Elementary Algebra and Geometry*. Report R-109, second revised edn.. Santa Monica, CA: The Rand Corporation.

J. Symbolic Computation (1988) **5**, 261–266

Quantifier Elimination:
Optimal Solution for Two Classical Examples

DANIEL LAZARD†

L.I.T.P. Université Paris VI, Tour 55–65,
4 place Jussieu, 75252 Paris Cedex 05, France

(*Received 5 March* 1986)

Optimal solutions are given for the two following problems: the condition for a degree 4 polynomial to have only positive values and the condition for an ellipse to be inside the unit circle.

The aim of this paper is to give optimal solutions for two classical quantifier elimination problems. These solutions are not obtained by existing general algorithms. However, the design of efficient algorithms must be oriented towards the production of simple solutions, and not just any odd solution. I do not know of any general algorithm whose solution has similar complexity; it would be interesting to design such an algorithm.

By complexity of a solution, we mean the size and the degree of the polynomials appearing in it, as well as the length of the logical formula. Rather surprisingly, all these complexities may be simultaneously optimised in both examples.

1. Positive Polynomials

Let $P(x) := ax^4 + bx^3 + cx^2 + dx + e$; the problem is to solve

$$(\forall\, x)P(x) \geqslant 0. \tag{1}$$

Suppose that $a > 0$. The change of variable $x \to x - b/4a$ gives the polynomial

$$Q(x) := x^4 + px^2 + qx + r,$$

and our problem is equivalent to

$$(\forall\, x)Q(x) \geqslant 0. \tag{2}$$

The first remark is the following:

LEMMA 1. *Let $Q(x)$ be a monic polynomial of degree d; the sign of its discriminant is $s = (-1)^{(d^2-r)/2}$, where r is the number of its real roots. Thus $r \equiv d^2 + s + 3 \bmod 4$ alinea: This is the proof.* See Weiss (1963, p. 166).

The second remark is that if $Q(x)$ has four real roots, they cannot be all of the same sign (their sum is 0). Thus, if $r > 0$, the product of the roots is positive and the curves

† L.I.T.P. and Greco de Calcul Formel (CNRS).

0747–7171/88/010261+06 $03.00/0

$y = x^4 + px^2 + r$ and $y = -qx$ have two intersection points with $x > 0$ and two with $x < 0$. The symmetry of $y = x^4 + px^2 + r$ shows easily that if $r > 0$ and if this curve has 0 or four common points with $y = -qx$ for some q, the same is true for all q' with $|q'| \leqslant |q|$. Property (2) implies that *discriminant* $(Q) \geqslant 0$, and putting $q' = 0$ in the last remark we get the equivalence of (2) with

$$\text{discriminant } (Q) \geqslant 0 \quad \text{and} \quad (\forall\, x)x^4 + px^2 + r \geqslant 0$$

(this implies $r \geqslant 0$; the limit cases $r = 0$ or *discriminant* $(Q) = 0$ have to be handled separately). Thus we get the following:

PROPOSITION 1. *The condition*

$$(\forall\, x)x^4 + px^2 + qx + r \geqslant 0$$

is equivalent to

$$\text{discriminant } (x^4 + px^2 + qx + r) \geqslant 0 \quad \text{and} \quad (p \geqslant 0 \text{ or } r \geqslant p^2/4).$$

THEOREM 1. *The condition*

$$(\forall\, x)ax^4 + bx^3 + cx^2 + dx + e \geqslant 0$$

is equivalent to

$$(a > 0 \text{ and } D \geqslant 0 \text{ and } (8ac - 3b^2 \geqslant 0 \text{ or } R \geqslant 0))$$

$$\text{or } (a = 0 \text{ and } b = 0 \text{ and } c > 0 \text{ and } d^2 - 4ce \leqslant 0)$$

$$\text{or } (a = 0 \text{ and } b = 0 \text{ and } c = 0 \text{ and } d = 0 \text{ and } e \geqslant 0)$$

where

$$R = 64a^3e - 16a^2bd - 16a^2c^2 + 16ab^2c - 3b^4$$

and

$$\begin{aligned}
D = {} & 256a^3e^3 - 192a^2bde^2 - 128a^2c^2e^2 + 144ab^2ce^2 - 27b^4e^2 \\
& + 144a^2cd^2e - 6ab^2d^2e - 80abc^2de + 18b^3cde + 16ac^4e \\
& - 4b^2c^3e - 27a^2d^4 + 18abcd^3 - 4b^3d^3 - 4ac^3d^2 + b^2c^2d^2.
\end{aligned}$$

It suffices to substitute in the conditions of Proposition 1 the values of p, q, r obtained by the change of variables of the beginning; the cases with $a = 0$ are easy.

REMARK 1. If we substitute (e, d, c, b, a) for (a, b, c, d, e), we obtain clearly an equivalent solution.

PROPOSITION 2. *The condition* $(\forall\, x)P(x) > 0$ *is equivalent with*

$$(e > 0 \text{ and } D > 0 \text{ and } (8ce - 3d^2 \geqslant 0 \text{ or } R' \geqslant 0))$$

$$\text{or } (e > 0 \text{ and } D = 0 \text{ and } 8ce - 3d^2 > 0 \text{ and } R' = 0)$$

where

$$R' = 64e^3a - 16e^2db - 16e^2c^2 + 16ed^2c - 3d^4.$$

We have used Remark 1 to avoid the cases $a = 0$ of Theorem 1; if $(\forall\, x)P(x) \geqslant 0$ and $(\exists\, x)P(x) = 0$, the last x has to be a double root and $D = 0$. However, if $P(x)$ has two imaginary double roots we have $D = 0$, but $(\forall\, x)P(x) > 0$. The second condition of the proposition takes this case into account; it corresponds to the condition $p < 0$ and $p^2 - 4r = 0$, the nullity of the discriminant implying $q = 0$.

REMARK 2. The Sturm theorem between $-\infty$ and $+\infty$ gives similar conditions with R replaced by

$$1/8 \det \begin{bmatrix} b & 0 & 0 & 4a \\ 2c & b & 4a & 3b \\ 3d & 2c & 3b & 2c \\ 4e & 3d & 2c & d \end{bmatrix} = \begin{aligned} &16a^2ce - 6ab^2e - 18a^2d^2 \\ &+ 14abcd - 3b^3d - 4ac^3 + b^2c^2 \end{aligned}$$

which is slightly more complicated than R. Moreover, when D and this determinant are null we have to add a condition which separates the case of two double roots from a triple root. This solution was given by Arnon (1985). Thus the polynomial and the logical formula appearing in this solution are both more complicated than in our.

REMARK 3. In the five-dimensional space of the parameters (a, b, c, d, e), the set of solutions of (1) is of dimension 5 and its boundary is of dimension 4; this boundary contains an open subset of the hypersurface $D = 0$. Thus the polynomial D or some multiple of it must appear in every solution. The fact that the polynomials R and $8ac - 3b^2$ are as simple as possible is clear in Proposition 1, where they are p and $p^2 - 4r$. Thus, the solution of Theorem 1 is the simplest with respect to the complexity of the polynomials which appear in it.

2. The Ellipse Problem

This problem, also called the Kahan problem, consists in writing down conditions such that the ellipse $E(x, y) = 0$ with

$$E(x, y) := \frac{(x-c)^2}{a^2} + \frac{(y-d)^2}{b^2} - 1$$

be inside the circle $C(x, y) = 0$ with

$$C(x, y) := x^2 + y^2 - 1.$$

This problem can be stated as

$$(\forall x)(\forall y)\, E(x, y) = 0 \Rightarrow C(x, y) \leqslant 0. \tag{3}$$

A first approach consists in a parametrisation of the ellipse

$$x = c + \frac{2at}{1+t^2}, \qquad y = d + b\frac{1-t^2}{1+t^2},$$

which gives the equivalent condition

$$(\forall t)(1+t^2)^2 C\left(c + \frac{2at}{1+t^2},\, d + b\frac{1-t^2}{1+t^2}\right) \leqslant 0.$$

The numerator of the left-hand side of this inequality is a polynomial in t of degree 4. Thus the formula of Theorem 2 gives a solution of the problem. Unfortunately, this solution is not the simplest. However, applying Lemma 1, we get the polynomial T in a, b, c, d such that $T = 0$ is equivalent to "the ellipse and the circle are tangent" and $T > 0$ is equivalent to "the ellipse and the circle have 0 or four simple common points". As above for D, an open subset of the hypersurface $T = 0$ is a piece of the boundary of the set of solutions and the polynomial T must appear in every solution. This irreducible polynomial is

$$
\begin{aligned}
T:= {}& a^4 d^8 + ((2a^2 b^2 + 2a^4)c^2 + (-4a^4 + 2a^2)b^2 + 2a^6 - 4a^4)d^6 \\
&+ ((b^4 + 4a^2 b^2 + a^4)c^4 + ((-6a^2 - 2)b^4 + (2a^4 + 2a^2)b^2 - 2a^6 - 6a^4)c^2 \\
&+ (6a^4 - 6a^2 + 1)b^4 + (-6a^6 + 10a^4 - 6a^2)b^2 + a^8 - 6a^6 + 6a^4)d^4 \\
&+ ((2b^4 + 2a^2 b^2)c^6 + (-2b^6 + (2a^2 - 6)b^4 + (-6a^4 + 2a^2)b^2 - 2a^4)c^4 \\
&+ ((6a^2 + 4)b^6 + (-10a^4 - 6a^2 + 6)b^4 + (6a^6 - 6a^4 - 10a^2)b^2 + 4a^6 \\
&+ 6a^4)c^2 + (-4a^4 + 6a^2 - 2)b^6 + (6a^6 - 8a^4 + 4a^2 - 2)b^4 \\
&+ (-2a^8 + 4a^6 - 8a^4 + 6a^2)b^2 - 2a^8 + 6a^6 - 4a^4)d^2 + b^4 c^8 \\
&+ (2b^6 + (-4a^2 - 4)b^4 + 2a^2 b^2)c^6 + (b^8 + (-6a^2 - 6)b^6 \\
&+ (6a^4 + 10a^2 + 6)b^4 + (-6a^4 - 6a^2)b^2 + a^4)c^4 + ((-2a^2 - 2)b^8 \\
&+ (6a^4 + 4a^2 + 6)b^6 + (-4a^6 - 8a^4 - 8a^2 - 4)b^4 + (6a^6 + 4a^4 + 6a^2)b^2 \\
&- 2a^6 - 2a^4)c^2 + (a^4 - 2a^2 + 1)b^8 + (-2a^6 + 2a^4 + 2a^2 - 2)b^6 \\
&+ (a^8 + 2a^6 - 6a^4 + 2a^2 + 1)b^4 + (-2a^8 + 2a^6 + 2a^4 - 2a^2)b^2 + a^8 - 2a^6 + a^4.
\end{aligned}
$$

In view of a simple solution of our problem, we need the following.

LEMMA 2. *Suppose $a > b$.*

(a) *If the ellipse is inside the circle, the same is true if c is replaced by zero.*

(b) *If, when c is replaced by zero, the new ellipse is inside the circle, then the ellipse and the circle have at most two real intersection points.*

Suppose by symmetry that $c \geqslant 0$.

(a) If c decreases, the half of the ellipse such that $x \geqslant c$ remains clearly inside the circle. When $c = 0$, the other half is also inside the circle, by symmetry.

(b) When c increases from zero, the number of intersection points changes only when the curves are tangent. We can compute the corresponding points: Let $x = \sin u$, $y = \cos u$ be a parametrisation of the circle, $x = a \sin t$, $y = d + b \cos t$ a parametrisation of the ellipse. The points which become tangent after a horizontal translation satisfy

$$
\cos u = d + b \cos t \quad \text{and} \quad \frac{\cos u}{\sin u} = \frac{a \cos t}{b \sin t}
$$

we get by eliminating u:

$$
b^2(a^2 - b^2)(\cos t)^4 + 2db(a^2 - b^2)(\cos t)^3
$$
$$
+ (d^2(a^2 - b^2) + b^4 - a^2)(\cos t)^2 + 2db^3 \cos t + d^2 b^2.
$$

This polynomial in $\cos t$ has exactly one root in each interval $]-\infty, -1[$, $]-1, 0[$, $]0, 1[$ and $]1, +\infty[$. This gives four values for t and two positive and two negative values of c for which the curves are tangent. The first positive value corresponds to the moment where the number of intersection points passes from 0 to two. The second is the value of c when the ellipse becomes outside of the circle. There is no possibility for passing from two to four intersection points.

COROLLARY. *If $a > b$, the ellipse is inside the circle, iff $T \geqslant 0$, the ellipse is inside the circle when c is replaced by 0 and the ellipse has a point strictly inside the circle.*

It remains to solve the case where $c = 0$. For that we suppose $d \geqslant 0$. Leaving a and b fixed, let d be varying. If the ellipse is inside the circle, the same is true when d decreases. When d increases, we find a value d_0 such that the ellipse is not inside the circle for $d > d_0$, but is in it for $d < d_0$. If the minimum curvature of the ellipse is less than 1, d_0 corresponds to the value of d for which the curves are tangent at $x = 0$, $y = 1$. In the other case, the value of d_0 corresponds to the case where the curves are bitangent. Thus we get

PROPOSITION 3. *If $a > b$ and $c = 0$, the ellipse is inside the circle if and only if*

$$b^2 + d^2 \leqslant 1 \quad \text{and}$$

$$(a^2 \leqslant b \text{ or } d^2 a^2 \leqslant (1 - a^2)(a^2 - b^2)).$$

THEOREM 2. *In the general case, the ellipse is inside the circle iff*

$$(a > b \quad \text{and} \quad T \geqslant 0 \quad \text{and} \quad c^2 + (b + |d|)^2 - 1 \leqslant 0$$

$$\text{and} \quad (a^2 \leqslant b \quad \text{or} \quad a^2 d^2 \leqslant (1 - a^2)(a^2 - b^2))$$

$$\text{or} \quad (a = b \quad \text{and} \quad c^2 + d^2 \leqslant (1 - a)^2 \quad \text{and} \quad a \leqslant 1)$$

$$\text{or} \quad (a < b \quad \text{and} \quad T \geqslant 0 \quad \text{and} \quad d^2 + (a + |c|)^2 - 1 \leqslant 0$$

$$\text{and} \quad (b^2 \leqslant a \quad \text{or} \quad b^2 c^2 \leqslant (1 - b^2)(b^2 - a^2)).$$

This follows immediately from all the preceding results.

REMARK 4. In all our proofs we have used some convexity argument. It seems that it is necessary for obtaining simple results in quantifier elimination. How can we put such convexity argument in a general procedure?

REMARK 5. A formulation which is logically simpler but where more polynomials have to be evaluated in each test is

$$T \geqslant 0 \quad \text{and} \quad c^2 + (b + |d|)^2 \leqslant 1 \quad \text{and} \quad d^2 + (a + |c|)^2 \leqslant 1$$

$$\text{and} \quad ((b^2 \leqslant a \quad \text{and} \quad a^2 \leqslant b)$$

$$\text{or} \quad (a > b \quad \text{and} \quad a^2 d^2 \leqslant (1 - a^2)(a^2 - b2))$$

$$\text{or} \quad (b < a \quad \text{and} \quad b^2 c^2 \leqslant (1 - b^2)(b^2 - a^2))).$$

REMARK 6. Clearly the above calculations are not purely hand made. The polynomials T, D and R were computed with MACSYMA which was also used for various experimentations.

REMARK 7. Among the polynomials which appear in Lauer's solution are

$$d^2(U - c^2(a^2 + b^2)) \quad \text{with} \quad U = a^2 d^2 - (1 - a^2)(a^2 - b^2)$$

$$-V + d^2(a^2 + b^2) \quad \text{with} \quad V = b^2 c^2 - (1 - b^2)(b^2 - a^2)$$

$$d^2 T,$$

which have to be compared with those appearing in our solution.

References

Arnon, D. S. (1985). On mechanical quantifier elimination for elementary algebra and geometry: solution of a non-trivial problem. EUROCAL 85. *Springer Lec. Notes Comp. Sci.* **204**, 270–271.

Arnon, D. S., Smith, S. F. (1983). Toward mechanical solution of the Kahan ellipse problem I. EUROCAL 83. *Springer Lec. Notes Comp. Sci.* **162**, 36–44.

Kahan, W. (1975). Problem #9: An ellipse problem. *SIGSAM Bull. ACM* **9**, 11.

Lauer, M. (1977). A solution to Kahan's problem (SIGSAM problem no. 9). *SIGSAM Bull. ACM* **11**, 16–20.

Mignotte, M. (1985). *Computer versus paper and pencil.* (CALSYF no. 4.)

Weiss, E. (1963). *Algebraic Number Theory.* New York: McGraw-Hill.

J. Symbolic Computation (1988) **5**, 267–274

A Bibliography of Quantifier Elimination for Real Closed Fields*

DENNIS S. ARNON

Xerox PARC, 3333 Coyote Hill Road, Palo Alto, California 94304, U.S.A.

(Received 15 November 1987)

A basic collection of literature relating to algorithmic quantifier elimination for real closed fields is assembled.

1 Introduction

This bibliography seeks to provide a basic collection of literature on algorithmic quantifier elimination for real closed fields. We do not try to be exhaustive, but rather to include a balanced selection of writings that provide entry points to theory, algorithms, and applications. Works that are exclusively devoted to supporting algorithms, e.g. polynomial factorization, have been excluded. A comprehensive source of information on such subsidiary algorithms is the volume of Buchberger *et. al.* (1982). It goes without saying that this bibliography reflects the interests, knowledge, and ignorance of the preparer. Nonetheless it is hoped that no essential papers have been omitted.

Each entry in the bibliography is followed by one or more sets of descriptors, plus possibly some notes. There is no rigid format for these annotations, however several major groups of descriptors are used.

"Presentation" descriptors refer to the nature of the presentation. These include: *Survey, Tutorial, Exposition, ResearchResult, ComplexityResult, Implementation* (i.e. reports on computer programs), and *Examples* (i.e. discusses actual examples in some significant way).

"Theory" descriptors attempt to classify the subject area of the theoretical contents of the paper. These include: *RealAlgebraicGeometry, RealClosedFields* (papers with this descriptor often discuss decision procedures), *LogicalTheories*, and *Algebra*.

"MainAlgorithm" descriptors are applied to papers that actually contain a quantifier elimination algorithm. These include: *TarskiAlgorithm, SeidenbergAlgorithm, CohenAlgorithm, MonkAlgorithm, CollinsAlgorithm, WuethrichAlgorithm, GrigorievAlgorithm,* and *Ben-Or/Kozen/ReifAlgorithm*. The arbitrary and simplistic nature of these categories is regrettable, but it is hoped that they are useful nonetheless.

*This work was supported by the Xerox Corporation. This paper was typeset at Xerox PARC using TeX in the Cedar environment.

0747–1717/88/010267+08 $03.00/0

"SubAlgorithm" descriptors are applied to papers that contain an important subalgorithm of a quantifier elimination algorithm, or an important related algorithm. These include: *CellAdjacencyAlgorithm*, *CellDecompositionAlgorithm*, and *ClusteringAlgorithm*.

"Application" descriptors are applied to papers that discuss some important application area. These include: *GeneralApplications*, *MotionPlanning*, *GeometryTheoremProving*, *CurveDisplay*, *HilbertSixteenthProblem*, *GeometricSearching*, and *RootClassification*.

I am indebted to B. Buchberger, J. Davenport, J. Heintz, and D. Kozen for valuable comments on preliminary versions of this bibliography.

2   References

A'Campo, N. (1979). Sur la première partie du seizième problème de Hilbert. Séminaire Bourbaki. *Springer Lec. Notes Math.* **770**, 207-227. (HilbertSixteenthProblem)

•Arnborg, S., Feng, H.-C. (1988). Algebraic decomposition of regular curves. *J. Symb. Comp.* **5**, (this issue). (ComplexityResult, ResearchResult)

Arnol'd, V. I., Oleinik, O. A. (1979). Topology of real algebraic manifolds. *Vestnik Moskovskogo Univ. Matematika* (Moscow University Mathematics Bulletin), **34**, 7-17. (RealAlgebraicGeometry)

Arnon, D. S. (1979). A cellular decomposition algorithm for semi-algebraic sets. Proceedings of an International Symposium on Symbolic and Algebraic Manipulation (EUROSAM '79). *Springer Lec. Notes Comp. Sci.* **72**, 301-315. (CollinsAlgorithm, Exposition; CellAdjacencyAlgorithm, CollinsAlgorithm, ResearchResult)

Arnon, D. S. (1981). *Algorithms for the Geometry of Semi-Algebraic Sets* (PhD thesis). Tech. Rept. #436, Comp. Sci. Dept., Univ. Wisconsin–Madison. (CollinsAlgorithm, Exposition; CellAdjacencyAlgorithm, ClusteringAlgorithm, CollinsAlgorithm, ResearchResult, Implementation, Examples)

Arnon, D. S., McCallum, S. (1982). Cylindrical algebraic decomposition by quantifier elimination. Proceedings of the European Computer Algebra Conference (EUROCAM '82). *Springer Lec. Notes Comp. Sci.* **144**, 215-222. (CollinsAlgorithm, ResearchResult, Examples)

Arnon, D. S. (1983). Topologically reliable display of algebraic curves. *Proc. ACM SIGGRAPH '83*. Detroit, MI (ed. by Peter Tanner), 219-227. (CurveDisplay, Implementation, CollinsAlgorithm)

Arnon, D. S., Collins, G. E., McCallum, S. (1984a). Cylindrical algebraic decomposition I: the basic algorithm, *SIAM J. Comp.* **13/4**, 865–877. (CollinsAlgorithm, Exposition, Implementation, Examples)

Arnon, D. S., Collins, G. E., McCallum, S. (1984b). Cylindrical algebraic decomposition II: an adjacency algorithm for the plane, *SIAM J. Comp.* **13/4**, 878–889. (CellAdjacencyAlgorithm, CollinsAlgorithm, ResearchResult, Implementation, Examples)

•Arnon, D. S. (1988a). A cluster-based cylindrical algebraic decomposition algorithm. *J. Symb. Comp.* **5**, (this issue). (CellAdjacencyAlgorithm, ClusteringAlgorithm, CollinsAlgorithm, ResearchResult)

Arnon, D. S. (1988b). Geometric reasoning with logic and algebra. *Artificial Intelligence J.* (special issue on Geometric Reasoning and Artificial Intelligence; to appear). (Exposition, GeometryTheoremProving, RootClassification, CollinsAlgorithm, Implementation, Examples, ResearchResult)

•Arnon, D. S., McCallum, S. (1988). A polynomial-time algorithm for the topological type of a real algebraic curve. *J. Symb. Comp.* **5**, (this issue). (CellAdjacencyAlgorithm, CollinsAlgorithm, HilbertSixteenthProblem, ResearchResult, Implementation, Examples)

Arnon, D. S., Mignotte, M. (1988). On mechanical quantifier elimination for elementary algebra and geometry. *J. Symb. Comp.* **5,** (this issue). (RealClosedFields, Examples, CollinsAlgorithm, Implementation, ResearchResult)

Arnon, D. S., Collins, G. E., McCallum, S. (1988). An adjacency algorithm for cylindrical algebraic decompositions of three-dimensional space. *J. Symb. Comp.* **5,** (this issue). (CellAdjacencyAlgorithm, CollinsAlgorithm, ResearchResult, Examples)

Bajaj, C. (1987). Compliant motion planning with geometric models. *Proc. Third ACM Symp. Computational Geometry,* 171-180. (ResearchResult, MotionPlanning)

Becker, E. (1986). On the real spectrum of a ring and its application to semialgebraic geometry. *Bull. (new series) American Math. Soc.,* **15,** pp. 19-60. (RealAlgebraicGeometry, Tutorial, ResearchResult)

Ben-Or, M., Kozen, D., Reif, J. (1986). The complexity of elementary algebra and geometry. *J. Comp. Sys. Sci.,* **32,** 251-264. (Ben-Or/Kozen/ReifAlgorithm, ComplexityResult, ResearchResult)

Berman, L. (1980). The complexity of logical theories. *Theor. Comp. Sci.,* **11,** 71-77. (ComplexityResult, ResearchResult)

Bochnak, J., Coste, M., Roy, M. F. (1987). *Geometrie Algébrique Réelle.* Berlin: Springer-Verlag. (Ergebnisse der Mathematik) (RealAlgebraicGeometry, Exposition, ResearchResult)

Böge, W. (1986). Quantifier elimination for real closed fields. Proc. AAECC-3. *Springer Lec. Notes Comp. Sci.*

Brumfiel, G. (1979). *Partially Ordered Rings and Semi-Algebraic Geometry* (London Math. Soc. Lect. Notes 37). Cambridge: Cambridge Univ. Press. (RealAlgebraicGeometry, Tutorial, ResearchResult ; CohenAlgorithm, Exposition)

Buchberger, B., Loos, R., Collins, G. E. (eds.) (1982). *Computer Algebra - Symbolic and Algebraic Computation* (Computing Supplementum 4). Vienna and New York: Springer-Verlag. (Survey, Tutorial, Exposition)

Canny, J. (1987). *The Complexity of Robot Motion Planning* (PhD thesis). Dept. of Electr. Engin. & Comp. Sci., Massachusetts Inst. of Tech. (ResearchResult, ComplexityResult, MotionPlanning, RealClosedFields)

Canny, J. (1987). A new algebraic method for robot motion planning and real geometry. *Proc. 1987 IEEE Conf. on Foundations of Comp. Sci.* (FOCS). (ResearchResult, ComplexityResult, MotionPlanning, RealClosedFields)

Chazelle, B. (1985). Fast searching in a real algebraic manifold with applications to geometric complexity. Proc. CAAP '85. *Springer Lec. Notes Comp. Sci.* (GeometricSearching, ResearchResult, Survey ; CollinsAlgorithm, Exposition)

Chou, Y. S. (1984). Proving elementary geometry theorems using Wu's method. In *Automated Theorem Proving after 25 Years* (AMS Contemporary Mathematics, 29), 243-286. Providence, RI: American Mathematical Society. (GeometryTheoremProving)

Chou, Y. S. (1988). *Mechanical Geometry Theorem Proving.* New York: D. Reidel. (GeometryTheoremProving)

Cohen, P. J. (1969). Decision procedures for real and p-adic fields. *Comm. Pure Appl. Math,* **22,** 131-151. (CohenAlgorithm, ResearchResult)

Collins, G. E. (1956). The Tarski decision procedure. *Proc. ACM Natl. Conf.* (TarskiAlgorithm, Exposition)

Collins, G. E. (1957). Tarski's decision method for elementary algebra. In *Proceedings of the Summer Inst. Symbolic Logic,* **1,** pp. 64-70. Cornell Univ. (TarskiAlgorithm, Exposition)

Collins, G. E. (1975). Quantifier elimination for real closed fields by cylindrical algebraic decomposition. Proceedings of the Second GI Conference on Automata Theory and Formal Languages. *Springer Lec. Notes Comp. Sci.* **33**, 515–532. (CollinsAlgorithm, ResearchResult)

Collins, G. E. (1976). Quantifier elimination for real closed fields by cylindrical algebraic decomposition - a synopsis. *ACM SIGSAM Bull.* **10/1**, 10-12. (CollinsAlgorithm, Exposition)

Collins, G. E. (1982). Factorization in cylindrical algebraic decomposition. Proceedings of the European Computer Algebra Conference (EUROCAM '82). *Springer Lec. Notes Comp. Sci.* **144**, 212-214. (CollinsAlgorithm, ResearchResult)

Collins, G. E. (1982). Quantifier elimination for real closed fields: a guide to the literature. In (Buchberger, B., Loos, R., Collins, G. E., eds.) *Computer Algebra - Symbolic and Algebraic Computation* (Computing Supplementum 4), pp. 79-81. Vienna and New York: Springer-Verlag. (RealClosedFields, CollinsAlgorithm, Survey)

* Coste, M., Roy, M. F. (1988). Thom's lemma, the coding of real algebraic numbers and the computation of the topology of semi-algebraic sets. *J. Symb. Comp.* **5**, (this issue). (RealAlgebraicGeometry, CellAdjacencyAlgorithm, ResearchResult)

Davenport, J. (1985). *Computer Algebra for Cylindrical Algebraic Decomposition*, Tech Report TRITA-NA-8511, Royal Inst. of Tech., Dept. of Numerical Analysis and Computer Science, Stockholm, Sweden. (CollinsAlgorithm, Survey)

Davenport, J. (1986). A piano movers problem. *ACM SIGSAM Bull.* **20/76**, 15-17. (MotionPlanning)

Davenport, J., Siret, Y., Tournier, E. (1987). *Calcul Formel: Systèmes et Algorithmes de Manipulations Algébriques.* Paris: Masson. (Exposition)

* Davenport, J., Heintz, J. (1988). Real quantifier elimination is doubly exponential. *J. Symb. Comp.* **5**, (this issue). (ComplexityResult, ResearchResult)

Delzell, C. (1982). A finiteness theorem for open semi-algebraic sets, with applications to Hilbert's 17th problem. In *Ordered Fields and Real Algebraic Geometry* (AMS Contemporary Mathematics, 8), 79-97. Providence, RI: American Mathematical Society. (RealAlgebraicGeometry, ResearchResult)

Delzell, C. (1984). A continuous, constructive solution to Hilbert's 17th problem. *Invent. math.*, **76**,, 365-384. (RealAlgebraicGeometry, ResearchResult)

Dershowitz, N. (1979). A note on simplification orderings. *Info. Proc. Letters*, **9/5**, 212-215. (GeneralApplications)

Dickmann, M. A. (1985). Applications of model theory to real algebraic geometry: a survey. Methods in Mathematical Logic Proceedings 1983. *Springer Lec. Notes Math.* **1130**, 76-150. (RealAlgebraicGeometry, Survey)

Dubois, D. W. (1969). A nullstellensatz for ordered fields. *Ark. Math.* **8**, 111-114. (RealAlgebraicGeometry, ResearchResult)

Dubois, D. W., Efroymson, G. (1970). Algebraic theory of real varieties. In *Studies and Essays Presented to Y.H. Chen for his 60th Birthday*, pp. 107-135. Taipei: Math. Res. Center Nat. Taiwan Univ. (RealAlgebraicGeometry, ResearchResult, Exposition)

Dubois, D. W. (ed.) (1982). *Ordered Fields and Real Algebraic Geometry* (AMS Contemporary Mathematics, 8). Providence, RI: American Mathematical Society. (RealAlgebraicGeometry, ResearchResult, Exposition)

Dubois, D. W. (ed.) (1984). *Ordered Fields and Real Algebraic Geometry* (Proceedings of the AMS Boulder Conference). *Rocky Mountain J. Math.*, **14/4**. (RealAlgebraicGeometry, ResearchResult, Exposition)

Ferrante, J., Rackoff, C. (1975). A decision procedure for the first order theory of real addition with order. *SIAM J. Comp.* **4**, 69-76. (ComplexityResult, ResearchResult)

Ferrante, J., Rackoff, C. (1979). *The Complexity of Decision Procedures for Logical Theories.* Springer Lec. Notes Math., **718**. (ComplexityResult, ResearchResult, Exposition)

Fischer, M., Rabin, M. (1974). Super-exponential complexity of Presburger Arithmetic. In *Complexity of Computation*, (AMS-SIAM Proceedings 7), pp. 27-41. (ComplexityResult, ResearchResult)

Fitchas, N., Galligo, A., Morgenstern, J. (1987). Algorithmes rapides en sequentiel et en parallele pour l'elimination de quantificateurs en geometrique elementaire. *Seminaire Structures Algebriques Ordonnees.* UER de Mathematiques, Univ. Paris VII. (to appear) (ComplexityResult, ResearchResult)

Galligo, A., Heintz, J., Morgenstern, J. (1987). Parallelism and fast quantifier elimination over algebraically (and real) closed fields. Proceedings of the FCT '87 Conf, Kazan USSR. *Springer Lec. Notes Comp. Sci.* (ComplexityResult, ResearchResult)

Grigor'ev, D. Y. (1988). The complexity of deciding Tarski algebra. *J. Symb. Comp.* **5**, (this issue). (ComplexityResult, ResearchResult)

Grigor'ev, D. Y., Vorobjov, N. N. (1988). Solving systems of polynomial inequalities in subexponential time. *J. Symb. Comp.* **5**, (this issue). (ComplexityResult, ResearchResult)

Gudkov, D. A. (1974). The topology of real projective algebraic varieties. *Russian Math. Surveys* **29**, 1-79. (RealAlgebraicGeometry, Survey)

Heintz, J., Wüthrich, H. (1975). An efficient quantifier elimination procedure for algebraically closed fields. *ACM SIGSAM Bull.* **9**, 11. (ComplexityResult, ResearchResult)

Heintz, J. (1983). Definability and fast quantifier elimination in algebraically closed fields. *Theoretical Comput. Sci.* **24**, 239-277. (ComplexityResult, ResearchResult)

Hironaka, H. (1975). Triangulations of algebraic sets. *AMS Symp. in Pure Math.*, **29**, 165-185. (RealAlgebraicGeometry)

Holthusen, C. (1974). *Vereinfachungen für Tarski's Entscheidungsverfahren der elementaren reeleen Algebra.* Diplomarbeit, Univ. Heidelberg. (TarskiAlgorithm, ResearchResult)

Jacobson, N. (1964). *Lectures in Abstract Algebra.* Princeton, NJ: D. Van Nostrand. (TarskiAlgorithm, SeidenbergAlgorithm, Algebra, Tutorial)

Jacobson, N. (1974). *Basic Algebra.* San Francisco, CA: W. H. Freeman. (TarskiAlgorithm, SeidenbergAlgorithm, Algebra, Tutorial, GeneralApplications)

Kahn, P. J. (1979). Counting types of rigid frameworks. *Inventiones Math.*, **55**, 297-308. (GeneralApplications, CollinsAlgorithm)

Kaplansky, I. (1977). *Hilbert's Problems - Preliminary Edition.* Manuscript, Math. Dept., Univ. Chicago. (HilbertSixteenthProblem, Exposition)

Kapur, D. (1986). Geometry theorem proving using Hilbert's Nullstellensatz. *Proc. 1986 ACM Symp. Symbolic and Algebraic Computation.* Waterloo, Ontario (B. Char, ed.), 202-208. (GeometryTheoremProving)

Kapur, D. (1988). A refutational approach to geometry theorem proving. *Artificial Intelligence J.* (special issue on geometric reasoning and artificial intelligence; to appear). (GeometryTheoremProving)

Kozen, D., Yap., C. K. (1985). Algebraic cell decomposition in NC. *Proc. IEEE Conf. on Foundations of Comp. Sci.* (FOCS), 515–521. (CellDecompositionAlgorithm, CellAdjacencyAlgorithm, BenOr/Kozen/ReifAlgorithm, ResearchResult)

Kreisel, G., Krivine, J. L. (1967). *Elements of Mathematical Logic (Model Theory).* Amsterdam: North-Holland. (RealClosedFields, Tutorial)

Kreisel, G. (1975). Some uses of proof theory for finding computer programs. *Colloques Internationeaux du CNRS*, **249**, 123-134. (RealClosedFields, Examples)

Kreisel, G. (1977). From foundations to science; justifying and unwinding proofs. In Symposium: Set theory. Foundations of Mathematics. *Recueil des travaux de l'Institut Mathématique, Nouvelle serié*, **2/10**, 63-72. (RealClosedFields, Examples)

Kutzler, B., Stifter, S. (1988). Automated geometry theorem proving using Buchberger's algorithm. *Proc. 1986 ACM Symp. Symbolic and Algebraic Computation.* Waterloo, Ontario (B. Char, ed.), 209-214. (GeometryTheoremProving)

Lam, T. Y. (1984). Introduction to real algebra. *Rocky Mountain J. Math.*, **14**, 767-814. (Survey, Tutorial, Exposition, RealAlgebraicGeometry)

Lazard, D. (1988). Quantifier elimination: optimal solution for two classical examples. *J. Symb. Comp.* **5**, (this issue). (RealClosedFields, Examples, ResearchResult)

MacIntyre, A., McKenna, K., van den Dries, L. (1983). Elimination of quantifiers in algebraic structures. *Advances in Math.*, **47**, 74-87. (LogicalTheories)

Massey W. S. (1978). *Homology and Cohomology Theory.* New York, NY: Marcel Dekker. (CellComplexes)

McCallum, S. (1979). *Constructive Triangulation of Real Curves and Surfaces* (MSc thesis). Math. Dept., Univ. of Sydney, Australia. (RealAlgebraicGeometry, CellAdjacencyAlgorithm, ResearchResult)

McCallum, S. (1985). *An improved projection operator for cylindrical algebraic decomposition* (Ph.D. thesis). Tech. Rept. #578, Comp. Sci. Dept., Univ. Wisconsin–Madison. (CollinsAlgorithm, ResearchResult, Implementation)

ᵇ McCallum, S. (1988). An improved projection operation for cylindrical algebraic decomposition of three-dimensional space. *J. Symb. Comp.* **5**, (this issue). (CollinsAlgorithm, ResearchResult, Implementation)

Meserve, B. (1955). Decision methods for elementary algebra. *Amer. Math. Monthly*, **62**, 1-8. (TarskiAlgorithm, Tutorial)

Monk, L. (1974). *An Elementary Recursive Decision Procedure for Th(R,+,*).* Manuscript, Math. Dept., Univ. California, Berkeley. (MonkAlgorithm, ResearchResult)

Monk, L. (1975). *Elementary Recursive Decision Procedures* (Ph.D. thesis). Math. Dept., Univ. California, Berkeley. (MonkAlgorithm, ResearchResult)

Müller, F. (1978). *Ein exakter Algorithmus zur nichtlinearen Optimierung für beliebige Polynome mit mehreren Veranderlichen.* Meisenheim am Glan: Verlag Anton Hain. (GeneralApplications, Implementation, Examples, ResearchResult)

Mundy, J., Kapur, D. (1988). Wu's method: an informal introduction. *Artificial Intelligence J.* (special issue on geometric reasoning and artificial intelligence; to appear). (GeometryTheoremProving)

O'Rourke, J. (1982). *The Complexity of Computing Minimum Convex Covers for Polygons.* Manscrpt, Dept. of Electr. Engin. and Comp. Sci., Johns Hopkins Univ. (GeneralApplications)

Paugam, A. (1986). *Algorithmes d'Elimination des Quantificateurs.* Manuscript, Math. Dept., Univ. Rennes. (SeidenbergAlgorithm, CollinsAlgorithm, Exposition)

Prestel, A. (1984). Lecture on formally real fields. Rio de Janeiro: IMPA 1975. *Springer Lec. Notes Math.* **1093.** (RealClosedFields)

Prill, D. (1986). On approximation and incidence in cylindrical algebraic decompositions. *SIAM J. Comp.* **15**, 972–993. (CellAdjacencyAlgorithm, CollinsAlgorithm, ResearchResult, ComplexityResult)

Rabin, M. (1977). Decidable theories. In (J. Barwise, ed.), , *Handbook of Mathematical Logic*, pp. 595-630. Amsterdam: North-Holland. (RealClosedFields, Tutorial)

Risler, J.-J. (1980). Sur le 16 problème de Hilbert: un résumé et quelques questions. In Séminaire sur la géométrie algébrique réelle. *Publ. Math. Univ. Paris VII*, **9**, 11-25. (HilbertSixteenthProblem, Survey)

Risler, J.-J. (1988). Some aspects of complexity in real algebraic geometry. *J. Symb. Comp.* **5**, (this issue). (RealAlgebraicGeometry, Survey)

Robinson, A. (1965). *Introduction to model theory and the metamathematics of algebra.* Amsterdam: North-Holland. (RealClosedFields, Tutorial)

Robinson, A. (1974). A decision method for elementary algebra and geometry - revisited. *AMS Symp. in Pure Math.*, **25**, 139-152. (TarskiAlgorithm, Survey)

Schwartz, J. T., Sharir, M. (1983). On the 'piano movers' problem II. General techniques for computing topological properties of real algebraic manifolds. *Adv. Applied Math.* **4**, 298–351. (MotionPlanning, ResearchResult; CollinsAlgorithm, Exposition)

Schwartz, J., Sharir, M. (1988). A survey of motion planning and related geometric algorithms. *Artificial Intelligence J.* (special issue on Geometric Reasoning and Artificial Intelligence; to appear). (MotionPlanning, Survey)

Seidenberg, A. (1954). A new decision method for elementary algebra. *Annals of Math.*, **60**, 365-374. (SeidenbergAlgorithm, ResearchResult)

Sharir, M., Schorr, A. (1984). On shortest paths in polyhedral spaces. *Proc. 16th ACM Symp. on Theory of Computing* (STOC), 144-153. (GeneralApplications, GeometricSearching)

Smoryński, C. (1981). Skolem's solution to a problem of Frobenius. *Math. Intelligencer*, **3/3**, 123-132. (LogicalTheories, RealClosedFields, CollinsAlgorithm, Tutorial, Examples,)

Sontag, E. (1985). Real addition and the polynomial hierarchy. *Info. Proc. Letters*, **20**, 115-120. (ResearchResult, ComplexityResult,)

Struth, M. (1980). *Vorzeichenbestimmung für Polynome auf der Zellenzerlegung nach Collins.* Diplomarbeit, Univ. Heidelberg. (CollinsAlgorithm, ResearchResult,)

Tarski, A. (1951). *A Decision Method for Elementary Algebra and Geometry.* Report R-109, second revised edn. Santa Monica, CA: The Rand Corporation. (TarskiAlgorithm, ResearchResult, Exposition, Examples)

Tarski, A. (1959). What is elementary geometry? In (L. Henkin, P. Suppes, A. Tarski, eds.) *Proceedings of an International Symposium on the Axiomatic Method, with Special Reference to Geometry and Physics* (Studies in Logic and the Foundations of Mathematics), pp. 16-29. Amsterdam: North Holland. (TarskiAlgorithm, GeometryTheoremProving)

Thom, R. (1972). *Structural Stability and Morphogenesis.* New York: Benjamin. (RealAlgebraicGeometry, ResearchResult, Exposition)

van der Waerden, B. L., Schütte, K. (1952). Das Problem der Dreizehn Kugeln. *Math. Ann.* **125**, 325-334. (RealClosedFields, ResearchResult, Examples)

van der Waerden, B. L. (1970). *Algebra.* New York: Frederick Ungar. (RealClosedFields, Algebra, Exposition, Tutorial)

Wang, H. (1981). *Popular lectures on mathematical logic.* Beijing: Science Press. New York: Van Nostrand Reihnold. (RealClosedFields, Tutorial)

Weispfenning, V. (1988). The complexity of linear problems in fields. *J. Symb. Comp.* **5**, (this issue). (ComplexityResult, ResearchResult)

Whitney, H. (1957). Elementary structure of real algebraic varieties. *Ann. of Math.* **66**, 545-556. (RealAlgebraicGeometry, ResearchResult)

Wilson, G. (1978). Hilbert's sixteenth problem. *Topology*, **17**, 53-73. (HilbertSixteenthProblem, Survey)

Wu, W.-T. (1984). Basic Principles of Mechanical Theorem proving in Elementary Geometries. *J. Syst. Sci. & Math. Sci.* **4**, (Beijing), 207-235. Reprinted in *J. Automated Reasoning* (1986), **2**, 221-252. (GeometryTheoremProving)

Wüthrich, H. (1976). Ein Entscheidungsverfahren für die Theorie der reell-abgeschlossenen Körper. *Springer Lec. Notes Comp. Sci.* **43**,, 138-162. (WuethrichAlgorithm, ResearchResult)

Wüthrich, H. (1977). *Ein schnelles Quantoreneliminationsverfahren für die Theorie der algebraisch abgeschlossenen Körper* (Ph.D. thesis). Philisophischen Fakultät II, Univ. Zurich. (ResearchResult)